Texts and Monographs in
Symbolic Computation

A Series of the
Research Institute for Symbolic Computation,
Johannes-Kepler-University, Linz, Austria

Edited by
B. Buchberger and G. E. Collins

D. Wang

Elimination Methods

SpringerWienNewYork

Dr. Dongming Wang
Laboratoire d'Informatique de Paris 6
Université Pierre et Marie Curie, Paris, France

Data conversion by Thomson Press (India) Ltd., New Delhi, India
Printed by Novographic Druck G.m.b.H., A-1230 Wien
Graphic design: Ecke Bonk
Printed on acid-free and chlorine-free bleached paper
SPIN 10637914

With 12 Figures

Library of Congress Cataloging-in-Publication Data

Wang, Dongming.
Elimination methods / D. Wang.
p. cm. — (Texts and monographs in symbolic computation, ISSN 0943-853X)
Includes bibliographical references and index.
ISBN 3211832416 (alk. paper)
1. Elimination. I. Title. II. Series.

QA192.W36 2000
512.9′434 — dc21 00-035777

ISSN 0943-853X
ISBN 3-211-83241-6 Springer-Verlag Wien New York

To my parents and
to Xiaofan, Simon, and Louise

Preface

The development of polynomial-elimination techniques from classical theory to modern algorithms has undergone a tortuous and rugged path. This can be observed from B. L. van der Waerden's elimination of the "elimination theory" chapter from his classic *Modern Algebra* in later editions, A. Weil's hope to eliminate "from algebraic geometry the last traces of elimination theory," and S. Abhyankar's suggestion to "eliminate the eliminators of elimination theory." The renaissance and recognition of polynomial elimination owe much to the advent and advance of modern computing technology, based on which effective algorithms are implemented and applied to diverse problems in science and engineering. In the last decade, both theorists and practitioners have more and more realized the significance and power of elimination methods and their underlying theories. Active and extensive research has contributed a great deal of new developments on algorithms and software tools to the subject, that have been widely acknowledged. Their applications have taken place from pure and applied mathematics to geometric modeling and robotics, and to artificial neural networks.

This book provides a systematic and uniform treatment of elimination algorithms that compute various zero decompositions for systems of multivariate polynomials. The central concepts are triangular sets and systems of different kinds, in terms of which the decompositions are represented. The prerequisites for the concepts and algorithms are results from basic algebra and some knowledge of algorithmic mathematics. Some of the operations and results on multivariate polynomials which are used throughout the book are collected in the first chapter. Chapters 2 to 5 are devoted to the description of the algorithms of zero decomposition. We start by presenting algorithms that decompose arbitrary polynomial systems into triangular systems; the latter are not guaranteed to have zeros. These algorithms are modified in Chap. 3 by incorporating the projection process and GCD computation so that the computed triangular systems always have zeros. Then, we elaborate how to make use of polynomial factorization in order to compute triangular systems that are irreducible. The proposed algorithms and their underlying theories are based on the previous work of J. F. Ritt, W.-t. Wu, A. Seidenberg, and J. M. Thomas and its further development by the author. A brief review of some relevant algorithms including those based on resultants and Gröbner bases is given in Chap. 5. Elimination methods play a special role in constructive algebraic geometry and polynomial-ideal theory. Chapter 6 contains investigations on a few problems from these two areas. The book ends with an introduction to several selected applications of symbolic elimination methods.

Most of the algorithms presented in the book have been implemented by the author in the Maple system, and they are among the most efficient elimination algorithms available by this time. The algorithms are described formally so that the reader can easily work out his own implementation. Nevertheless, both theoretical complexity and practical implementation issues are not addressed in the book.

This book can be used as a textbook for a graduate course in elimination theory and methods. Some of the material was taught by the author at RISC-Linz, Johannes Kepler University a few times from 1989 to 1998.

I am very grateful to Professor Wen-tsün Wu who introduced me to the fascinating subject of polynomial elimination, taught me his method of characteristic sets, and has kept advising me for more than a decade. His work and thoughts have been so influential in my research that I have referred to them in most of my relevant publications.

I am greatly indebted to Professor Bruno Buchberger from whom I have learned so much beyond Gröbner bases. His generous support and help of numerous forms have made me easy at work and life for years.

Many colleagues and students have kindly helped me in different ways, like inviting me for a talk, a visit, or simply a dinner, being available to help when my languages run short, and giving me a hand when my computer gets stuck. It is impossible to mention all the names; I wish to thank all of them sincerely.

The members of the ATINF group led by Professor Ricardo Caferra at Laboratoire Leibniz, Institut d'Informatique et Mathématiques Appliquées de Grenoble deserve special thanks. They have created an ideal working environment, where I could enjoy thinking, writing, and programming. It is my pleasure to thank Mrs. Silvia Schilgerius and Mr. Thomas Redl at Springer-Verlag Wien, with whom I have worked for publishing this and two previous books.

Dongming Wang

Contents

List of symbols

$\triangleq$	"defined to be"
$\prec$, $\succ$	order for variables, terms, polynomials, and triangular sets
$\precsim$, $\succsim$, $\sim$	order for polynomials and triangular sets
$\backsim$	similarity of polynomials
$\sqrt{\ }$	radical (of an ideal)
$\Longleftrightarrow$	"if and only if"
$\leadsto$	"simplified to"
$\Rightarrow$, $\vee$, $\wedge$	logical "imply," "or," "and"
$\mathbf{A}^n_{\tilde{\boldsymbol{K}}}$	n-dimensional affine space over $\tilde{\boldsymbol{K}}$
$\mathbf{C}$	field of complex numbers
cls	class of a polynomial
coef	coefficient of a polynomial in a term
cont	content of a polynomial with respect to a variable
deg	degree of a polynomial in a variable
det	determinant of a square matrix
Dim	dimension of an algebraic variety or of a polynomial set or system
dim	dimension of a perfect triangular set or system
GB	reduced Gröbner basis of a polynomial set
gcd	greatest common divisor of a set of polynomials or of two polynomials with respect to a variable
Ideal	ideal generated by a set of polynomials
ini	initial of a polynomial; or the set of initials of the polynomials in a set
ITS	irreducible triangular series of a polynomial set or system
lc	leading coefficient of a polynomial (in a variable)
ldeg	leading degree of a polynomial
level	level of a polynomial set or system
lm	leading monomial of a polynomial
lt	leading term of a polynomial
lv	leading variable of a polynomial
$\boldsymbol{K}$	field of characteristic 0
$\tilde{\boldsymbol{K}}$	extension field of $\boldsymbol{K}$
$\tilde{\boldsymbol{K}}$-Zero	set of all zeros in $\tilde{\boldsymbol{K}}$ of a polynomial set or system
$\bar{\boldsymbol{K}}$	algebraic closure of $\boldsymbol{K}$
$\boldsymbol{K}(\theta)$	extension field obtained from $\boldsymbol{K}$ by adjoining θ

op	ith element of a tuple or an (ordered) set
$\mathbb{P}$	polynomial set (i.e., a finite set of nonzero polynomials)
$\mathbb{P}^{(i)}$	$\mathbb{P} \cap \boldsymbol{K}[x_1, \ldots, x_i]$
$\mathbb{P}^{[i]}$	$\mathbb{P} \setminus \mathbb{P}^{(i)}$
$\mathbb{P}^{\langle i \rangle}$	$\mathbb{P}^{(i)} \setminus \mathbb{P}^{(i-1)}$
$\mathbb{P}^{\langle \bar{x}, i \rangle}$	$\mathbb{P}\vert_{x_1=\bar{x}_1, \ldots, x_i=\bar{x}_i}$
$[\mathbb{P}, \mathbb{Q}]$, $\mathfrak{P}$	polynomial system (i.e., a pair of polynomial sets)
$\mathfrak{P}^{(i)}$	$[\mathbb{P}^{(i)}, \mathbb{Q}^{(i)}]$ if $\mathfrak{P} = [\mathbb{P}, \mathbb{Q}]$
$\mathfrak{P}^{\langle i \rangle}$	$[\mathbb{P}^{\langle i \rangle}, \mathbb{Q}^{\langle i \rangle}]$ if $\mathfrak{P} = [\mathbb{P}, \mathbb{Q}]$
$\mathfrak{P}^{\langle \bar{x}, i \rangle}$	$[\mathbb{P}^{\langle \bar{x}, i \rangle}, \mathbb{Q}^{\langle \bar{x}, i \rangle}]$ if $\mathfrak{P} = [\mathbb{P}, \mathbb{Q}]$
$\breve{\mathfrak{P}}$	$\mathbb{P} \cup \mathbb{Q}$ if $\mathfrak{P} = [\mathbb{P}, \mathbb{Q}]$
PB	prime basis of an irreducible triangular set
pp	primitive part of a polynomial with respect to a variable
pquo	pseudo-quotient of a polynomial with respect to a nonzero polynomial in a variable
prem	pseudo-remainder of a polynomial with respect to a nonzero polynomial (in a variable) or with respect to a triangular set; or set of pseudo-remainders of the polynomials in a set with respect to a polynomial or with respect to a triangular set
$\mathbf{Q}$	field of rational numbers
$\mathbf{R}$	field of real numbers
$\boldsymbol{R}$	ring
$\boldsymbol{R}[\boldsymbol{x}]$	ring of polynomials in $\boldsymbol{x}$ with coefficients in $\boldsymbol{R}$
Rad	radical of an ideal
red	reductum of a polynomial (with respect to a variable)
RegZero	set of regular zeros of a regular set, a triangular system, or a polynomial system
rem	remainder of a polynomial with respect to a polynomial set; or set of remainders of the polynomials in a set with respect to another polynomial set
res	resultant of two polynomials with respect to a variable or of a polynomial with respect to a triangular set
RS	regular series of a polynomial set or system
sat	saturation of a triangular set
sqfr	greatest squarefree divisor of a polynomial
SS	simple series of a polynomial set or system
$\mathbb{T}$	triangular set
$\mathbb{T}^{\{i\}}$	$[T_1, \ldots, T_i]$ if $\mathbb{T} = [T_1, \ldots, T_r]$
$[\mathbb{T}, \tilde{\mathbb{T}}]$, $\mathfrak{S}$	simple system
$[\mathbb{T}, \mathbb{U}]$, $\mathfrak{T}$	triangular system
tdeg	total degree of a polynomial
$\boldsymbol{u}$	$(u_1, \ldots, u_d)$, or $u_1, \ldots, u_d$
$\mathcal{V}$	algebraic variety
$\boldsymbol{x}$	$(x_1, \ldots, x_n)$, or $x_1, \ldots, x_n$
$\boldsymbol{x}^{\{i\}}$	$(x_1, \ldots, x_i)$, or $x_1, \ldots, x_i$
$\boldsymbol{\xi}$	$(\xi_1, \ldots, \xi_n)$, or $\xi_1, \ldots, \xi_n$, or $(\boldsymbol{u}, \eta_1, \ldots, \eta_r)$, or $\boldsymbol{u}, \eta_1, \ldots, \eta_r$

$\boldsymbol{\xi}^{\{i\}}$	$(\xi_1, \ldots, \xi_i)$, or $\xi_1, \ldots, \xi_i$, or $(\boldsymbol{u}, \eta_1, \ldots, \eta_i)$, or $\boldsymbol{u}, \eta_1, \ldots, \eta_i$
$\mathbf{Z}$	ring of integers
$\boldsymbol{z}$	$(\boldsymbol{u}, y_1, \ldots, y_r)$, or $\boldsymbol{u}, y_1, \ldots, y_r$
$\boldsymbol{z}^{\{i\}}$	$(\boldsymbol{u}, y_1, \ldots, y_i)$, or $\boldsymbol{u}, y_1, \ldots, y_i$
Zero	set of all zeros of a polynomial set or system

1 Polynomial arithmetic and zeros

We start by collecting some concepts, operations, and properties on multivariate polynomials, which are fundamental and will be used throughout the following chapters. Most of the results presented here are not proved formally; their proofs may be found in standard textbooks on algebra. Wherever no reference is given, the reader is advised to look up them in van der Waerden (1950, 1953) and Knuth (1981).

1.1 Polynomials

Let $\boldsymbol{R}$ be a ring and $x_1, x_2, \ldots, x_n$ be n distinct symbols, not in $\boldsymbol{R}$, called *indeterminates, unknowns* or *variables*. We often write $\boldsymbol{x}$ for $x_1, x_2, \ldots, x_n$ or $(x_1, x_2, \ldots, x_n)$. For n nonnegative integers $i_1, i_2, \ldots, i_n$, one can form a power product

$$\mu = x_1^{i_1} x_2^{i_2} \ldots x_n^{i_n}.$$

It is called a *term*.

Let a be an element of $\boldsymbol{R}$, i.e., $a \in \boldsymbol{R}$. The formal expression

$$\alpha = a\mu = a x_1^{i_1} x_2^{i_2} \ldots x_n^{i_n}$$

is called a *monomial* and written sometimes as $\alpha = a\boldsymbol{x}^{\boldsymbol{i}}$, where

$$\boldsymbol{x} = (x_1, \ldots, x_n), \quad \boldsymbol{i} = (i_1, \ldots, i_n).$$

The above a is called the *coefficient* of α. The monomial α is said to be *nonzero* if $a \neq 0$.

For an n-tuple $\boldsymbol{i} = (i_1, \ldots, i_n)$, the lth element i_l is denoted $\mathrm{op}(l, \boldsymbol{i})$. Sometimes we write $\boldsymbol{i}^{\{l\}}$ for $(i_1, \ldots, i_l)$. Any two n-tuples $\boldsymbol{i}$ and $\boldsymbol{j}$ of nonnegative integers are said to be *distinct* if there is an l $(1 \leq l \leq n)$ such that $\mathrm{op}(l, \boldsymbol{i}) \neq \mathrm{op}(l, \boldsymbol{j})$. Two terms $\boldsymbol{x}^{\boldsymbol{i}}$ and $\boldsymbol{x}^{\boldsymbol{j}}$ are *distinct* if so are $\boldsymbol{i}$ and $\boldsymbol{j}$. Let $a_1, \ldots, a_t \in \boldsymbol{R}$ and $\boldsymbol{i}_1, \ldots, \boldsymbol{i}_t$ be t pairwise distinct n-tuples of nonnegative integers. The finite sum

$$P = \sum_{l=1}^{t} a_l \boldsymbol{x}^{\boldsymbol{i}_l} \tag{1.1.1}$$

is called a *polynomial* in the indeterminates $\boldsymbol{x}$ with coefficients $a_1, \ldots, a_t$ in $\boldsymbol{R}$. A polynomial P is 0 if all the monomials of P are 0, i.e., $a_1 = \cdots = a_t = 0$. Since the monomial 0 can be arbitrarily added to and deleted from a polynomial, we assume that in any nonzero polynomial P all monomials are nonzero, i.e., $a_1 \neq 0, \ldots, a_t \neq 0$, and call t the *number of monomials* of P. P is said to be a *constant* if $P \in \boldsymbol{R}$. Let $\boldsymbol{x}^{\boldsymbol{i}}$ be a term. If there is an $a \in \boldsymbol{R}$ and $a \neq 0$ such that the monomial $a\boldsymbol{x}^{\boldsymbol{i}}$ appears in P, then a is called the *coefficient* of P in $\boldsymbol{x}^{\boldsymbol{i}}$, denoted by $\mathrm{coef}(P, \boldsymbol{x}^{\boldsymbol{i}})$. Otherwise, $\mathrm{coef}(P, \boldsymbol{x}^{\boldsymbol{i}})$ is defined to be 0.

Let P be a nonzero polynomial as in (1.1.1) and x_k an arbitrary indeterminate. We define the *degree* of P in x_k as

$$\deg(P, x_k) \triangleq \max_{1 \leq l \leq t} \mathrm{op}(k, \boldsymbol{i}_l),$$

where $\triangleq$ reads "is defined to be." For convenience, we define $\deg(0, x_k) = -1$. The *total degree* of P is defined by

$$\mathrm{tdeg}(P) \triangleq \max_{1 \leq l \leq t} \sum_{k=1}^{n} \mathrm{op}(k, \boldsymbol{i}_l).$$

A polynomial is said to be *homogeneous* if all its terms have the same total degree.

Example 1.1.1. The following is a polynomial in $x_1, \ldots, x_4$ with integer coefficients

$$F_1 = x_4^2 + x_1 x_4^2 - x_2 x_4 - x_1 x_2 x_4 + x_1 x_2 + 3x_2.$$

One sees that

$$\begin{aligned}
&\mathrm{coef}(F_1, x_1 x_2 x_4) = -1, \ \mathrm{coef}(F_1, x_2 x_4^3) = 0, \\
&\deg(F_1, x_2) = 1, \ \deg(F_1, x_4) = 2, \\
&\mathrm{tdeg}(F_1) = 3,
\end{aligned}$$

and F_1 is not homogeneous.

Let $Q = \sum_{l=1}^{s} b_l \boldsymbol{x}^{\boldsymbol{j}_l}$ be any other polynomial. The *sum* of P and Q is defined as

$$P + Q \triangleq \sum_{l=1}^{r} c_l \boldsymbol{x}^{\boldsymbol{k}_l},$$

where $\boldsymbol{k}_1, \ldots, \boldsymbol{k}_r$ are all the distinct n-tuples among $\boldsymbol{i}_1, \ldots, \boldsymbol{i}_t, \boldsymbol{j}_1, \ldots, \boldsymbol{j}_s$ and

$$c_l = \mathrm{coef}(P, \boldsymbol{x}^{\boldsymbol{k}_l}) + \mathrm{coef}(Q, \boldsymbol{x}^{\boldsymbol{k}_l}), \quad l = 1, \ldots, r.$$

Form the n-tuples $\boldsymbol{k}_{\boldsymbol{i}_u \boldsymbol{j}_v} = (\mathrm{op}(1, \boldsymbol{i}_u) + \mathrm{op}(1, \boldsymbol{j}_v), \ldots, \mathrm{op}(n, \boldsymbol{i}_u) + \mathrm{op}(n, \boldsymbol{j}_v))$, with $u = 1, \ldots, t$ and $v = 1, \ldots, s$, and let $\boldsymbol{k}_1, \ldots, \boldsymbol{k}_r$ be all the distinct ones among them. The *product* of P and Q is defined as

$$PQ \triangleq \sum_{l=1}^{r} c_l \boldsymbol{x}^{\boldsymbol{k}_l},$$

where

$$c_l = \sum_{k_{i_u j_v} = k_l} a_u b_v, \quad l = 1, \ldots, r.$$

Theorem 1.1.1. Under the above definition of addition and multiplication, all the polynomials in $\boldsymbol{x}$ with coefficients in $\boldsymbol{R}$ form a ring.

The ring of polynomials in the n indeterminates $x_1, \ldots, x_n$ with coefficients in $\boldsymbol{R}$ is denoted by $\boldsymbol{R}[x_1, \ldots, x_n]$, or $\boldsymbol{R}[\boldsymbol{x}]$ for short. It is also known as a *polynomial ring* derived from $\boldsymbol{R}$ by *adjoining* $\boldsymbol{x}$. If $\boldsymbol{R}$ is commutative, then so is $\boldsymbol{R}[\boldsymbol{x}]$. If, in particular, $\boldsymbol{R}$ is the integral ring $\mathbf{Z}$, then $\boldsymbol{R}[\boldsymbol{x}]$ is a ring of polynomials with integer coefficients.

Theorem 1.1.2. If $\boldsymbol{R}$ is an integral domain, then so is $\boldsymbol{R}[\boldsymbol{x}]$.

Remember that n is the number of variables $\boldsymbol{x}$. We say that the polynomials are *univariate* if $n = 1$, *bivariate* if $n = 2$, and *multivariate* if $n \geq 2$. Accordingly, the polynomial ring $\boldsymbol{R}[\boldsymbol{x}]$ is said to be *univariate*, *bivariate*, or *multivariate* respectively, depending on whether n is 1, 2, or ≥ 2. The multivariate polynomial ring $\boldsymbol{R}[\boldsymbol{x}]$ derived from $\boldsymbol{R}$ by adjoining the indeterminates $\boldsymbol{x}$ can also be considered as the ring $\boldsymbol{R}[x_1][x_2]\ldots[x_n]$ derived from $\boldsymbol{R}$ by successively adjoining the indeterminates $x_1, x_2, \ldots, x_n$.

Theorem 1.1.3. The polynomial rings $\boldsymbol{R}[x_1]\ldots[x_n]$, $\boldsymbol{R}[x_{q_1}]\ldots[x_{q_n}]$, and $\boldsymbol{R}[\boldsymbol{x}]$ are isomorphic, where $q_1 \ldots q_n$ is an arbitrary permutation of $1 \ldots n$.

Therefore, a multivariate polynomial $P \in \boldsymbol{R}[\boldsymbol{x}]$ can also be understood as a univariate polynomial in a fixed indeterminate, for example, in x_n with coefficients in $\boldsymbol{R}[x_1, \ldots, x_{n-1}]$. In other words, P may be considered as an element of $\boldsymbol{R}[\boldsymbol{x}^{\{n-1\}}][x_n]$.

By a *polynomial set* we mean a finite set of nonzero polynomials in $\boldsymbol{R}[\boldsymbol{x}]$. While speaking about a *polynomial system*, we refer to a pair $[\mathbb{P}, \mathbb{Q}]$ of polynomial sets. As a general convention, in this book we denote polynomials by capital letters like P, Q, F, polynomial sets by blackboard bold letters like $\mathbb{P}, \mathbb{Q}, \mathbb{T}$, polynomial systems by Gothic (Fraktur) letters like $\mathfrak{P}, \mathfrak{T}, \mathfrak{S}$, and sets or sequences of polynomial systems by Greek letters like Ψ.

In what follows, let us fix an ordering for the indeterminates $x_1 \prec \cdots \prec x_n$.

Definition 1.1.1. For any two distinct terms $\boldsymbol{x}^{\boldsymbol{i}}$ and $\boldsymbol{x}^{\boldsymbol{j}}$ with

$$\boldsymbol{i} = (i_1, \ldots, i_n), \quad \boldsymbol{j} = (j_1, \ldots, j_n),$$

we say that $\boldsymbol{x}^{\boldsymbol{i}}$ *precedes* $\boldsymbol{x}^{\boldsymbol{j}}$ or $\boldsymbol{x}^{\boldsymbol{j}}$ *follows* $\boldsymbol{x}^{\boldsymbol{i}}$, denoted as

$$\boldsymbol{x}^{\boldsymbol{i}} \prec \boldsymbol{x}^{\boldsymbol{j}} \text{ or } \boldsymbol{x}^{\boldsymbol{j}} \succ \boldsymbol{x}^{\boldsymbol{i}},$$

if there is a k ($1 \leq k \leq n$) such that

$$i_k < j_k \text{ and } i_l = j_l, \quad \text{for } k < l \leq n.$$

Under "$\prec$" all the terms in $\boldsymbol{x}$ may be ordered, and so may the monomials of any nonzero polynomial in $\boldsymbol{R}[\boldsymbol{x}]$. We call "$\prec$" the *purely lexicographical ordering* of terms or monomials.

In fact, any nonzero polynomial in $\boldsymbol{R}[\boldsymbol{x}]$ can be written in the form (1.1.1) with

$$a_1 \neq 0, \ldots, a_t \neq 0, \quad a_i \in \boldsymbol{R},$$
$$\boldsymbol{x}^{\boldsymbol{i}_1} \succ \cdots \succ \boldsymbol{x}^{\boldsymbol{i}_t}.$$

In this case, $\boldsymbol{x}^{\boldsymbol{i}_1}$ is called the *leading term*, $a_1\boldsymbol{x}^{\boldsymbol{i}_1}$ the *leading monomial*, and a_1 the *leading coefficient* of P, denoted by $\mathrm{lt}(P)$, $\mathrm{lm}(P)$, and $\mathrm{lc}(P)$, respectively. When $P \notin \boldsymbol{K}$, the biggest index p such that

$$\deg(P, x_p) > 0 \text{ or equivalently } \deg(\boldsymbol{x}^{\boldsymbol{i}_1}, x_p) > 0$$

is called the *class*, x_p the *leading variable*, and $\deg(P, x_p)$ the *leading degree* of P, denoted by $\mathrm{cls}(P)$, $\mathrm{lv}(P)$, and $\mathrm{ldeg}(P)$, respectively. Symbolically,

$$\mathrm{lv}(P) = x_{\mathrm{cls}(P)}, \quad \mathrm{ldeg}(P) = \deg(P, \mathrm{lv}(P)).$$

For any $P \in \boldsymbol{K}$ and $P \neq 0$, we define the *class*, the *leading variable*, and the *leading degree* of P to be 0, x_0, and 0, respectively, where x_0 is a new variable ordered to be $\prec x_1$.

Let P be a polynomial with $\mathrm{cls}(P) = p > 0$, which may also be considered as one in x_p. Any other polynomial $Q \in \boldsymbol{R}[\boldsymbol{x}]$ is said to be *reduced* with respect to P if $\deg(Q, x_p) < \mathrm{ldeg}(P)$. The leading coefficient $\mathrm{lc}(P, x_p)$ of P in x_p is called the *initial* of P, denoted by $\mathrm{ini}(P)$, which is a polynomial in $x_1, \ldots, x_{p-1}$. The *initial* of any $P \in \boldsymbol{K}$ is defined to be itself. For any polynomial set $\mathbb{P}$, we define

$$\mathrm{ini}(\mathbb{P}) \triangleq \{\mathrm{ini}(P)\colon \ P \in \mathbb{P}\}.$$

Example 1.1.2. With $x_1 \prec \cdots \prec x_4$, the polynomial F_1 in Example 1.1.1 may be rewritten as

$$\begin{aligned} F_1 &= x_1x_4^2 + x_4^2 - x_1x_2x_4 - x_2x_4 + x_1x_2 + 3x_2 \\ &= (x_1 + 1)x_4^2 + (-x_1x_2 - x_2)x_4 + x_1x_2 + 3x_2. \end{aligned}$$

We have

$$\begin{aligned} &\mathrm{lc}(F_1) = 1, \\ &\mathrm{lm}(F_1) = \mathrm{lt}(F_1) = x_1x_4^2, \\ &\mathrm{cls}(F_1) = 4, \quad \mathrm{lv}(F_1) = x_4, \\ &\mathrm{ldeg}(F_1) = 2, \quad \mathrm{ini}(F_1) = x_1 + 1. \end{aligned}$$

The polynomial $F_2 = x_1x_4 + x_3 - x_1x_2$ is reduced with respect to F_1, but F_1 is not with respect to F_2.

1.2 Greatest common divisor, pseudo-division, and polynomial remainder sequences

Let the ring $\boldsymbol{R}$ be restricted to a *unique factorization domain* (UFD), i.e., a commutative ring with identity. In this case, $ab \neq 0$ whenever a and b are nonzero elements of $\boldsymbol{R}$, and every $a \in \boldsymbol{R}$ either is a "unit" or has a "unique" representation of the form

$$a = up_1 \cdots p_t, \quad t \geq 1,$$

where $p_1, \ldots, p_t$ are "primes" and u is a unit. Every field in which each nonzero element is a unit and there is no prime is a UFD. When $\boldsymbol{R}$ is assumed to be a UFD, by Theorem 1.1.2 $\boldsymbol{R}[\boldsymbol{x}]$ is also a UFD.

Let F and G be two polynomials in $\boldsymbol{R}[\boldsymbol{x}]$, with $G \neq 0$. We say that G *divides* F or F is *divisible* by G, denoted as $G \mid F$, if there exists a quotient polynomial $Q \in \boldsymbol{R}[\boldsymbol{x}]$ such that

$$F = QG.$$

In this case, G is called a *divisor* of F, and F is called a *multiple* of G.

Definition 1.2.1. Let $P_1, \ldots, P_s$ be polynomials in $\boldsymbol{R}[\boldsymbol{x}]$ which are not all 0. A polynomial $G \in \boldsymbol{R}[\boldsymbol{x}]$ is called a *greatest common divisor* (GCD) of $P_1, \ldots, P_s$ if G divides $P_1, \ldots, P_s$ and every common divisor of $P_1, \ldots, P_s$ divides G.

A polynomial $L \in \boldsymbol{R}[\boldsymbol{x}]$ is called a *least common multiple* of $P_1, \ldots, P_s$ if all $P_1, \ldots, P_s$ divide L and L divides every common multiple of $P_1, \ldots, P_s$.

The polynomial G in this definition is not unique: For any unit a, aG is also a GCD. However, by the UFD property any two GCDs are different only by a unit factor. Hence, all the GCDs of $P_1, \ldots, P_s$ will be considered identical. It is so also for the least common multiples. Let $\mathbb{P} = \{P_1, \ldots, P_s\}$.

$$\gcd(\mathbb{P}) = \gcd(P_1, \ldots, P_s) \text{ and } \operatorname{lcm}(\mathbb{P}) = \operatorname{lcm}(P_1, \ldots, P_s)$$

stand for any GCD and least common multiple of $P_1, \ldots, P_s$, respectively.

Example 1.2.1. Consider the polynomials

$$\begin{aligned} G_1 &= 3x_4^2 - 3x_2x_4 + 6x_1x_4 - 3x_3x_4 + 3x_2x_3 - 6x_1x_3, \\ G_2 &= 6x_4^2 + 15x_1x_2x_4 - 6x_3x_4 - 15x_1x_2x_3. \end{aligned}$$

One can verify that $3x_3 - 3x_4$ divides both G_1 and G_2. Actually, $x_4 - x_3$ (multiplied by any constant) is a GCD of G_1 and G_2 in $\mathbf{Q}[x_1, \ldots, x_4]$, where $\mathbf{Q}$ denotes the field of rational numbers.

Let F be a polynomial in $\boldsymbol{R}[\boldsymbol{x}]$ and x_k a fixed variable. While considered as a polynomial in x_k, F can be written as

$$\begin{aligned} F &= F_0x_k^m + F_1x_k^{m-1} + \cdots + F_m, \\ F_i &\in \boldsymbol{R}[x_1, \ldots, x_{k-1}, x_{k+1}, \ldots, x_n], \end{aligned}$$

where $m = \deg(F, x_k)$. In this expression, F_{m-i} is called the *coefficient* of F in x_k^i and denoted by $\operatorname{coef}(F, x_k^i)$ for each i. In particular, F_0 is the *leading coefficient* of F in x_k, denoted by $\operatorname{lc}(F, x_k)$. Namely,

$$\operatorname{lc}(F, x_k) = \operatorname{coef}(F, x_k^{\deg(F, x_k)}).$$

The polynomial $F - F_0x_k^m$ is called the *reductum* of F with respect to x_k and denoted by $\operatorname{red}(F, x_k)$. When $x_k = \operatorname{lv}(F)$, it is omitted in $\operatorname{red}(F, x_k)$. Symbolically,

$$\begin{aligned} &\operatorname{lc}(F, x_k) \triangleq F_0, \\ &\operatorname{red}(F, x_k) \triangleq F_1x_k^{m-1} + \cdots + F_m, \\ &\operatorname{red}(F) \triangleq \operatorname{red}(F, \operatorname{lv}(F)). \end{aligned}$$

Any GCD of $F_0, \ldots, F_m$ as polynomials in $\boldsymbol{R}[x_1, \ldots, x_{k-1}, x_{k+1}, \ldots, x_n]$ is called the *content* of F with respect to x_k, denoted by $\operatorname{cont}(F, x_k)$. If $\operatorname{cont}(F, x_k)$ is a unit of $\boldsymbol{R}$, then F is said to be *primitive* with respect to x_k. For any nonzero polynomial F, $F/\operatorname{cont}(F, x_k)$ is called the *primitive part* of F with respect to x_k, denoted by $\operatorname{pp}(F, x_k)$; therefore, F may be written as

$$F = \operatorname{cont}(F, x_k) \cdot \operatorname{pp}(F, x_k).$$

Lemma 1.2.1 (Gauss' lemma). The product of primitive polynomials over a UFD is primitive.

Let $F \neq 0$, $m = \deg(F, x_k)$ as above and G be any other polynomial of degree l in x_k. For pseudo-dividing G by F, considered as polynomials in x_k, we have a division algorithm as follows. Let $R = G$; repeat the following process until $r = \deg(R, x_k) < m$:

$$R \leftarrow F_0 R - R_0 x_k^{r-m} F,$$

where $R_0 = \operatorname{lc}(R, x_k)$. As r strictly decreases for each iteration, the procedure must terminate. Finally, one obtains two polynomials Q and R in $\boldsymbol{R}[\boldsymbol{x}]$ satisfying the relation

$$I^q G = QF + R, \qquad (1.2.1)$$

where

$$I = \operatorname{lc}(F, x_k), \quad q = \max(l - m + 1, 0),$$
$$\deg(R, x_k) < m, \quad \deg(Q, x_k) = \max(l - m, -1).$$

In case $m = 0$, $R = 0$ and $Q = F^l G$.

The expression (1.2.1) is called a *pseudo-remainder formula*; Q is called the *pseudo-quotient* and R the *pseudo-remainder* of G with respect to F in x_k, denoted by $\operatorname{pquo}(G, F, x_k)$ and $\operatorname{prem}(G, F, x_k)$, respectively. Actually, the polynomials Q and R in (1.2.1) are uniquely determined by F and G. This fact is stated as follows for later use.

Proposition 1.2.2. Let the polynomials F, G, I, Q, R and integer q be as above. If Q' and R' are two polynomials in $\boldsymbol{R}[\boldsymbol{x}]$ such that $I^q G = Q'F + R'$, then $Q' = Q$ and $R' = R$.

Proof. Knuth (1981, pp. 402 and 407). □

The process of acquiring Q and R in pseudo-dividing G by F is called a *pseudo-reduction* (with respect to x_k). It is a fundamental operation underlying many of the algorithms described in this book and thus will play a key role in the following chapters. For this reason, let us describe the computational process of a pseudo-remainder in the form of the following algorithm.

Algorithm prem: $R \leftarrow \operatorname{prem}(G, F, x)$. Given two polynomials $G, F \in \boldsymbol{R}[\boldsymbol{x}]$ and a variable $x \in \{\boldsymbol{x}\}$, this algorithm computes a pseudo-remainder R of G with respect to F in x.

P1. Set $R \leftarrow G, r \leftarrow \deg(R, x), H \leftarrow F, h \leftarrow \deg(H, x), d \leftarrow r - h + 1$.
P2. If $h \leq r$, then set $L \leftarrow \mathrm{lc}(H, x), H \leftarrow \mathrm{red}(H, x)$; else set $L \leftarrow 1$.
P3. While $h \leq r$ and $R \neq 0$, do:
  P3.1. Compute $T \leftarrow x^{r-h}\mathrm{lc}(R, x)H$.
  P3.2. If $r = 0$, then set $R \leftarrow 0$; else set $R \leftarrow \mathrm{red}(R, x)$.
  P3.3. Compute $R \leftarrow LR - T$ and set $r \leftarrow \deg(R, x), d \leftarrow d - 1$.
P4. Return $R \leftarrow L^d R$.

When $x_k = \mathrm{lv}(F)$, it is omitted in $\mathrm{prem}(G, F, x_k)$. For a polynomial set $\mathbb{Q}$, $\mathrm{prem}(\mathbb{Q}, F)$ stands for $\{\mathrm{prem}(Q, F)\colon\ Q \in \mathbb{Q}\}$. The following simple example illustrates the division process. More complicated calculations will be given in the next example.

Example 1.2.2. Let $F = xy^2 + 1$ and $G = 2y^3 - y^2 + x^2y$. With respect to y, the corresponding R and Q can be calculated as follows

$$\begin{array}{rll}
 & 2xy - x & = Q \\
xy^2 + 1 & \sqrt{\ 2y^3 - y^2 + x^2y} & G \\
 & 2xy^3 - xy^2 + x^3y & xG \\
 & \underline{-(2xy^3 + 2y)\qquad\quad} & -2yF \\
 & -xy^2 + x^3y - 2y & \bar{R} \\
 & -x^2y^2 + x^4y - 2xy & x\bar{R} \\
 & \underline{-(-x^2y^2 - x)\qquad\quad} & xF \\
 & x^4y - 2xy + x & = R.
\end{array}$$

This implies that

$$x^2G = (2xy - x)F + x^4y - 2xy + x. \tag{1.2.2}$$

The integer q in (1.2.1) may be determined as small as possible, provided that the division process does not introduce fractions into Q and R. For example, the multiplier L^d in step P4 of algorithm prem may be omitted (for some applications). One can take $q = 1$ instead of 2 in (1.2.2) so that it simplifies to

$$xG = (2y - 1)F + x^3y - 2y + 1.$$

Taking the smallest q is rather crucial for control of the size expansion of the pseudo-remainder in practical computation. Moreover, one can modify formula (1.2.1) by replacing I^q with $I_1^{q_1} \cdots I_e^{q_e}$, where $I_1, \ldots, I_e$ are all the distinct irreducible factors of I (see Sect. 1.4 for the definition of irreducibility), and choosing the smallest $q_1, \ldots, q_e$ so that the corresponding pseudo-remainder formula still holds. For this modification the determination of R requires additional computation and thus takes more time at every individual step. However, the modified division may avoid some redundant factors so that the subsequent computation profits.

Example 1.2.3. Refer to the polynomials F_1, F_2, G_1, G_2 given in Examples 1.1.1, 1.1.2, and 1.2.1. Pseudo-dividing F_1 by F_2 in x_4, we get the following pseudo-remainder formula

$$x_1^2 F_1 = QF_2 + R,$$

where

$$\begin{aligned} Q &= x_1^2 x_4 + x_1 x_4 - x_1 x_3 - x_3, \\ R &= \mathrm{prem}(F_1, F_2) = x_1 x_3^2 + x_3^2 - x_1^2 x_2 x_3 - x_1 x_2 x_3 + x_1^3 x_2 + 3x_1^2 x_2. \end{aligned}$$

One can also verify that

$$\begin{aligned} G_3 &= \mathrm{prem}(G_1, G_2, x_4) \\ &= -45x_1x_2x_4 - 18x_2x_4 + 36x_1x_4 + 45x_1x_2x_3 + 18x_2x_3 - 36x_1x_3, \\ G_3' &= \mathrm{prem}(F_1, G_2, x_4) \\ &= 6x_1x_3x_4 + 6x_3x_4 - 15x_1^2x_2x_4 - 21x_1x_2x_4 - 6x_2x_4 + 15x_1^2x_2x_3 \\ &\quad + 15x_1x_2x_3 + 6x_1x_2 + 18x_2, \end{aligned}$$

and

$$\begin{aligned} &\mathrm{cont}(F_1, x_4) = 1, \\ &\mathrm{cont}(G_1, x_4) = \mathrm{cont}(G_2, x_4) = \mathrm{cont}(G_3', x_4) = 3, \\ &\mathrm{cont}(G_3, x_4) = 45x_1x_2 + 18x_2 - 36x_1, \\ &\mathrm{pp}(G_3, x_4) = x_3 - x_4. \end{aligned}$$

Two polynomials $F, G \in \boldsymbol{R}[x]$ are said to be *similar*, denoted as $F \backsim G$, if there exist $a, b \in \boldsymbol{R}$, $ab \neq 0$, such that $aF = bG$.

Let the polynomials G and F be renamed P_1 and P_2, and assume that $\deg(P_1, x_k) \geq \deg(P_2, x_k)$. We form a sequence of polynomials $P_1, P_2, P_3, \ldots, P_r$ such that

$$P_i \backsim \mathrm{prem}(P_{i-2}, P_{i-1}, x_k), \quad i = 3, \ldots, r$$

and

$$\mathrm{prem}(P_{r-1}, P_r, x_k) = 0.$$

Such a sequence is called a *polynomial remainder sequence* (PRS) of G and F with respect to x_k.

From the pseudo-remainder formula and the formation of PRS one may see that $\gcd(P_1, P_2), \gcd(P_2, P_3), \ldots, \gcd(P_{r-1}, P_r)$, and P_r differ from each other only by factors of polynomials in $\boldsymbol{R}[x_1, \ldots, x_{k-1}, x_{k+1}, \ldots, x_n]$. If P_1 and P_2 are both primitive with respect to x_k, then

$$\gcd(G, F) = \gcd(P_1, P_2) = \mathrm{pp}(P_r, x_k).$$

It is easy to see, on the other hand, that

$$\gcd(G, F) = \gcd(\mathrm{cont}(G, x_k), \mathrm{cont}(F, x_k)) \gcd(\mathrm{pp}(G, x_k), \mathrm{pp}(F, x_k))$$

for any polynomials G and F. It follows that the formation of PRS provides a means for determining the GCD of two polynomials; while the determination of GCDs of more polynomials can be easily reduced to the case of two polynomials.

Example 1.2.4. Consider the polynomials in Example 1.2.1. Calculations using algorithm prem show that

$$\begin{aligned}
&\mathrm{prem}(G_2, G_3, x_4) = 0,\\
G_4' &= \mathrm{prem}(G_2, G_3', x_4)\\
&= 2430x_1^2x_2^2x_3^2 + 3240x_1^3x_2^2x_3^2 - 2430x_1^2x_2^3x_3 + 864x_1x_2x_3^2 - 540x_1x_2^3x_3\\
&\quad + 216x_1^2x_2x_3^2 + 1350x_1^4x_2^2x_3^2 - 216x_1^2x_2^2x_3 - 3240x_1^3x_2^3x_3\\
&\quad - 1350x_1^4x_2^3x_3 + 540x_1x_2^2x_3^2 - 864x_1x_2^2x_3 + 1296x_1x_2^2 + 216x_1^2x_2^2\\
&\quad + 6210x_1^2x_2^3 + 5940x_1^3x_2^3 + 1350x_1^4x_2^3 + 1620x_1x_2^3 - 648x_2^2x_3\\
&\quad + 648x_2x_3^2 + 1944x_2^2,\\
&\mathrm{prem}(G_3', G_4', x_4) = 0.
\end{aligned}$$

Thus, G_1, G_2, G_3 and F_1, G_2, G_3', G_4' are both PRS. It follows that

$$\begin{aligned}
\gcd(G_1, G_2) &= \mathrm{pp}(G_3, x_4) = x_3 - x_4,\\
\gcd(F_1, G_2) &= \mathrm{pp}(G_4', x_4) = 1.
\end{aligned}$$

Definition 1.2.2. A sequence of nonzero polynomials $P_1, P_2, \ldots, P_r$ in $\boldsymbol{R}[x]$ with

$$r \geq 2,\ \ d_i = \deg(P_i, x),\ \ d_1 \geq d_2,\ \ I_i = \mathrm{lc}(P_i, x)$$

is called the *subresultant polynomial remainder sequence* (subresultant PRS) of P_1 and P_2 with respect to x if

$$\begin{aligned}
P_{i+2} &= \mathrm{prem}(P_i, P_{i+1}, x)/Q_{i+2}, \quad 1 \leq i \leq r-2,\\
&\mathrm{prem}(P_{r-1}, P_r, x) = 0,
\end{aligned}$$

where

$$\begin{aligned}
Q_3 &= (-1)^{d_1-d_2+1}, \quad H_3 = -1;\\
Q_i &= -I_{i-2}H_i^{d_{i-2}-d_{i-1}},\\
H_i &= (-I_{i-2})^{d_{i-3}-d_{i-2}}H_{i-1}^{1-d_{i-3}+d_{i-2}}, \quad i = 4, \ldots, r.
\end{aligned}$$

In the following section we shall present several known results about *subresultants*. They ensure that subresultant PRS above is well-defined, i.e., $P_i \in \boldsymbol{R}[x]$ for all $i \geq 3$ so long as $P_1, P_2 \in \boldsymbol{R}[x]$.

1.3 Resultants and subresultants

The *resultant* of two univariate polynomials $F, G \in \boldsymbol{R}[x]$ is a form in the coefficients of F and G whose vanishing provides certain conditions for these two polynomials to have common zeros for x. A common *zero* $\bar{x}$ of F and G means a number in some extension of the quotient field of $\boldsymbol{R}$ such that $F(\bar{x}) = G(\bar{x}) = 0$. It will be defined formally in Sect. 1.5. An ideal reference for this section is Mishra (1993, chap. 7).

Let F and G be of respective degrees m and l in x with $m \geq l > 0$, written as

$$\begin{aligned} F &= a_0x^m + a_1x^{m-1} + \cdots + a_{m-1}x + a_m, \\ G &= b_0x^l + b_1x^{l-1} + \cdots + b_{l-1}x + b_l. \end{aligned} \tag{1.3.1}$$

We form a matrix of dimension $m + l$ by $m + l$, called the *Sylvester matrix* of F and G with respect to x, as follows

$$\mathbf{M} = \begin{pmatrix} a_0 & a_1 & \dots & a_m & & & \\ & a_0 & a_1 & \dots & a_m & & \\ & & \dots & \dots & \dots & \dots & \\ & & & a_0 & a_1 & \dots & a_m \\ b_0 & b_1 & \dots & b_l & & & \\ & b_0 & b_1 & \dots & b_l & & \\ & & \dots & \dots & \dots & \dots & \\ & & & b_0 & b_1 & \dots & b_l \end{pmatrix},$$

where the blank spaces are filled with 0 as usual.

Definition 1.3.1. The determinant of the Sylvester matrix $\mathbf{M}$ is called the *Sylvester resultant* or *eliminant* of F and G with respect to x, denoted $\operatorname{res}(F, G, x)$.

As usual we use $\det(\square)$ to denote the *determinant* of any square matrix $\square$. The resultant $\operatorname{res}(F, G, x) = \det(\mathbf{M})$ is homogeneous of degree l in the a_i and of degree m in the b_i.

Example 1.3.1. Consider the cubic polynomial $F = ax^3 + bx^2 + cx + d$ in x. The resultant R of F and its derivative

$$\mathrm{d}F/\mathrm{d}x = 3ax^2 + 2bx + c$$

is also called the *discriminant* of F. When $a \neq 0$, a necessary and sufficient condition for F to have multiple zeros is $R = 0$.

The 5×5 Sylvester matrix $\mathbf{M}$ of F and $\mathrm{d}F/\mathrm{d}x$ with respect to x is shown below

$$\mathbf{M} = \begin{pmatrix} a & b & c & d & 0 \\ 0 & a & b & c & d \\ 3a & 2b & c & 0 & 0 \\ 0 & 3a & 2b & c & 0 \\ 0 & 0 & 3a & 2b & c \end{pmatrix}.$$

Thus, the resultant of F and $\mathrm{d}F/\mathrm{d}x$ with respect to x is

$$\operatorname{res}(F, \mathrm{d}F/\mathrm{d}x, x) = \det(\mathbf{M}) = a(27a^2d^2 - 18abcd + 4b^3d + 4ac^3 - b^2c^2).$$

Lemma 1.3.1. Let F and G be as in (1.3.1). Then there exist polynomials $A, B \in \boldsymbol{R}[x]$ such that $AF + BG = \operatorname{res}(F, G, x)$, where $\deg(A, x) < \deg(G, x)$ and $\deg(B, x) < \deg(F, x)$.

A proof of this lemma can be found, for example, in van der Waerden (1953, p. 85) or Mishra (1993, pp. 228f). As a consequence of the above lemma and definition, we have the sufficiency in the following theorem.

Theorem 1.3.2. Let F and G be as in (1.3.1). Then $\operatorname{res}(F, G, x) = 0$ if and only if either F and G have a common zero for x or $a_0 = b_0 = 0$.

The necessity can be proved without much difficulty (see, e.g., van der Waerden 1953, pp. 83f). Therefore, if one of a_0 and b_0 is nonzero, $\operatorname{res}(F, G, x) = 0$ is a necessary and sufficient condition for F and G to have a common zero.

Now let $\mathbf{M}_{ij}$ be the submatrix of $\mathbf{M}$ obtained by deleting the last j of the l rows of F coefficients, the last j of the m rows of G coefficients, and the last $2j + 1$ columns, excepting column $m + l - i - j$, for $0 \leq i \leq j < l$.

Definition 1.3.2. The polynomial

$$S_j(x) = \sum_{i=0}^{j} \det(\mathbf{M}_{ij})x^i$$

is called the jth *subresultant* of F and G with respect to x, for $0 \leq j < l$. Here $\deg(S_j, x) \leq j$, and $R_j = \det(\mathbf{M}_{jj})$ is called the jth *principal subresultant coefficient* (PSC) or the jth *resultant* of F and G with respect to x.

If $m > l + 1$, the definition of the jth subresultant $S_j(x)$ and PSC R_j of F and G with respect to x is extended as follows:

$$S_l(x) = b_0^{m-l-1}G, \quad R_l = b_0^{m-l}; \quad S_j(x) = R_j = 0, \; l < j < m - 1.$$

S_j is said to be *defective* of degree r if $\deg(S_j, x) = r < j$, and *regular* otherwise.

It is easy to see that $S_0 = R_0$ is the *resultant* of F and G with respect to x.

Theorem 1.3.3. Let F and G be two polynomials in $\boldsymbol{R}[x]$ with $m = \deg(F, x) \geq \deg(G, x) = l > 0$ and S_j be the jth subresultant of F and G with respect to x, for $0 \leq j < m - 1$. Then there exist polynomials $A_j, B_j \in \boldsymbol{R}[x]$ such that $A_jF + B_jG = S_j$, where $\deg(A_j, x) < l - j$ and $\deg(B_j, x) < m - j$.

Proof. Mishra (1993, pp. 255f).

Definition 1.3.3. Let F and G be two polynomials in $\boldsymbol{R}[x]$ with $m = \deg(F, x) \geq \deg(G, x) = l > 0$ and set

$$\mu = \begin{cases} m - 1 & \text{if } m > l, \\ l & \text{otherwise.} \end{cases}$$

Let $S_{\mu+1} = F$, $S_\mu = G$, and S_j be the jth subresultant of F and G with respect to x for $0 \leq j < \mu$. The sequence of polynomials $S_{\mu+1}, S_\mu, S_{\mu-1}, \ldots, S_0$ in $\boldsymbol{R}[x]$ is called the *subresultant chain* of F and G with respect to x. It is said to be *regular* if all S_j are regular, and *defective* otherwise.

Let

$$R_{\mu+1} = 1 \text{ and } R_j = \begin{cases} \mathrm{lc}(S_j, x) & \text{if } S_j \text{ is regular,} \\ 0 & \text{otherwise; for } 0 \leq j \leq \mu. \end{cases}$$

The sequence of polynomials $R_{\mu+1}, R_\mu, \ldots, R_0$ is called the *PSC chain* of F and G with respect to x.

The PSC chain defined here is consistent with the PSCs in Definition 1.3.2. In fact, for $1 \leq j < \mu$, R_j above is the jth PSC, which vanishes when S_j is defective.

Theorem 1.3.4 (Subresultant chain). Let $S_{\mu+1}$ and S_μ be two polynomials in $\boldsymbol{R}[x]$ with $\deg(S_{\mu+1}, x) \geq \deg(S_\mu, x) > 0$ and let $S_{\mu+1}, S_\mu, \ldots, S_0$ be the subresultant chain of $S_{\mu+1}$ and S_μ with respect to x, with PSC chain $R_{\mu+1}, R_\mu, \ldots, R_0$. If both S_{j+1} and S_j are regular, then

$$R_{j+1}^2 S_{j-1} = \mathrm{prem}(S_{j+1}, S_j, x), \quad 1 \leq j \leq \mu.$$

If S_{j+1} is regular and S_j is defective of degree $r < j$, then

$$\begin{aligned} S_{j-1} = S_{j-2} = \cdots = S_{r+1} = 0, &\quad -1 \leq r < j < \mu; \\ R_{j+1}^{j-r} S_r = \mathrm{lc}(S_j, x)^{j-r} S_j, &\quad 0 \leq r \leq j < \mu; \\ (-1)^{j-r} R_{j+1}^{j-r+2} S_{r-1} = \mathrm{prem}(S_{j+1}, S_j, x), &\quad 0 < r \leq j < \mu. \end{aligned}$$

Proof. Loos (1983, pp. 122f) or Mishra (1993, pp. 268 and 274–283). □

Theorem 1.3.4 provides an effective algorithm for constructing subresultant chains by means of pseudo-division. However, in the case $\deg(S_{\mu+1}, x) = \deg(S_\mu, x)$, $S_{\mu+1}$ is defective and thus how to obtain $S_{\mu-1}$ is not covered by the theorem. To deal with this special case, we need the following result, which will also be used later.

Proposition 1.3.5. Let ϕ denote a ring homomorphism of $\boldsymbol{R}$ into another UFD $\tilde{\boldsymbol{R}}$ as well as its induced ring homomorphism of $\boldsymbol{R}[x]$ into $\tilde{\boldsymbol{R}}[x]$, F, G, m, l be as in (1.3.1), and

$$\begin{aligned} &\tilde{a}_0 = \phi(a_0), \ \tilde{b}_0 = \phi(b_0), \\ &\tilde{m} = \deg(\phi(F), x), \ \tilde{l} = \deg(\phi(G), x). \end{aligned}$$

Then with respect to x the jth subresultant $\tilde{S}_j$ of $\phi(F)$ and $\phi(G)$ is equal to the jth subresultant S_j of F and G multiplied by δ, i.e., $\tilde{S}_j = \delta S_j$, for $0 \leq j < \max(\tilde{m}, \tilde{l}) - 1$, where

$$\delta = \begin{cases} 1 & \text{if } \tilde{a}_0 \tilde{b}_0 \neq 0, \\ \tilde{a}_0^{l-\tilde{l}} & \text{if } \tilde{a}_0 \neq 0 \text{ and } \tilde{b}_0 = 0, \\ \tilde{b}_0^{m-\tilde{m}} & \text{if } \tilde{a}_0 = 0 \text{ and } \tilde{b}_0 \neq 0, \\ 0 & \text{if } \tilde{a}_0 = \tilde{b}_0 = 0. \end{cases}$$

Proof. Corollary 7.8.2 in Mishra (1993, pp. 264 f). □

We turn back to the subresultant chain as before and consider $S_{\mu+1}$ as obtained from a generic polynomial S of degree $\mu+1$ in x with indeterminate coefficients by specializing $\mathrm{lc}(S,x)$ to 0 and $\mathrm{coef}(S,x^i)$ to $\mathrm{coef}(S_{\mu+1},x^i)$ for $i=\mu,\ldots,0$. According to Proposition 1.3.5, $S_{\mu-1}$ is identical to the $(\mu-1)$th subresultant of S and S_μ with respect to x multiplied by $\mathrm{lc}(S_\mu,x)$. It follows that

$$S_{\mu-1}=\mathrm{lc}(S_\mu,x)\mathrm{prem}(S_{\mu+1},S_\mu,x).$$

From Theorem 1.3.4 and the above discussions, we derive the following algorithm for computing subresultant chains.

Algorithm SubresChain: $\mathfrak{S} \leftarrow$ SubresChain(F,G). Given two polynomials $F,G\in \boldsymbol{R}[x]$ with $\deg(F,x)\geq\deg(G,x)>0$, this algorithm computes the subresultant chain $\mathfrak{S}$ of F and G with respect to x.

S1. Set $m\leftarrow\deg(F,x), l\leftarrow\deg(G,x)$. If $l<m$, then set $j\leftarrow m-1$; else set $j\leftarrow l$. Set $S_{j+1}\leftarrow F$, $S_j\leftarrow G$, $R_{j+1}\leftarrow 1$, $\mu\leftarrow j$.

S2. If $S_j=0$, then set $r\leftarrow -1$; else set $r\leftarrow\deg(S_j,x)$. Set $S_k\leftarrow 0$ for $k=j-1,j-2,\ldots,r+1$.

S3. If $0\leq r<j$, then compute $S_r\leftarrow \mathrm{lc}(S_j,x)^{j-r}S_j/R_{j+1}^{j-r}$.
If $r\leq 0$, then return $\mathfrak{S}\leftarrow[S_{\mu+1},S_\mu,\ldots,S_0]$ and the algorithm terminates.

S4. If $r=m=l$, then set $I\leftarrow\mathrm{lc}(G,x)$; else set $I\leftarrow 1$. Compute

$$S_{r-1}\leftarrow I\,\mathrm{prem}(S_{j+1},S_j,x)/(-R_{j+1})^{j-r+2}.$$

Set $j\leftarrow r-1$, $R_{j+1}\leftarrow\mathrm{lc}(S_{j+1},x)$ and go back to S2.

Example 1.3.2. Let

$$F=-x^4-z^3x^2+x^2-z^4+2z^2-1,$$
$$G=x^4+z^2x^2-r^2x^2+z^4-2z^2+1.$$

Application of SubresChain yields the following subresultant chain of F and G with respect to x:

$$F,\ G,\ -Hx^2,\ H^2x^2,\ (z^4-2z^2+1)H^3,\ (z^4-2z^2+1)^2H^4,$$

where $H=z^3-z^2+r^2-1$. Now, $\mu=4$; S_4, S_2, and S_0 are regular and S_5, S_3, and S_1 are defective of degrees 4, 2, and 0, respectively.

Definition 1.3.4. Let $S_{\mu+1}$ and S_μ be two polynomials in $\boldsymbol{R}[x]$ with $\deg(S_{\mu+1},x)\geq\deg(S_\mu,x)>0$ and let $\mathfrak{S}$: $S_{\mu+1},S_\mu,\ldots,S_0$ be the subresultant chain of $S_{\mu+1}$ and S_μ with respect to x. A finite sequence $d_1,d_2,\ldots,d_r$ of steadily decreasing nonnegative integers is called the *block indices* of $\mathfrak{S}$ if $d_1=\mu+1$, each S_{d_i} is regular for $2\leq i\leq r$, and for any $0\leq j\leq\mu$ and $j\notin\{d_2,\ldots,d_r\}$ S_j is defective.

The sequence of regular subresultants $S_{d_2},\ldots,S_{d_r}$ is called the *subresultant regular subchain* (SRS) of $S_{\mu+1}$ and S_μ with respect to x.

The subresultant chain $\mathfrak{S}$ possesses interesting block structures characterized by its block indices $d_1,\ldots,d_r$. The first block consists of the single term $S_{\mu+1}$. For any $2\leq i\leq r$, we have

$$S_{d_i}\neq 0,\quad S_{d_i}\backsim S_{d_{i-1}-1}\ \text{ and }\ S_{d_{i-1}-2}=\cdots=S_{d_i+1}=0.$$

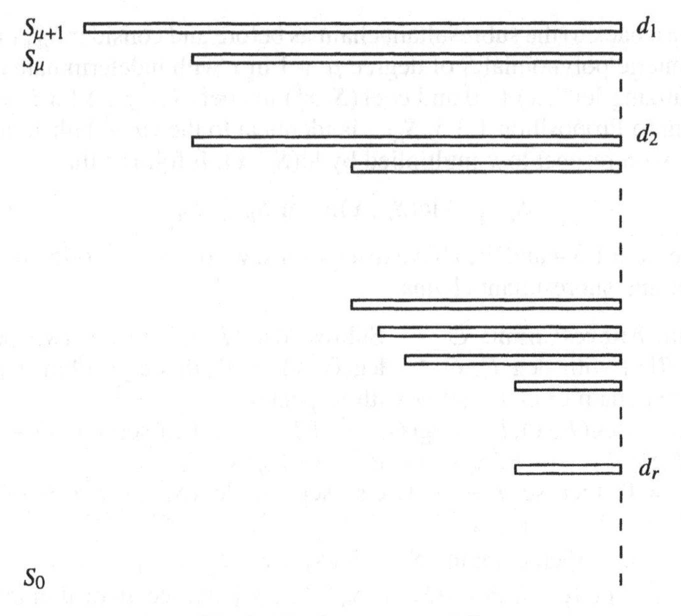

Fig. 1. Block structure of $\mathfrak{S}$

Namely, the ith *nonzero block* of $\mathfrak{S}$ can be put in the form

$$S_{d_{i-1}-1}, 0, \ldots, 0, S_{d_i},$$

where $S_{d_{i-1}-1} \backsim S_{d_i}$ and $d_{i-1} - 1 \geq d_i$. If $d_r > 0$, then

$$S_{d_r-1} = \cdots = S_0 = 0;$$

this is the last block, called the *zero block*, of $\mathfrak{S}$. The block structure of $\mathfrak{S}$ is illustrated in Fig. 1.

The following theorem establishes the relationship between subresultant PRS and subresultant chains and shows that subresultant PRS is well-defined (see Definition 1.2.2).

Theorem 1.3.6. Let $S_{\mu+1}, S_\mu, \ldots, S_0$ and $d_1, d_2, \ldots, d_r$ be as in Definition 1.3.4. Then the sequence of polynomials $S_{d_1}, S_{d_1-1}, S_{d_2-1}, \ldots, S_{d_{r-1}-1}$ is the subresultant PRS of $S_{\mu+1}$ and S_μ with respect to x.

Proof. Collins (1967) or Mishra (1993, pp. 272f). □

It is easy to see that $S_{\mu+1}, S_\mu, S_{d_3}, \ldots, S_{d_r}$ is also a PRS of $S_{\mu+1}$ and S_μ with respect to x. Thus, the SubresChain algorithm may be modified to compute PRS, subresultant PRS and resultants of polynomials.

Example 1.3.3. As a more complicated example, consider

$$\begin{aligned}P_1 = {} & 729y^6 - 1458x^3y^4 + 729x^2y^4 - 4158xy^4 - 1685y^4 + 729x^6y^2 \\ & - 1458x^5y^2 - 2619x^4y^2 - 4892x^3y^2 - 297x^2y^2 + 5814xy^2 \\ & + 427y^2 + 729x^8 + 216x^7 - 2900x^6 - 2376x^5 + 3870x^4 + 4072x^3 \\ & - 1188x^2 - 1656x + 529,\end{aligned}$$

$$P_2 = 2187y^4 - 4374x^3y^2 - 972x^2y^2 - 12474xy^2 - 2868y^2 + 2187x^6 - 1944x^5 - 10125x^4 - 4800x^3 + 2501x^2 + 4968x - 1587.$$

The subresultant chain $\mathfrak{S}$ of P_1 and P_2 with respect to y is

$$\begin{aligned}
S_6 &= P_1,\\
S_5 &= P_2,\\
S_4 &= 2187P_2,\\
S_3 &= 1549681956x^2(-8748x^3y^2 - 8262x^2y^2 - 8478xy^2 + 498y^2 + 2187x^6\\
&\quad - 7776x^5 - 18252x^4 + 4812x^3 + 4787x^2 - 540x - 2766),\\
S_2 &= -1944x^2F_1F_2S_3,\\
S_1 &= 12050326889856x^6F_1F_2F_3^2F_4^2,\\
S_0 &= 8033551259904x^8F_3^4F_4^4,
\end{aligned}$$

where

$$\begin{aligned}
F_1 &= 18x - 1,\\
F_2 &= 81x^2 + 81x + 83,\\
F_3 &= 81x^2 + 18x + 28,\\
F_4 &= 729x^4 + 972x^3 - 1026x^2 + 1684x + 765.
\end{aligned}$$

Hence, the block indices of $\mathfrak{S}$ are 6, 4, 2, 0, and S_6, S_5, S_3, S_1 is a subresultant PRS of P_1 and P_2 with respect to y. The polynomials above are written in factorized form for brevity and readability.

If, for instance, x is specialized to $1/18$, then F_1 becomes 0. Let

$$\bar{S}_j = S_j|_{x=1/18}, \quad j = 6, \dots, 0.$$

Then, $\bar{S}_1 = \bar{S}_2 = 0$ and $\bar{S}_0, \bar{S}_3$ are both constants. Thus the block indices of the specialized subresultant chain are 6, 4, 0. An application of Proposition 1.3.5 ensures that the jth subresultant of $\bar{S}_6$ and $\bar{S}_5$ with respect to y is identical to $\bar{S}_j$ for each j. Hence $\bar{S}_6, \bar{S}_5, \bar{S}_3$ is a subresultant PRS of $\bar{S}_6$ and $\bar{S}_5$ with respect to y.

Resultant-based elimination theory is one of the classical in constructive algebra and has wide applications in modern computer algebra and geometry. The idea and its development owe to L. Euler, É. Bézout, A. L. Dixon, A. Cayley, and J. J. Sylvester, among others. Two easy references are van der Waerden (1950, 1953) and Mishra (1993, chap. 7). In Sect. 5.4, we shall explain another formulation of univariate resultants and introduce multivariate resultants as well as various related elimination techniques.

The often-mentioned modern references to the concept, theory, and algorithms of subresultants include Collins (1967, 1971), Brown and Traub (1971), Knuth (1981), Loos (1983) and the early approach of W. Habicht. Here we want to point out the earlier work by Thomas (1937, 1946), in which the concept was also introduced.

1.4 Field extension and factorization

Let $\boldsymbol{R}$ be a UFD. A polynomial $F \in \boldsymbol{R}[\boldsymbol{x}]$ is said to be *irreducible* over $\tilde{\boldsymbol{R}} \supset \boldsymbol{R}$ if it cannot be written as the product of two nonconstant polynomials in $\tilde{\boldsymbol{R}}[\boldsymbol{x}]$. Otherwise, F is said to be *reducible* over $\tilde{\boldsymbol{R}}$. Over $\boldsymbol{R}$, any polynomial can be factorized as the product of irreducible polynomials uniquely up to a constant factor.

Now let $\boldsymbol{K}$ be the quotient field of $\boldsymbol{R}$. One simple, concrete example of $\boldsymbol{R}$ is the ring $\mathbf{Z}$ of integers, where $\boldsymbol{K}$ becomes the rational-number field $\mathbf{Q}$. According to a lemma of Gauss (see van der Waerden 1953, p. 73), if a polynomial in $\boldsymbol{R}[\boldsymbol{x}]$ factors over $\boldsymbol{K}$, so does it over $\boldsymbol{R}$. It is therefore appropriate to deal with factorization over $\boldsymbol{K}$ instead of $\boldsymbol{R}$. A very fundamental problem is to factorize a given polynomial in $\boldsymbol{K}[\boldsymbol{x}]$ as the product of irreducible polynomials in $\boldsymbol{K}[\boldsymbol{x}]$. This conceptually simple problem is by no means trivial as far as practical computation is of concern. Nevertheless, powerful algorithms have been well developed (see, e.g., Knuth 1981, pp. 420–441) and implemented in popular computer algebra systems. We shall feel free to use such algorithms and software systems when polynomial factorization over $\boldsymbol{K}$ is necessary.

In Chap. 4 factorization of polynomials in $\boldsymbol{K}[\boldsymbol{x}]$ over *algebraic-extension fields* of $\boldsymbol{K}$ is also needed. Let us explain this precisely as follows.

Let θ be an element in some extension field $\tilde{\boldsymbol{K}}$ of $\boldsymbol{K}$, but not in $\boldsymbol{K}$. Denote by $\boldsymbol{K}(\theta)$ the set of all rational functions $F(\theta)/G(\theta)$, where F and G are both polynomials in θ with coefficients in $\boldsymbol{K}$ and $G(\theta)$ is nonzero in $\tilde{\boldsymbol{K}}$. Then under the operations of $\tilde{\boldsymbol{K}}$, $\boldsymbol{K}(\theta)$ constitutes a field containing $\boldsymbol{K}$, called a *simple-extension field* obtained from $\boldsymbol{K}$ by adjoining θ. If, for any univariate polynomial $A \in \boldsymbol{K}[y]$, $A(\theta) \neq 0$, then θ is a *transcendental* number over $\boldsymbol{K}$ and $\boldsymbol{K}(\theta)$ is called a *transcendental-extension field* obtained from $\boldsymbol{K}$ by adjoining θ. In this case, $\boldsymbol{K}(\theta)$ is also called a *rational-function field* of $\boldsymbol{K}$.

Next we turn to the case when there exist polynomials $A \in \boldsymbol{K}[y]$ such that $A(\theta) = 0$. Let A be one of such polynomials which have minimal degree m in y. Now, θ is an *algebraic* number over $\boldsymbol{K}$, $\boldsymbol{K}(\theta)$ is called an *algebraic-extension field* obtained from $\boldsymbol{K}$ by adjoining θ, and m is called the *degree* of θ or $\boldsymbol{K}(\theta)$ over $\boldsymbol{K}$. The polynomial A is obviously irreducible over $\boldsymbol{K}$. It is called an *adjoining polynomial* of θ.

Let $F(\theta)/G(\theta)$ be an arbitrary number in $\boldsymbol{K}(\theta)$. Since $G(\theta) \neq 0$ and $A \in \boldsymbol{K}[y]$ is irreducible over $\boldsymbol{K}$, G and A do not have any common zero. This implies that $\operatorname{res}(G, A, y) \in \boldsymbol{K}$ is nonzero. By Lemma 1.3.1, there are polynomials $K, L \in \boldsymbol{K}[y]$ such that

$$KG + LA = 1, \tag{1.4.1}$$

where $\deg(L, y) < \deg(G, y)$ and $\deg(K, y) < \deg(A, y) = m$. Dividing FK by A leads to the following remainder formula

$$FK = QA + R, \tag{1.4.2}$$

where $Q, R \in \boldsymbol{K}[y]$ and $\deg(K, y) < m$. From the expressions (1.4.1) and (1.4.2), one gets

$$\frac{F}{G} = R + \left(\frac{FL}{G} - Q\right) A.$$

As $A(\theta) = 0$, it follows that $F(\theta)/G(\theta) = R(\theta)$. Therefore, an arbitrary number in $\boldsymbol{K}(\theta)$ can be represented as a polynomial of θ whose degree is less than or equal to $m - 1$. The representation is unique and can be constructively determined via algebraic operations.

Note that θ is only a symbol and in general it cannot be given explicitly. What we are usually given is the irreducible polynomial A, by means of which θ is defined. In view of this, we shall denote $\boldsymbol{K}(\theta)$ simply by $\boldsymbol{K}(y)$ when the adjoining polynomial A is mentioned.

A field $\tilde{\boldsymbol{K}} \supset \boldsymbol{K}$ is said to be *algebraically closed* if for every nonconstant polynomial $P \in \tilde{\boldsymbol{K}}[x]$ there exists an $\bar{x} \in \tilde{\boldsymbol{K}}$ such that $P(\bar{x}) = 0$. Any algebraically closed algebraic-extension field of $\boldsymbol{K}$ is called an *algebraic closure* of $\boldsymbol{K}$. For example, the field $\mathbf{C}$ of complex numbers is an algebraic closure of $\mathbf{Q}$.

Now consider a sequence of r (>1) polynomials

$$A_1(y_1), A_2(y_1, y_2), \ldots, A_r(y_1, \ldots, y_r),$$

in which $A_i \in \boldsymbol{K}[y_1, \ldots, y_i]$ and $\deg(A_i, y_i) \geq 1$ for each i. Such a sequence satisfies the property that each A_i, considered as a polynomial in y_i, is irreducible over the algebraic-extension field

$$\boldsymbol{K}_{i-1} = \boldsymbol{K}(y_1) \ldots (y_{i-1}) = \boldsymbol{K}(y_1, \ldots, y_{i-1})$$

with $A_1, \ldots, A_{i-1}$ as adjoining polynomials, respectively. Therefore, we have a sequence of algebraic-extension fields $\boldsymbol{K}_1, \ldots, \boldsymbol{K}_r$. For each i the ordered set

$$\mathbb{A}_i = [A_1, \ldots, A_i]$$

of adjoining polynomials will be called an *irreducible ascending set*, and $\boldsymbol{K}_i$ an *algebraic-extension field* of $\boldsymbol{K}$ with *adjoining ascending set* $\mathbb{A}_i$.

Let $\mathbb{A}_r$ and $\boldsymbol{K}_r$ be as before and a polynomial $F \in \boldsymbol{K}[y_1, \ldots, y_r, y]$, considered as $\bar{F} \in \boldsymbol{K}_r[y]$, be reducible over $\boldsymbol{K}_r$. Then an irreducible factorization of $\bar{F}$ is of the form

$$\bar{F} = \bar{F}_1 \cdots \bar{F}_t,$$

in which each $\bar{F}_i \in \boldsymbol{K}_r[y]$ is irreducible over $\boldsymbol{K}_r$, and $t \geq 2$. We shall see in Sect. 4.1 that there are polynomials $F_1, \ldots, F_t, Q_1, \ldots, Q_r \in \boldsymbol{K}[y_1, \ldots, y_r, y]$ and $D \in \boldsymbol{K}[y_1, \ldots, y_r]$ such that

$$I(DF - F_1 \cdots F_t) = \sum_{i=1}^{r} Q_i A_i,$$

where I is a power product of $\mathrm{lc}(A_i, y_i)$. Alternatively the factorization of F is written as $DF \doteq F_1 \cdots F_t$ over the extension field $\boldsymbol{K}_r$. The problem of *algebraic factorization* amounts to constructing the polynomials $F_1, \ldots, F_t$ from F and $\mathbb{A}_r$, for which several algorithms are available. Two of them will be explained in Sect. 7.5.

Example 1.4.1. Refer to the polynomials in Examples 1.1.1, 1.2.1, 1.2.3, and 1.2.4. Over $\mathbf{Q}$, F_1 and G_3' are both irreducible, and G_1, G_2, G_3, G_4' are all reducible and have the following factorizations

$$\begin{aligned}
G_1 &= 3(x_4 - x_3)(x_4 - x_2 + 2x_1),\\
G_2 &= 3(x_4 - x_3)(2x_4 + 5x_1x_2),\\
G_3 &= -9(x_4 - x_3)(5x_1x_2 + 2x_2 - 4x_1),\\
G_4' &= -54x_2(25x_1^3x_2 + 35x_1^2x_2 + 10x_1x_2 + 4x_1 + 12)\\
&\quad \cdot (-x_1x_3^2 - x_3^2 + x_1x_2x_3 + x_2x_3 - x_1x_2 - 3x_2).
\end{aligned}$$

Let

$$\begin{aligned}
A &= 2x_1^2x_2^2 + 2x_1x_2^2 - 2x_1^2x_2,\\
F &= x_1x_3^2 + x_3^2 - x_1^2x_2x_3 - x_1x_2x_3 + x_1^3x_2 + 3x_1^2x_2.
\end{aligned}$$

Both A and F are irreducible over $\mathbf{Q}$. Over the extension field $\mathbf{Q}(x_1, x_2)$, where x_1 is a transcendental element and x_2 an algebraic element with adjoining polynomial A, the polynomial F can be factorized as

$$F \doteq (x_1 + 1)(x_3 - 2x_1x_2 + x_1)(x_3 + x_1x_2 - x_1).$$

1.5 Zeros and ideals

Let $\boldsymbol{K}$ be an arbitrary field of characteristic 0 and $\boldsymbol{K}[\boldsymbol{x}]$ the ring of polynomials in the indeterminates $\boldsymbol{x} = (x_1, \ldots, x_n)$ with coefficients in $\boldsymbol{K}$. Let $\tilde{\boldsymbol{K}}$ be an arbitrary extension field of $\boldsymbol{K}$. Any n-tuple $\bar{\boldsymbol{x}} = (\bar{x}_1, \ldots, \bar{x}_n)$ of numbers in $\tilde{\boldsymbol{K}}$ is called a *point* of the *affine n-space* $\mathbf{A}^n$ over $\tilde{\boldsymbol{K}}$. Let $P \in \boldsymbol{K}[\boldsymbol{x}]$ be a polynomial. The point $\bar{\boldsymbol{x}}$ is called a *zero* of P or alternatively a *solution* of the polynomial equation $P = 0$ if $P(\bar{\boldsymbol{x}}) = 0$, that is, P vanishes when $\bar{x}_1, \ldots, \bar{x}_n$ are substituted for $x_1, \ldots, x_n$.

Let $\mathfrak{P} = [\mathbb{P}, \mathbb{Q}]$ be a polynomial system. If an n-tuple of numbers in $\tilde{\boldsymbol{K}}$ is a common zero of all the polynomials in $\mathbb{P}$ but not a zero of any polynomial in $\mathbb{Q}$, it is called a *zero* of $\mathfrak{P}$ or a *solution* of the system of polynomial equations $\mathbb{P} = 0$ and inequations $\mathbb{Q} \neq 0$. We may speak about the set of all zeros of $\mathfrak{P}$, which is denoted by $\mathrm{Zero}(\mathfrak{P})$ or $\mathrm{Zero}(\mathbb{P}/\mathbb{Q})$. Symbolically, it is defined as

$$\mathrm{Zero}(\mathbb{P}/\mathbb{Q}) \triangleq \left\{\bar{\boldsymbol{x}} \in \tilde{\boldsymbol{K}}^n : P(\bar{\boldsymbol{x}}) = 0, Q(\bar{\boldsymbol{x}}) \neq 0, \forall P \in \mathbb{P}, Q \in \mathbb{Q}\right\}.$$

We simply write $\mathrm{Zero}(\mathbb{P})$ for $\mathrm{Zero}(\mathbb{P}/\mathbb{Q})$ when $\mathbb{Q} \subset \boldsymbol{K} \setminus \{0\}$. In this case, $\mathrm{Zero}(\mathbb{P})$ is the set of all common zeros of the polynomials in $\mathbb{P}$. Sometimes, we write $\mathrm{Zero}(\mathbb{P}/Q)$ for $\mathrm{Zero}(\mathbb{P}/\{Q\})$ and $\mathrm{Zero}(P/\mathbb{Q})$ for $\mathrm{Zero}(\{P\}/\mathbb{Q})$, etc. It is easy to see that

$$\mathrm{Zero}(\mathbb{P}/\mathbb{Q}) = \mathrm{Zero}\left(\mathbb{P}/\prod_{Q\in\mathbb{Q}} Q\right) = \mathrm{Zero}(\mathbb{P}) \setminus \mathrm{Zero}\left(\prod_{Q\in\mathbb{Q}} Q\right).$$

And, for any polynomial sets $\mathbb{H}$, $\mathbb{P}_i$, $\mathbb{Q}_i$,

$$\operatorname{Zero}(\mathbb{P}/\mathbb{Q}) = \bigcup_i \operatorname{Zero}(\mathbb{P}_i/\mathbb{Q}_i)$$

implies that

$$\operatorname{Zero}(\mathbb{P} \cup \mathbb{H}/\mathbb{Q}) = \bigcup_i \operatorname{Zero}(\mathbb{P}_i \cup \mathbb{H}/\mathbb{Q}_i),$$
$$\operatorname{Zero}(\mathbb{P}/\mathbb{Q} \cup \mathbb{H}) = \bigcup_i \operatorname{Zero}(\mathbb{P}_i/\mathbb{Q}_i \cup \mathbb{H}).$$

The components $\bar{x}_i$ of a zero of a polynomial, a polynomial set, or a polynomial system – which are numbers of $\tilde{\boldsymbol{K}}$ – may be still in $\boldsymbol{K}$. In order to make the involved field $\tilde{\boldsymbol{K}}$ explicit, we shall sometimes call the zero (solution) defined above a *$\tilde{\boldsymbol{K}}$-zero* ($\tilde{\boldsymbol{K}}$-solution) or an *extended zero* (extended solution). Accordingly, we use the notations $\tilde{\boldsymbol{K}}$-Zero($\mathbb{P}$), $\tilde{\boldsymbol{K}}$-Zero($\mathbb{P}/\mathbb{Q}$), etc.

Unless specified otherwise, $\operatorname{Zero}(\mathfrak{P}) = \emptyset$ is always meant in *any* extension of the ground field $\boldsymbol{K}$, and so is $\operatorname{Zero}(\mathfrak{P}) \neq \emptyset$ in *some* extension field of $\boldsymbol{K}$.

Let $\mathbb{P} = \{P_1, \ldots, P_s\} \subset \boldsymbol{K}[\boldsymbol{x}]$ be a (nonempty) polynomial set. Form the following infinite set of polynomials:

$$\mathfrak{I} = \left\{ \sum_{i=1}^{s} Q_i P_i : \; Q_1, \ldots, Q_s \in \boldsymbol{K}[\boldsymbol{x}] \right\}.$$

Theorem 1.5.1. $\mathfrak{I}$ is an *ideal* in $\boldsymbol{K}[\boldsymbol{x}]$.

The ideal $\mathfrak{I}$ formed above is called a *polynomial ideal* generated by $P_1, \ldots, P_s$ or simply by $\mathbb{P}$, denoted by Ideal($\mathbb{P}$). $P_1, \ldots, P_s$ and $\mathbb{P}$ are called the *generators* and *generating set* for $\mathfrak{I}$, respectively, and are said to form a finite *basis* for $\mathfrak{I}$. Let the definition of zeros be extended naturally to infinite sets of polynomials. It is also easy to see that

$$\operatorname{Zero}(\operatorname{Ideal}(\mathbb{P})) = \operatorname{Zero}(\mathbb{P}).$$

According to Hilbert's finite-basis theorem, one knows that for any subset $\mathfrak{I}$ of $\boldsymbol{K}[\boldsymbol{x}]$, if it is an ideal, then there is a finite nonempty set $\mathbb{P}$ of polynomials such that $\mathfrak{I} = \operatorname{Ideal}(\mathbb{P})$.

Let $\mathfrak{I}$ be any ideal in $\boldsymbol{K}[\boldsymbol{x}]$. The set of polynomials

$$\{F \in \boldsymbol{K}[\boldsymbol{x}] : \; F^m \in \mathfrak{I} \text{ for some integer } m \geq 1\}$$

forms an ideal, called the *radical ideal* of $\mathfrak{I}$ and denoted by Rad($\mathfrak{I}$) or sometimes by $\sqrt{\mathfrak{I}}$. It is also easy to see that

$$\operatorname{Zero}(\sqrt{\mathfrak{I}}) = \operatorname{Zero}(\mathfrak{I}).$$

1.6 Hilbert's Nullstellensatz

A polynomial ideal $\mathfrak{J}$ is called a *unit ideal* if it can be generated by the constant polynomial 1.

Theorem 1.6.1. Every polynomial ideal $\mathfrak{J} \subset \boldsymbol{K}[\boldsymbol{x}]$ which has no zero, i.e., $\mathrm{Zero}(\mathfrak{J}) = \emptyset$, in any extension field of $\boldsymbol{K}$ is a unit ideal.

This theorem may be restated as

Theorem 1.6.2. If the polynomials $P_1, \ldots, P_s \in \boldsymbol{K}[\boldsymbol{x}]$ have no common zero, i.e., $\mathrm{Zero}(\{P_1, \ldots, P_s\}) = \emptyset$, in an algebraically closed extension field of $\boldsymbol{K}$, then there exist polynomials $Q_1, \ldots, Q_s \in \boldsymbol{K}[\boldsymbol{x}]$ such that the following identity holds: $1 = Q_1 P_1 + \cdots + Q_s P_s$.

Proof. Van der Waerden (1950, p. 5). □

Theorem 1.6.2 may be regarded as a special case of Hilbert's Nullstellensatz.

Theorem 1.6.3 (Nullstellensatz). Let $\mathbb{P} = \{P_1, \ldots, P_s\}$ be a polynomial set and P a polynomial in $\boldsymbol{K}[\boldsymbol{x}]$. If $\mathrm{Zero}(\mathbb{P}) \subset \mathrm{Zero}(P)$ in an algebraically closed extension field of $\boldsymbol{K}$, then there exist polynomials $Q_1, \ldots, Q_s \in \boldsymbol{K}[\boldsymbol{x}]$ such that

$$P^q = Q_1 P_1 + \cdots + Q_s P_s$$

holds for some integer $q > 0$.

For a proof of this theorem, one uses the well-known trick of Rabinowitsch by reducing it to the case of Theorem 1.6.2 (see van der Waerden 1950, p. 6). In detail, under the hypothesis of the theorem, $P_1, \ldots, P_s, Pz - 1$ have no common zero, where z is a new variable. By Theorem 1.6.2 there are polynomials $H_1, \ldots, H_s, H \in \boldsymbol{K}[\boldsymbol{x}, z]$, such that

$$1 = H_1 P_1 + \cdots + H_s P_s + H(Pz - 1).$$

Replacing z in this equality by $1/P$ and multiplying it by some power of P to clean out the denominators, one immediately gets the identity in Theorem 1.6.3.

The containment relation $\mathrm{Zero}(\mathbb{P}) \subset \mathrm{Zero}(P)$, which means that P vanishes at every common zero of $P_1, \ldots, P_s$, is written sometimes as

$$P|_{\mathrm{Zero}(\mathbb{P})} = 0. \tag{1.6.1}$$

By Theorem 1.6.3 and the definition of radical ideals, (1.6.1) is equivalent to

$$P \in \sqrt{\mathrm{Ideal}(\mathbb{P})}.$$

Let $\iff$ stand for "if and only if." The following theorem is a consequence of the above results.

Theorem 1.6.4. Let $\mathbb{P}$ be a polynomial set in $\boldsymbol{K}[\boldsymbol{x}]$ and $\mathfrak{J} = \mathrm{Ideal}(\mathbb{P})$. Then

$$\begin{aligned} P \in \sqrt{\mathfrak{J}} &\iff 1 \in \mathrm{Ideal}(\mathbb{P} \cup \{Pz - 1\}) \\ &\iff \mathrm{Zero}(\mathbb{P} \cup \{Pz - 1\}) = \emptyset, \end{aligned}$$

where z is a new variable.

2 Zero decomposition of polynomial systems

From now on we come to describe elimination algorithms that decompose arbitrary systems of multivariate polynomials into special systems of triangular form – the theme of this book. Meanwhile, various zero relations between the given and the constructed systems will be established. In this chapter three kinds of different yet related algorithms are presented which compute such decompositions of relatively coarse form.

2.1 Triangular systems

Let $\boldsymbol{K}$ be a computable field of characteristic 0. The field $\mathbf{Q}$ of rational numbers is a concrete example for $\boldsymbol{K}$. A polynomial set is a finite set of nonzero polynomials in $\boldsymbol{K}[\boldsymbol{x}]$. By a polynomial system in $\boldsymbol{K}[\boldsymbol{x}]$ we mean a pair $[\mathbb{P}, \mathbb{Q}]$ of polynomial sets with which the set $\mathrm{Zero}(\mathbb{P}/\mathbb{Q})$ is of concern. In other words, we are concerned with the solutions of a system of polynomial equations $\mathbb{P} = 0$ and inequations $\mathbb{Q} \neq 0$.

In what follows, the number of elements of a finite set $\mathbb{S}$ is denoted $|\mathbb{S}|$. It is also called the *length* of $\mathbb{S}$. An *ordered set* is written by enclosing its elements in a pair of square brackets. For any nonempty ordered set $\mathbb{T} = [T_1, \ldots, T_r]$ and $1 \leq i \leq r$, the following symbols are often used:

$$\mathrm{op}(i, \mathbb{T}) \triangleq T_i, \quad \mathbb{T}^{\{i\}} \triangleq [T_1, \ldots, T_i].$$

If $\mathbb{S} = [S_1, \ldots, S_s]$ is another ordered set which has no intersection with $\mathbb{T}$, we define

$$\mathbb{S} \cup \mathbb{T} \triangleq [S_1, \ldots, S_s, T_1, \ldots, T_r].$$

$\mathbb{S} \cup \mathbb{T}$ and $\mathbb{T} \cup \mathbb{S}$ are distinguished when they are considered as ordered sets. In other words, the ordering is preserved for union of nonintersecting ordered sets. If one or both of $\mathbb{S}$ and $\mathbb{T}$ are usual sets, then so is $\mathbb{S} \cup \mathbb{T} = \mathbb{T} \cup \mathbb{S}$.

Definition 2.1.1. A finite nonempty ordered set of nonconstant polynomials in $\boldsymbol{K}[\boldsymbol{x}]$

$$\mathbb{T} = [T_1, T_2, \ldots, T_r]$$

is called a *triangular set* or a *noncontradictory quasi-ascending set* if

$$\mathrm{cls}(T_1) < \mathrm{cls}(T_2) < \cdots < \mathrm{cls}(T_r).$$

Any triangular set can be written in the following form

$$\mathbb{T} = \begin{bmatrix} T_1(x_1, \ldots, x_{p_1}), & \\ T_2(x_1, \ldots, x_{p_1}, \ldots, x_{p_2}), & \\ \cdots\cdots & \\ T_r(x_1, \ldots, x_{p_1}, \ldots, x_{p_2}, \ldots, x_{p_r}) & \end{bmatrix}, \tag{2.1.1}$$

where

$$0 < p_1 < p_2 < \cdots < p_r \leq n,$$
$$p_i = \text{cls}(T_i), \quad x_{p_i} = \text{lv}(T_i), \quad i = 1, \ldots, r.$$

Let $\mathbb{T}$ be a triangular set as in (2.1.1) and P any polynomial. P is said to be *reduced* with respect to $\mathbb{T}$ if P is reduced with respect to every $T \in \mathbb{T}$, i.e., $\deg(P, x_{p_i}) < \text{ldeg}(T_i)$ for all i. The polynomial

$$R = \text{prem}(\ldots \text{prem}(P, T_r), \ldots, T_1),$$

denoted simply by $\text{prem}(P, \mathbb{T})$, is called the *pseudo-remainder* of P with respect to $\mathbb{T}$. From the expression (1.2.1), one can easily deduce the following *pseudo-remainder formula*

$$I_1^{q_1} \cdots I_r^{q_r} P = \sum_{i=1}^{r} Q_i T_i + R, \tag{2.1.2}$$

where each q_i is a nonnegative integer and

$$I_i = \text{ini}(T_i), \quad Q_i \in \boldsymbol{K}[\boldsymbol{x}], \quad i = 1, \ldots, r.$$

Apparently, $\text{prem}(P, \mathbb{T}) = P$ when P is reduced with respect to $\mathbb{T}$. For any polynomial set $\mathbb{P}$, $\text{prem}(\mathbb{P}, \mathbb{T})$ stands for $\{\text{prem}(P, \mathbb{T}) : P \in \mathbb{P}\}$.

Example 2.1.1. Recall F_1, F_2 in Example 1.1.2 and let

$$F_3 = x_3 x_4 - 2x_2^2 - x_1 x_2 - 1, \quad F_4 = \text{prem}(F_1, F_2).$$

F_4 has been calculated in Example 1.2.3. F_3 is reduced with respect to F_1, but F_1 is not with respect to F_3. Also, no one of F_2 and F_3 is reduced with respect to the other. With respect to $x_1 \prec \cdots \prec x_4$, $\mathbb{T}_1 = [F_4, F_2]$ is clearly a triangular set. Both F_1 and F_3 are not reduced with respect to $\mathbb{T}_1$. One can verify that

$$F_6 = \text{prem}(F_3, \mathbb{T}_1) = 2x_1 x_2^2 + 2x_1^2 x_2^2 - 2x_1^2 x_2 + x_1^2 + x_1,$$
$$\text{prem}(F_1, \mathbb{T}_1) = 0.$$

In the following definition and hereafter, the ordering is preserved for difference of ordered sets in the natural way. For example, $[a, b, c, d] \setminus [a, c] = [b, d]$.

Definition 2.1.2. A polynomial system $[\mathbb{T}, \mathbb{U}]$ in $\boldsymbol{K}[\boldsymbol{x}]$ is called a *triangular system* if $\mathbb{T}$ is a triangular set and $I(\bar{\boldsymbol{x}}) \neq 0$ for any $I \in \text{ini}(\mathbb{T})$ of class i and $\bar{\boldsymbol{x}} \in \text{Zero}(\mathbb{T}^{(i)}/\mathbb{U})$.

A triangular system $[\mathbb{T}, \mathbb{U}]$ is said to be *fine* if $0 \notin \text{prem}(\mathbb{U}, \mathbb{T})$. It is said to be *reduced* if every $T \in \mathbb{T} \cup \mathbb{U}$ is reduced with respect to $\mathbb{T} \setminus [T]$.

Lemma 2.1.1. For any triangular system $[\mathbb{T}, \mathbb{U}]$ and polynomial P in $\boldsymbol{K}[\boldsymbol{x}]$, if $\text{prem}(P, \mathbb{T}) = 0$, then $\text{Zero}(\mathbb{T}/\mathbb{U}) \subset \text{Zero}(P)$.

Proof. Let $\bar{\boldsymbol{x}} \in \text{Zero}(\mathbb{T}/\mathbb{U})$. By definition, $I(\bar{\boldsymbol{x}}) \neq 0$ for any $I \in \text{ini}(\mathbb{T})$. From the pseudo-remainder formula (2.1.2) one sees that $P(\bar{\boldsymbol{x}}) = 0$. □

Definition 2.1.3. A triangular set $\mathbb{T} \subset \boldsymbol{K}[\boldsymbol{x}]$ is said to be *fine* or *reduced* if $[\mathbb{T}, \text{ini}(\mathbb{T})]$ is fine or reduced, respectively.

A reduced triangular set is also called a *noncontradictory ascending set*.

A triangular set $\mathbb{T}$ is called a *noncontradictory weak-ascending set* if for every $T \in \mathbb{T}$, $\text{ini}(T)$ is reduced with respect to $\mathbb{T} \setminus [T]$.

Any set of a single nonzero constant is called a *contradictory (quasi-, weak-) ascending set*.

Note that the pseudo-remainder of any polynomial with respect to a contradictory ascending set is 0. A (quasi-, weak-) ascending set is either a contradictory one or a noncontradictory one.

Example 2.1.2. Let $x_1 \prec x_2 \prec x_3$ and $\mathbb{T} = [x_1 - 2, (x_1^2 - 4)x_3 + x_2]$. $\mathbb{T}$ is a triangular set, but it is not fine. $[\mathbb{T}, \{x_1, x_1 - 2\}]$ is a triangular system (not fine), but not so is $[\mathbb{T}, \{x_1 + 2\}]$. The triangular set

$$[x_1^2 - 2, x_2^2 - 2x_1x_2 + 2, (x_2 - x_1)x_3 + 1]$$

is both fine and reduced, so it is a noncontradictory ascending set.

It is easy to show that if $[\mathbb{T}, \mathbb{U}]$ is a fine triangular system, then either $\mathbb{T}$ is fine or $\text{Zero}(\mathbb{T}/\mathbb{U}) = \emptyset$.

Lemma 2.1.2. Let $F \in \boldsymbol{K}[x]$ and $G \in \boldsymbol{K}[x, y]$ be two polynomials. Then

$$\text{prem}(\text{coef}(G, y^k), F, x) \neq 0 \iff \text{coef}(\text{prem}(G, F, x), y^k) \neq 0 \tag{2.1.3}$$

for any $0 \leq k \leq \deg(G, y)$.

Proof. Let $I = \text{lc}(F, x), m = \deg(F, x), l = \deg(G, y)$ and G be written as

$$G = G_l y^l + G_{l-1} y^{l-1} + \cdots + G_0, \quad G_i \in \boldsymbol{K}[x].$$

Set

$$R_i = \text{prem}(G_i, F, x), \quad i = 0, 1, \ldots, l.$$

Corresponding to the pseudo-remainder formula (1.2.1), one has

$$I^{q_i} G_i = Q_i F + R_i, \quad q_i = \max(\deg(G_i, x) - m + 1, 0), \tag{2.1.4}$$

for each i. Let $q = \max(\deg(G, x) - m + 1, 0) = \max_{0 \leq i \leq l} q_i$. Multiplying the remainder formula in (2.1.4) by $y^i I^{q - q_i}$ for each i and adding the resulting formulae together, we obtain

$$I^q G = \left(\sum_{i=0}^{l} I^{q-q_i} Q_i y^i\right) F + \sum_{i=0}^{l} I^{q-q_i} R_i y^i.$$

By Proposition 1.2.2,

$$\sum_{i=0}^{l} I^{q-q_i} R_i y^i = \operatorname{prem}(G, F, x).$$

It follows that

$$\operatorname{coef}(\operatorname{prem}(G, F, x), y^k) = I^{q-q_k} R_k = I^{q-q_k} \operatorname{prem}(\operatorname{coef}(G, y^k), F, x)$$

for any $0 \leq k \leq l$. Clearly, $I \neq 0$; (2.1.3) is therefore proved. □

The following is an obvious consequence of Lemma 2.1.2.

Corollary 2.1.3. Let $\mathbb{T} \subset \boldsymbol{K}[\boldsymbol{x}]$ be a triangular set and $P \in \boldsymbol{K}[\boldsymbol{x}, y]$ be any polynomial, where y is a new indeterminate. Then

$$\operatorname{prem}(\operatorname{coef}(P, y^k), \mathbb{T}) \neq 0 \iff \operatorname{coef}(\operatorname{prem}(P, \mathbb{T}), y^k) \neq 0,$$

for any $0 \leq k \leq \deg(P, y)$,

Lemma 2.1.4. From any fine triangular set $\mathbb{T} \subset \boldsymbol{K}[\boldsymbol{x}]$ one can compute a reduced triangular set $\mathbb{T}^*$ such that

$$\operatorname{Zero}(\mathbb{T}^*/\operatorname{ini}(\mathbb{T}^*)) = \operatorname{Zero}(\mathbb{T}/\operatorname{ini}(\mathbb{T})). \tag{2.1.5}$$

Proof. Let $\mathbb{T} = [T_1, \ldots, T_r]$ with $p_i = \operatorname{cls}(T_i)$ and $I_i = \operatorname{ini}(T_i)$, for $i = 1, \ldots, r$. The case $r = 1$ is trivial, so we may assume $r > 1$ and set

$$\begin{aligned} \mathbb{T}^{\{i-1\}} &= [T_1, \ldots, T_{i-1}], \\ T_i^* &= \operatorname{prem}(T_i, \mathbb{T}^{\{i-1\}}), \\ \mathbb{T}^{*\{i\}} &= [T_1, T_2^*, \ldots, T_i^*]; \quad i = 2, \ldots, r. \end{aligned}$$

As $\mathbb{T}^{\{i-1\}}$ does not involve the variables $x_{p_i}, \ldots, x_n$, by Corollary 2.1.3 we have

$$\operatorname{cls}(T_i^*) = p_i, \ \operatorname{ldeg}(T_i^*) = \operatorname{ldeg}(T_i), \quad 2 \leq i \leq r.$$

Hence, $\mathbb{T}^*$ is a reduced triangular set.

To show (2.1.5), write down the following formula corresponding to (2.1.2)

$$T_i^* = I_1^{q_{i1}} \cdots I_{i-1}^{q_{i,i-1}} T_i + \sum_{j=1}^{i-1} Q_{ij} T_j, \quad 2 \leq i \leq r. \tag{2.1.6}$$

Let $\bar{\boldsymbol{x}}^{\{p_{i-1}\}} \in \operatorname{Zero}(\mathbb{T}^{\{i-1\}}/\operatorname{ini}(\mathbb{T}^{\{i-1\}}))$. By (2.1.6), we have

$$\bar{T}_i^* = I_1^{q_{i1}}(\bar{\boldsymbol{x}}^{\{p_{i-1}\}}) \cdots I_{i-1}^{q_{i,i-1}}(\bar{\boldsymbol{x}}^{\{p_{i-1}\}}) \bar{T}_i,$$

where

$$\bar{T}_i = T_i(\bar{\boldsymbol{x}}^{\{p_{i-1}\}}, x_{p_{i-1}+1}, \ldots, x_{p_i}),$$
$$\bar{T}_i^* = T_i^*(\bar{\boldsymbol{x}}^{\{p_{i-1}\}}, x_{p_{i-1}+1}, \ldots, x_{p_i}).$$

Thus, $\bar{T}_i^*$ and $\bar{T}_i$ have the same set of zeros for $x_{p_{i-1}+1}, \ldots, x_{p_i}$. As this is true for any $i \geq 2$, it follows that

$$\mathrm{Zero}(\bar{T}_i^*/\mathrm{ini}(\bar{T}_i^*)) = \mathrm{Zero}(\bar{T}_i/\mathrm{ini}(\bar{T}_i)),$$

and hence

$$\mathrm{Zero}(\mathbb{T}^{*\{i\}}/\mathrm{ini}(\mathbb{T}^{*\{i\}})) = \mathrm{Zero}(\mathbb{T}^{\{i\}}/\mathrm{ini}(\mathbb{T}^{\{i\}})).$$

With $i = r$, (2.1.5) is thereby established. □

Remark 2.1.1. Let $[\mathbb{T}, \mathbb{U}]$ be a fine triangular system with $\mathrm{Zero}(\mathbb{T}/\mathbb{U}) \neq \emptyset$. In this case, $\mathbb{T}$ is also fine as noted above. Therefore, we can compute a reduced triangular set $\mathbb{T}^*$ such that (2.1.5) holds. Let $\mathbb{U}^* = \mathrm{prem}(\mathbb{U}, \mathbb{T}^*)$; then

$$\begin{aligned}\mathrm{Zero}(\mathbb{T}^*/\mathbb{U}^*) &= \mathrm{Zero}(\mathbb{T}^*/\mathrm{ini}(\mathbb{T}^*) \cup \mathbb{U}^*)\\ &= \mathrm{Zero}(\mathbb{T}/\mathrm{ini}(\mathbb{T}) \cup \mathbb{U}) = \mathrm{Zero}(\mathbb{T}/\mathbb{U}).\end{aligned}$$

This is to say, one can compute from $[\mathbb{T}, \mathbb{U}]$ a reduced triangular system $[\mathbb{T}^*, \mathbb{U}^*]$ such that

$$\mathrm{Zero}(\mathbb{T}^*/\mathbb{U}^*) = \mathrm{Zero}(\mathbb{T}/\mathbb{U}). \tag{2.1.7}$$

The main objective of this chapter is to describe algorithms that decompose any given polynomial system $\mathfrak{P}$ into finitely many fine triangular systems $\mathfrak{T}_1, \ldots, \mathfrak{T}_e$ such that

$$\mathrm{Zero}(\mathfrak{P}) = \bigcup_{i=1}^{e} \mathrm{Zero}(\mathfrak{T}_i). \tag{2.1.8}$$

We assign $e = 0$ when $\mathrm{Zero}(\mathfrak{P}) = \emptyset$ is verified.

2.2 Characteristic-set-based algorithm

The concept of characteristic sets was introduced by Ritt (1932, 1950) for (differential) polynomial ideals in the context of his work on differential algebra. However, this concept and the algorithmic method proposed by Ritt drew little attention until 1978 when W.-t. Wu realized that the constructive algebraic tools underlying his method of mechanical theorem proving in geometry appeared already in Ritt's two books. Since then, Wu has considerably developed Ritt's work by removing his analytic arguments using continuity and limit, etc., by adapting the concept and method for polynomial sets instead of ideals, and by demonstrating its powerfulness in various geometric applications. For instance, Wu dropped *irreducibility*, a major requirement in Ritt's process, so that a characteristic set can be effectively constructed from an arbitrary polynomial set. Wu's insight and extensive work have stimulated a great deal of research interest and activity on the subject. These altogether have contributed to the theoretical development of the method and made it more efficient and appropriate for practical applications. The characteristic-set-based algorithms presented in this book owe much to Wu (1984, 1986a, 1987, 1989a, 1994).

Ritt–Wu's characteristic sets

Definition 2.2.1. For two nonzero polynomials F and G in $\boldsymbol{K}[\boldsymbol{x}]$, F is said to have a *lower rank* than G, which is denoted as $F \prec G$ or $G \succ F$, if either $\mathrm{cls}(F) < \mathrm{cls}(G)$, or $\mathrm{cls}(F) = \mathrm{cls}(G) > 0$ and $\mathrm{ldeg}(F) < \mathrm{ldeg}(G)$. In this case, G is said to have a *higher rank* than F.

If neither $F \prec G$ nor $G \prec F$, F and G are said to have the *same rank*, denoted as $F \sim G$.

We write $F \precsim G$ for "$F \prec G$ or $F \sim G$," and similarly for "$\succsim$."

Example 2.2.1. Recall F_1, F_2, F_3 in Examples 1.1.2 and 2.1.1. With $x_1 \prec \cdots \prec x_4$, we have

$$\mathrm{cls}(F_1) = \mathrm{cls}(F_2) = \mathrm{cls}(F_3) = 4,$$
$$\mathrm{ldeg}(F_1) = 2, \quad \mathrm{ldeg}(F_2) = \mathrm{ldeg}(F_3) = 1.$$

It follows that $F_3 \sim F_2$, $F_2 \prec F_1$.

Definition 2.2.2. For two triangular sets

$$\mathbb{T} = [T_1, \dots, T_r], \quad \mathbb{T}' = [T_1', \dots, T_{r'}'],$$

$\mathbb{T}$ is said to have a *higher rank* than $\mathbb{T}'$, which is denoted as $\mathbb{T} \succ \mathbb{T}'$ or $\mathbb{T}' \prec \mathbb{T}$, if either condition holds:

a. There exists a $j \leq \min(r, r')$ such that $T_1 \sim T_1', \dots, T_{j-1} \sim T_{j-1}'$, while $T_j \succ T_j'$;
b. $r' > r$ and $T_1 \sim T_1', \dots, T_r \sim T_r'$.

In this case, $\mathbb{T}'$ is said to have a *lower rank* than $\mathbb{T}$. If neither $\mathbb{T} \prec \mathbb{T}'$ nor $\mathbb{T}' \prec \mathbb{T}$, $\mathbb{T}$ and $\mathbb{T}'$ are said to have the *same rank*, denoted as $\mathbb{T} \sim \mathbb{T}'$. In this case,

$$r = r', \text{ and } T_1 \sim T_1', \dots, T_r \sim T_r'.$$

Example 2.2.2. Let the polynomials $F_1, \dots, F_4$ be as in Examples 1.1.2 and 2.1.1, and

$$F_5 = \mathrm{prem}(F_3, F_2) = -x_3^2 + x_1x_2x_3 - 2x_1x_2^2 - x_1^2x_2 - x_1.$$

Then

$$\mathbb{T}_1 = [F_4, F_2], \quad \mathbb{T}_2 = [F_5, F_2], \quad \mathbb{T}_3 = [F_4, F_1]$$

are reduced triangular sets. $\mathbb{T}_1$ and $\mathbb{T}_2$ have the same rank which is lower than that of $\mathbb{T}_3$, i.e., $\mathbb{T}_1 \sim \mathbb{T}_2 \prec \mathbb{T}_3$.

The above-defined "$\precsim$" is a partial order, under which the collection of all triangular sets is partially ordered. Thus, for any set of ascending sets one is free to talk about the notion of *minimal ascending set* if it exists.

Lemma 2.2.1. Let $\mathbb{T}_1 \succsim \mathbb{T}_2 \succsim \cdots \succsim \mathbb{T}_k \succsim \cdots$ be a sequence of triangular sets whose ranks never increase. Then there exists a k' such that $\mathbb{T}_k \sim \mathbb{T}_{k'}$ for all $k \geq k'$.

Proof. Let $T_k = \mathrm{op}(1, \mathbb{T}_k)$ and $r_k = |\mathbb{T}_k|$ for each k (recall that $\mathrm{op}(i, \mathbb{T}_k)$ denotes the ith element of $\mathbb{T}_k$). Then

$$T_1 \succsim T_2 \succsim \cdots \succsim T_k \succsim \cdots.$$

In other words, for any k either $\mathrm{cls}(T_{k+1}) < \mathrm{cls}(T_k)$, or

$$\mathrm{cls}(T_{k+1}) = \mathrm{cls}(T_k) > 0 \text{ and } \mathrm{ldeg}(T_{k+1}) \leq \mathrm{ldeg}(T_k).$$

As both class and degree are nonnegative integers, there exists an index k_1 such that $T_k \sim T_{k_1}$ for all $k \geq k_1$.

If there is a $k_1' \geq k_1$ such that $r_k = 1$ for all $k \geq k_1'$, then the lemma is clearly true. Otherwise, there exists a $k_1' \geq k_1$ such that $r_k \geq 2$ for all $k \geq k_1'$. Let $T_k' = \mathrm{op}(2, \mathbb{T}_k)$ for $k \geq k_1'$; then

$$T_{k_1'}' \succsim T_{k_1'+1}' \succsim \cdots \succsim T_k' \succsim \cdots.$$

As before there exists a $k_2 \geq k_1'$ such that $T_k' \sim T_{k_2}'$ for all $k \geq k_2$.

If $r_k \leq 2$ for all $k \geq k_2$, the lemma is already proved. Otherwise, there exists a $k_2' \geq k_2$ such that $r_k \geq 3$ for all $k \geq k_2'$. In this case, we may consider $T_k'' = \mathrm{op}(3, \mathbb{T}_k)$ and form a sequence of polynomials with nonincreasing ranks. As $r_k \leq n$ for all k, proceeding in this way one should stop at some r and k' such that

$$r_k = r, \;\; \mathrm{op}(r, \mathbb{T}_k) \sim \mathrm{op}(r, \mathbb{T}_{k'}), \quad \forall k \geq k'.$$

It follows that $\mathbb{T}_k \sim \mathbb{T}_{k'}$ for all $k \geq k'$, and the lemma is proved. □

Consider any nonempty polynomial set $\mathbb{P}$. Let Φ be the set of all ascending sets contained in $\mathbb{P}$. Since each single polynomial forms by itself an ascending set, $\Phi \neq \emptyset$. Any minimal ascending set of Φ is called a *basic set* of $\mathbb{P}$. Such a basic set exists and can be determined as follows.

Starting with $\mathbb{P} = \mathbb{F}_1$, one chooses a polynomial, say B_1, of lowest rank from $\mathbb{F}_1$. If $\mathrm{cls}(B_1) = 0$, then $[B_1]$ is already a basic set of $\mathbb{P}$. Otherwise, let

$$\mathbb{F}_2 = \{F \in \mathbb{F}_1 \setminus \{B_1\}\colon \; F \text{ is reduced w.r.t. } B_1\}.$$

If $\mathbb{F}_2 = \emptyset$, then $[B_1]$ is a basic set of $\mathbb{F}_1 = \mathbb{P}$. From the choice of B_1 all the polynomials in $\mathbb{F}_2$ have rank higher than that of B_1. Now, let B_2 be a polynomial in $\mathbb{F}_2$ of lowest rank and

$$\mathbb{F}_3 = \{F \in \mathbb{F}_2 \setminus \{B_2\}\colon \; F \text{ is reduced w.r.t. } B_2\}.$$

If $\mathbb{F}_3 = \emptyset$, then $[B_1, B_2]$ is a basic set of $\mathbb{P}$. Otherwise, choose from $\mathbb{F}_3$ a polynomial B_3 of lowest rank and proceed as before. As

$$\mathrm{cls}(B_1) < \mathrm{cls}(B_2) < \mathrm{cls}(B_3) < \cdots \leq n,$$

the procedure must terminate in a finite number of steps. Finally, a basic set of $\mathbb{P}$ is constructed.

The above process can be described as the following algorithm.

Algorithm BasSet: $\mathbb{B} \leftarrow \mathrm{BasSet}(\mathbb{P})$. Given a nonempty polynomial set $\mathbb{P} \subset \boldsymbol{K}[\boldsymbol{x}]$, this algorithm computes a basic set $\mathbb{B}$ of $\mathbb{P}$.
B1. Set $\mathbb{F} \leftarrow \mathbb{P}, \mathbb{B} \leftarrow \emptyset$.
B2. While $\mathbb{F} \neq \emptyset$, do:
  B2.1. Let B be an element of $\mathbb{F}$ with lowest rank.
  B2.2. Set $\mathbb{B} \leftarrow \mathbb{B} \cup [B]$.
  B2.3. If $\mathrm{cls}(B) = 0$, then set $\mathbb{F} \leftarrow \emptyset$; else set

$$\mathbb{F} \leftarrow \{F \in \mathbb{F} \setminus \{B\}: \ F \text{ is reduced w.r.t. } B\}.$$

A basic set of $\mathbb{P}$ is contradictory if and only if $\mathbb{P}$ contains a constant. In this case algorithm BasSet terminates at the first iteration of the while-loop. See Example 2.2.3 for examples of basic sets.

Definition 2.2.3. An ascending set $\mathbb{C}$ is called a *characteristic set* of a nonempty polynomial set $\mathbb{P} \subset \boldsymbol{K}[\boldsymbol{x}]$ if

$$\mathbb{C} \subset \mathrm{Ideal}(\mathbb{P}), \quad \mathrm{prem}(\mathbb{P}, \mathbb{C}) = \{0\}.$$

Here, a characteristic set of $\mathbb{P}$ is defined à la Wu. Ritt's definition of a characteristic set is for the ideal $\mathfrak{I}$ (generated by $\mathbb{P}$) and requires that $\mathrm{prem}(\mathfrak{I}, \mathbb{C}) = \{0\}$; thus for computing $\mathbb{C}$ one has to consider its *irreducibility* as in Sect. 4.1 or use alternative algorithms (see Mishra 1993, sect. 5.6).

Proposition 2.2.2. Let $\mathbb{C} = [C_1, \ldots, C_r]$ be a characteristic set of any polynomial set $\mathbb{P} \subset \boldsymbol{K}[\boldsymbol{x}]$ and

$$I_i = \mathrm{ini}(C_i), \ \mathbb{P}_i = \mathbb{P} \cup \{I_i\}, \quad i = 1, \ldots, r,$$
$$\mathbb{I} = \mathrm{ini}(\mathbb{C}) = \{I_1, \ldots, I_r\}.$$

Then

$$\mathrm{Zero}(\mathbb{C}/\mathbb{I}) \subset \mathrm{Zero}(\mathbb{P}) \subset \mathrm{Zero}(\mathbb{C}), \tag{2.2.1}$$

$$\mathrm{Zero}(\mathbb{P}) = \mathrm{Zero}(\mathbb{C}/\mathbb{I}) \cup \bigcup_{i=1}^{r} \mathrm{Zero}(\mathbb{P}_i) \tag{2.2.2}$$

in $\boldsymbol{K}$ or any extension field of $\boldsymbol{K}$.

Proof. Since $\mathbb{C} \subset \mathrm{Ideal}(\mathbb{P})$, $\mathrm{Zero}(\mathbb{P}) \subset \mathrm{Zero}(\mathbb{C})$.

On the other hand, for any $P \in \mathbb{P}$ there are nonnegative integers q_i and polynomials Q_i such that

$$I_1^{q_1} \cdots I_r^{q_r} P = \sum_{i=1}^{r} Q_i C_i.$$

It follows that

$$\text{Zero}(\mathbb{C}/\mathbb{I}) \subset \text{Zero}(\mathbb{P}).$$

This is true clearly for $\boldsymbol{K}$ or any extension field of $\boldsymbol{K}$. Thus, (2.2.1) is proved.

Note that the zeros of $\mathbb{P}$ which make the vanishing of some I_i are considered additionally as those of $\mathbb{P}_i$. The relation (2.2.2) is obtained with ease. □

Now we are ready to present the characteristic-set algorithm of Ritt and Wu, which points out how to construct a characteristic set from any given polynomial set.

Algorithm CharSet: $\mathbb{C} \leftarrow \text{CharSet}(\mathbb{P})$. Given a nonempty polynomial set $\mathbb{P} \subset \boldsymbol{K}[\boldsymbol{x}]$, this algorithm computes a characteristic set $\mathbb{C}$ of $\mathbb{P}$.

C1. Set $\mathbb{F} \leftarrow \mathbb{P}$, $\mathbb{R} \leftarrow \mathbb{P}$.

C2. While $\mathbb{R} \neq \emptyset$, do:

C2.1. Compute $\mathbb{C} \leftarrow \text{BasSet}(\mathbb{F})$.

C2.2. If $\mathbb{C}$ is contradictory, then set $\mathbb{R} \leftarrow \emptyset$; else compute

$$\mathbb{R} \leftarrow \text{prem}(\mathbb{F} \setminus \mathbb{C}, \mathbb{C}) \setminus \{0\}$$

and set $\mathbb{F} \leftarrow \mathbb{F} \cup \mathbb{R}$.

In order to show the termination of this algorithm, let us first prove the following lemma.

Lemma 2.2.3. Let $\mathbb{P} \subset \boldsymbol{K}[\boldsymbol{x}]$ be a nonempty polynomial set having a basic set $\mathbb{B} = [B_1, B_2, \ldots, B_r]$, where $\text{cls}(B_1) > 0$. If B is a nonzero polynomial reduced with respect to $\mathbb{B}$, then $\mathbb{P} \cup \{B\}$ has a basic set of rank lower than that of $\mathbb{B}$.

Proof. Let $\mathbb{P}^+ = \mathbb{P} \cup \{B\}$. If $\text{cls}(B) = 0$, then $[B]$ is a basic set of $\mathbb{P}^+$ and has rank lower than that of $\mathbb{B}$. Suppose otherwise $\text{cls}(B) = p > 0$. As B is reduced with respect to $\mathbb{B}$, there exists an i $(1 \leq i \leq r)$ such that $p \leq \text{cls}(B_i)$, and $p > \text{cls}(B_{i-1})$ when $i > 1$. Moreover, in the case $p = \text{cls}(B_i)$, $\deg(B, x_p) < \text{ldeg}(B_i)$. Hence $[B_1, B_2, \ldots, B_{i-1}, B]$ is an ascending set contained in $\mathbb{P}^+$ and has rank lower than that of $\mathbb{B}$. The basic set of $\mathbb{P}^+$ has therefore rank lower than that of $\mathbb{B}$. □

Proof of CharSet. Algorithm CharSet may be sketched as follows:

$$\begin{array}{cccc} \mathbb{P} = \mathbb{F}_1 \subset & \mathbb{F}_2 \subset & \cdots \subset & \mathbb{F}_m \\ \cup & \cup & & \cup \\ \mathbb{B}_1 & \mathbb{B}_2 & \cdots & \mathbb{B}_m = \mathbb{C} \\ \mathbb{R}_1 & \mathbb{R}_2 & & \mathbb{R}_m = \emptyset \end{array} \tag{2.2.3}$$

where

$$\mathbb{R}_i = \text{prem}(\mathbb{F}_i \setminus \mathbb{B}_i, \mathbb{B}_i) \setminus \{0\},$$
$$\mathbb{F}_{i+1} = \mathbb{F}_i \cup \mathbb{R}_i$$

and $\mathbb{B}_i$ is a basic set of $\mathbb{F}_i$ for each i.

Termination. We need to show that the while-loop has only finitely many iterations, i.e., to show the finiteness of m in sketch (2.2.3). If some $\mathbb{B}_i$ is contradictory, the algorithm terminates obviously. Otherwise, by Lemma 2.2.3 $\mathbb{B}_{i+1} \prec \mathbb{B}_i$ for all i. Hence, $\mathbb{B}_1 \succ \mathbb{B}_2 \succ \cdots$. By Lemma 2.2.1, such a sequence is composed of a finite number of terms. In other words, m is finite and thus the algorithm must terminate.

Correctness. From the formula (2.1.2) one knows that for any polynomial $F \in \mathbb{F}_i$, $\mathrm{prem}(F, \mathbb{B}_i) \in \mathrm{Ideal}(\mathbb{B}_i \cup \{F\})$. It follows that

$$\mathrm{Ideal}(\mathbb{F}_{i+1}) = \mathrm{Ideal}(\mathbb{F}_i) = \mathrm{Ideal}(\mathbb{P})$$

for each i. Therefore,

$$\mathbb{C} = \mathbb{B}_m \subset \mathbb{F}_m \subset \mathrm{Ideal}(\mathbb{P}).$$

As $\mathbb{R}_m = \emptyset$, we have

$$\mathrm{prem}(\mathbb{F}_m, \mathbb{C}) = \mathrm{prem}(\mathbb{F}_m \setminus \mathbb{C}, \mathbb{C}) \cup \mathrm{prem}(\mathbb{C}, \mathbb{C}) = \{0\}.$$

By definition, $\mathbb{C}$ is a characteristic set of $\mathbb{P}$. The proof is complete. □

The above procedure of acquiring a characteristic set $\mathbb{C}$ from $\mathbb{P}$ is called *well-ordering principle* and was attributed to Ritt by Wu (1984, 1986a).

Example 2.2.3. Let $\mathbb{P} = \{F_1, F_2, F_3\}$ with

$$\begin{aligned}
F_1 &= x_1x_4^2 + x_4^2 - x_1x_2x_4 - x_2x_4 + x_1x_2 + 3x_2,\\
F_2 &= x_1x_4 + x_3 - x_1x_2,\\
F_3 &= x_3x_4 - 2x_2^2 - x_1x_2 - 1.
\end{aligned}$$

These polynomials already appeared in Examples 1.1.2 and 2.1.1. The sequence of polynomial sets and their basic sets corresponding to those in sketch (2.2.3) are as follows:

$$\begin{array}{lll}
\mathbb{P} = \mathbb{F}_1 = \{F_1, F_2, F_3\} \subset & \mathbb{F}_2 = \{F_1, \ldots, F_5\} \subset & \mathbb{F}_3 = \{F_1, \ldots, F_6\}\\
\quad\cup & \cup & \cup\\
\quad\mathbb{B}_1 = [F_2] & \mathbb{B}_2 = [F_4, F_2] & \mathbb{B}_3 = [F_6, F_4, F_2] = \mathbb{C}\\
\quad\mathbb{R}_1 = \{F_4, F_5\} & \mathbb{R}_2 = \{F_6\} & \mathbb{R}_3 = \emptyset,
\end{array}$$

where F_4, F_5, F_6 are given in Examples 2.1.1 and 2.2.2. Hence, the last basic set $\mathbb{B}_3$ is a characteristic set $\mathbb{C}$ of $\mathbb{P}$. Let the polynomials F_6, F_4, F_2 be renamed C_1, C_2, C_3 and copied here for easy reference:

$$\mathbb{C} = [C_1, C_2, C_3] = \begin{bmatrix} x_1(2x_1x_2^2 + 2x_2^2 - 2x_1x_2 + x_1 + 1),\\ x_1x_3^2 + x_3^2 - x_1^2x_2x_3 - x_1x_2x_3 + x_1^3x_2 + 3x_1^2x_2,\\ x_1x_4 + x_3 - x_1x_2 \end{bmatrix}.$$

The initials of C_1, C_2, C_3 are

$$I_1 = 2x_1(x_1 + 1),\ \ I_2 = x_1 + 1,\ \ I_3 = x_1.$$

Clearly, $I_1 \neq 0$ implies that $I_1 I_2 I_3 \neq 0$, since both I_2 and I_3 are factors of I_1. So only the initial I_1 has to be further considered. Let $\mathbb{P}_1$ and $\mathbb{P}_2$ be the enlarged polynomial sets obtained from $\mathbb{P}$ by adjoining $x_1 + 1$ and x_1, respectively, i.e.,

$$\mathbb{P}_1 = \mathbb{P} \cup \{x_1 + 1\}, \;\; \mathbb{P}_2 = \mathbb{P} \cup \{x_1\}.$$

We have the following zero relation

$$\text{Zero}(\mathbb{P}) = \text{Zero}(\mathbb{C}/I_1) \cup \text{Zero}(\mathbb{P}_1) \cup \text{Zero}(\mathbb{P}_2). \tag{2.2.4}$$

It is important to remark that, during the computation of characteristic sets by CharSet, there appear inevitably some superfluous factors of initials. These factors should be removed in order to control the growth of polynomial size. The appearance of superfluous factors during the computation of polynomial remainder sequences was discovered by Collins (1967). Such factors appearing in the computation of characteristic sets were studied in Li (1989a).

Definition 2.2.4. An ascending set $\mathbb{C}$ is called a $\mathbb{Q}$-*modified characteristic set* of a nonempty polynomial set $\mathbb{P} \subset \boldsymbol{K}[\boldsymbol{x}]$ if

$$\text{Zero}(\mathbb{P}/\mathbb{Q}) \subset \text{Zero}(\mathbb{C}), \quad \text{prem}(\mathbb{P}, \mathbb{C}) = \{0\},$$

where $\mathbb{Q}$ is a polynomial set.

The prefix $\mathbb{Q}$ is omitted when $\mathbb{Q} \subset \boldsymbol{K}$.

Let algorithm CharSet be modified by allowing the removal of polynomial factors during the computation and designate the resulting algorithm by ModCharSet. Then the output of ModCharSet consists of an ascending set $\mathbb{C}$ and a set $\mathbb{F}$ of distinct removed factors $F_1, \ldots, F_t$. It is clear to see that $\mathbb{C}$ is an $\mathbb{F}$-modified characteristic set of the input polynomial set $\mathbb{P}$. Moreover, the zero relation (2.2.2) can be modified accordingly as

$$\text{Zero}(\mathbb{P}) = \text{Zero}(\mathbb{C}/\mathbb{I}) \cup \bigcup_{i=1}^{r} \text{Zero}(\mathbb{P}_i) \cup \bigcup_{j=1}^{t} \text{Zero}(\mathbb{Q}_j), \tag{2.2.5}$$

where $\mathbb{P}_i = \mathbb{P} \cup \{I_i\}$, $\mathbb{Q}_j = \mathbb{P} \cup \{F_j\}$. Furthermore, let $H_1, \ldots, H_q$ be any choice of polynomials such that $\text{Zero}(\emptyset/H_1 \cdots H_q) = \text{Zero}(\emptyset/\mathbb{I} \cup \mathbb{F})$. Then (2.2.5) can be replaced by

$$\text{Zero}(\mathbb{P}) = \text{Zero}(\mathbb{C}/\mathbb{I}) \cup \bigcup_{k=1}^{q} \text{Zero}(\mathbb{P} \cup \{H_k\}). \tag{2.2.6}$$

The inevitable occurrence of initial factors often renders the appearing polynomials too large to be manageable. The incessant trial for removing such factors often costs much computing time.

Remark 2.2.1. Weak-basic sets and quasi-basic sets may be defined similarly. The algorithms for computing a weak-basic set and a quasi-basic set $\mathbb{B}$ of any polynomial set $\mathbb{P}$ can be obtained from algorithm BasSet by replacing the last line with

$$\mathbb{F} \leftarrow \{F \in \mathbb{F} \setminus \{B\}: \ \mathrm{cls}(F) > \mathrm{cls}(B), \mathrm{ini}(F) \text{ is reduced w.r.t. } B\}$$

and

$$\mathbb{F} \leftarrow \{F \in \mathbb{F} \setminus \{B\}: \ \mathrm{cls}(F) > \mathrm{cls}(B)\}$$

respectively. Lemma 2.2.3 and the specification of algorithm CharSet are still true when the basic set is replaced by a weak-basic set or a quasi-basic set, and the corresponding weak-ascending set or quasi-ascending set $\mathbb{C}$ computed as in CharSet is called a *weak-* or a *quasi-characteristic set* of $\mathbb{P}$, respectively.

Let a fine triangular set also be called a *noncontradictory W-ascending set*. Any set comprising a single nonzero polynomial of class 0 is a *contradictory W-ascending set*. A W-ascending set is called an *ascending chain in weak sense* in Chou (1988) and Chou and Gao (1990b); the notion W-prem is also introduced therein. It is easy to see that algorithm CharSet can also be modified to compute the corresponding *W-characteristic sets* by replacing the ascending set and the basic set with the corresponding W-ascending and W-basic sets.

We shall see that the method of characteristic sets in the standard sense is theoretically more complete than that in the other senses.

Zero decomposition

Let us turn back to the zero relation (2.2.2). As each I_i is reduced with respect to $\mathbb{C}$, by Lemma 2.2.3 any basic set of the polynomial set $\mathbb{P}_i \cup \mathbb{C}$ has rank lower than that of $\mathbb{C}$. Note that $\mathrm{Zero}(\mathbb{P}_i \cup \mathbb{C}) = \mathrm{Zero}(\mathbb{P}_i)$. Therefore, in proceeding further with each $\mathbb{P}_i \cup \mathbb{C}$ as $\mathbb{P}$ by means of algorithm CharSet, one may arrive after a finite number of steps at a zero decomposition of the form

$$\mathrm{Zero}(\mathbb{P}) = \bigcup_{i=1}^{e} \mathrm{Zero}(\mathbb{C}_i/\mathbb{I}_i), \tag{2.2.7}$$

in which $\mathbb{C}_i$ is an ascending set and $\mathbb{I}_i = \mathrm{ini}(\mathbb{C}_i)$ for each i.

Definition 2.2.5. A finite set or sequence Ψ of (weak-) ascending sets $\mathbb{C}_1, \ldots, \mathbb{C}_e$ is called a (*weak-*) *characteristic series* of a polynomial set $\mathbb{P}$ in $\boldsymbol{K}[\boldsymbol{x}]$ if (2.2.7) holds and $\mathrm{prem}(\mathbb{P}, \mathbb{C}_i) = \{0\}$ for every i.

If $\Psi = \emptyset$, it is meant that $e = 0$ and thus $\mathrm{Zero}(\mathbb{P}) = \emptyset$.

Algorithm CharSer: $\Psi \leftarrow \mathrm{CharSer}(\mathbb{P})$. Given a nonempty polynomial set $\mathbb{P} \subset \boldsymbol{K}[\boldsymbol{x}]$, this algorithm computes a characteristic series Ψ of $\mathbb{P}$.

C1. Set $\Phi \leftarrow \{\mathbb{P}\}$, $\Psi \leftarrow \emptyset$.
C2. While $\Phi \neq \emptyset$, do:
  C2.1. Let $\mathbb{F}$ be an element of Φ and set $\Phi \leftarrow \Phi \setminus \{\mathbb{F}\}$.
  C2.2. Compute $\mathbb{C} \leftarrow \mathrm{CharSet}(\mathbb{F})$.
  C2.3. If $\mathbb{C}$ is noncontradictory, then set

$$\Psi \leftarrow \Psi \cup \{\mathbb{C}\},$$
$$\Phi \leftarrow \Phi \cup \{\mathbb{F} \cup \mathbb{C} \cup \{I\}: \ I \in \mathrm{ini}(\mathbb{C}) \setminus \boldsymbol{K}\}.$$

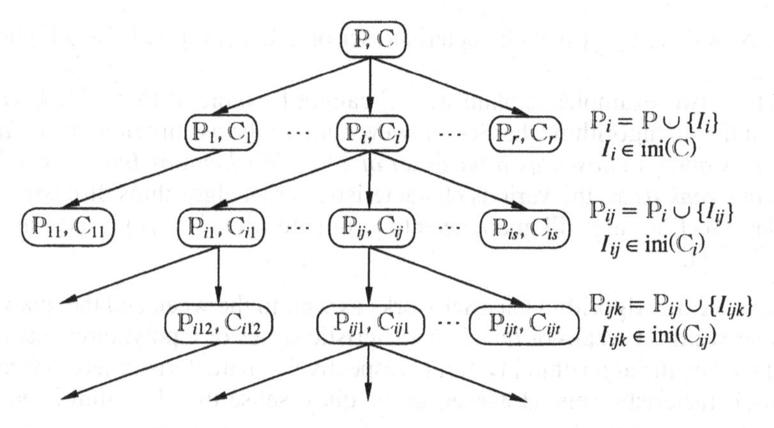

Fig. 2. Decomposition tree of polynomial set $\mathbb{P}$

Actually, this algorithm computes from $\mathbb{P}$ a multibranch tree, called a *decomposition tree* of $\mathbb{P}$. The tree has a root associated with $\mathbb{P}$ and its characteristic set $\mathbb{C}$ and is branched at each node by forming enlarged polynomial sets with adjunction of initials and their characteristic sets. Such a decomposition tree is shown in Fig. 2.

Example 2.2.4. Let $\mathbb{P} = \{F_1, F_2, F_3\}$ and $\mathbb{C}$ be the characteristic set of $\mathbb{P}$ as in Example 2.2.3. One can easily compute a characteristic set $\mathbb{C}_1$ of $\mathbb{P}_1 \cup \mathbb{C}$ and $\mathbb{C}_2$ of $\mathbb{P}_2 \cup \mathbb{C}$ as follows

$$\mathbb{C}_1 = [x_1 + 1, x_2, x_3^2 - 1, x_4 - x_3],$$
$$\mathbb{C}_2 = [x_1, 2x_2^2 + 1, x_3, x_4^2 - x_2x_4 + 3x_2].$$

Observe that all the initials of the polynomials in $\mathbb{C}_1$ and $\mathbb{C}_2$ are constant. We obtain therefore a characteristic series $\Psi = \{\mathbb{C}, \mathbb{C}_1, \mathbb{C}_2\}$ of $\mathbb{P}$ which furnishes a zero decomposition of the form

$$\mathrm{Zero}(\mathbb{P}) = \mathrm{Zero}(\mathbb{C}/I_1) \cup \mathrm{Zero}(\mathbb{C}_1) \cup \mathrm{Zero}(\mathbb{C}_2).$$

Remark 2.2.2. Let $\mathbb{C}$ be a characteristic set of $\mathbb{P} \subset \boldsymbol{K}[\boldsymbol{x}]$ and P any polynomial in $\boldsymbol{K}[\boldsymbol{x}]$ reduced with respect to $\mathbb{C}$. Neither the basic set nor the characteristic set of $\mathbb{P} \cup \{P\}$ necessarily has rank lower than that of $\mathbb{C}$. For example, let

$$\mathbb{P} = \{x_1^2, x_1^2 + x_1, x_1x_2, x_2x_3\}.$$

With $x_1 \prec x_2 \prec x_3$,

$$\mathbb{B} = [x_1^2, x_1x_2], \quad \mathbb{C} = [x_1, x_2x_3]$$

are a basic set and a characteristic set of $\mathbb{P}$, respectively. Now x_2 is reduced with respect to CharSet. However, the basic set of $\mathbb{P} \cup \{x_2\}$ has the same rank as $\mathbb{B}$.

As another example, consider the polynomial set $\mathbb{P} = \{x_1^2 - x_2^2, x_1^2 - 2x_2^2, x_2^2\}$. A characteristic set of $\mathbb{P}$ is $\mathbb{C} = [x_1^2, x_2^2]$. Clearly, x_1x_2 is reduced with respect to $\mathbb{C}$. Now, $[x_1^3, x_1x_2]$ is a characteristic set of $\mathbb{P} \cup \{x_1x_2\}$ and has a higher rank than $\mathbb{C}$.

These two examples explain why $\mathbb{C}$ cannot be omitted from $\mathbb{F} \cup \mathbb{C} \cup \{I\}$ in the last line of algorithm CharSer. However, under the assumption that *a basic set* $\mathbb{B}$ *of* $\mathbb{P}$ *is always chosen as a basic set of* $\mathbb{P}^* \supset \mathbb{P}$ *when any basic set of* $\mathbb{P}^*$ *has the same rank as* $\mathbb{B}$, the various characteristic-series algorithms discussed in this and later sections are still guaranteed to terminate when $\mathbb{F} \cup \{I\}$ is used instead of $\mathbb{F} \cup \mathbb{C} \cup \{I\}$.

Remark 2.2.3. Algorithm CharSer works as well in the weak and the quasi sense. In other words, a weak- or quasi-characteristic series of a polynomial set may be computed by the algorithm in altering respectively characteristic sets to weak- and quasi-characteristic sets. However, in the quasi sense the algorithm is no longer guaranteed to terminate.

During the computation of characteristic series, numerous branches of the decomposition tree may be produced due to the recursive generation of enlarged polynomial sets. Some of these branches are completely redundant and should be removed. Various techniques have been developed for controlling the expansion of branches (see Chou and Gao 1990b, Wang 1995a). For example, in Fig. 2, if the subtree with a root at some $\mathbb{P}_i$ is already computed, then any branch $\mathbb{P}_j$ which contains $\mathbb{P}_i$ as a subset need not be further considered.

Generalization and extensions

In algorithm CharSet, each enlarged polynomial set $\mathbb{F}_{i+1}$, as shown in sketch (2.2.3), is the union of $\mathbb{F}_i$ and $\mathbb{R}_i$. This results in rapid expansion of $\mathbb{F}_{i+1}$ as i increases. To reduce computational expenses, one strategy is to let $\mathbb{F}_{i+1}$ just be the union of $\mathbb{B}_i$ and $\mathbb{R}_i$ and check finally whether all the polynomials in $\mathbb{P}$ have pseudo-remainder 0 with respect to the last basic set. This strategy was proposed in Wu (1987, 1989a). In the first half of this section, we formulate this strategy as a generalized characteristic-set algorithm which may lead to several variants of the standard one.

Definition 2.2.6. Let $\mathbb{P}$ be a nonempty polynomial set in $\boldsymbol{K}[\boldsymbol{x}]$. Any ascending set which is contained in Ideal($\mathbb{P}$) and has rank not higher than that of any basic set of $\mathbb{P}$ is called a *medial set* of $\mathbb{P}$.

A medial set $\mathbb{M}$ of $\mathbb{P}$ is a *characteristic set* of $\mathbb{P}$ if $\mathrm{prem}(\mathbb{P}, \mathbb{M}) = \{0\}$.

Apparently, any basic set itself is a medial set of $\mathbb{P}$. The characteristic set mentioned here is consistent with that in Definition 2.2.3.

Lemma 2.2.4. Let a nonempty polynomial set $\mathbb{P} \subset \boldsymbol{K}[\boldsymbol{x}]$ have a medial set

$$\mathbb{M} = [M_1, M_2, \ldots, M_r],$$

where $\mathrm{cls}(M_1) > 0$. If M is a nonzero polynomial reduced with respect $\mathbb{M}$, then any medial set $\mathbb{M}^+$ of the polynomial set $\mathbb{P}^+ = \mathbb{P} \cup \mathbb{M} \cup \{M\}$ has rank lower than that of $\mathbb{M}$.

Proof. Let $\mathbb{B}^+$ and $\mathbb{B}^*$ be basic sets of $\mathbb{P}^+$ and $\mathbb{P} \cup \mathbb{M}$, respectively. Then $\mathbb{B}^* \precsim \mathbb{M}$.

If $\mathbb{B}^* \sim \mathbb{M}$, then M is reduced with respect to $\mathbb{B}^*$. Hence, by Definition 2.2.6 and Lemma 2.2.3 we have

$$\mathbb{M}^+ \precsim \mathbb{B}^+ \prec \mathbb{B}^* \sim \mathbb{M}.$$

If $\mathbb{B}^* \prec \mathbb{M}$, then

$$\mathbb{M}^+ \precsim \mathbb{B}^+ \precsim \mathbb{B}^* \prec \mathbb{M}$$

holds. Therefore, in either case $\mathbb{M}^+ \prec \mathbb{M}$. □

Let GenCharSet designate the algorithm obtained from CharSet by replacing step C2.1 therein with

C2.1. Compute a medial set $\mathbb{C}$ of $\mathbb{F}$.

Theorem 2.2.5. Algorithm GenCharSet terminates and its specification is correct; that is, it computes a characteristic set $\mathbb{C}$ of any given nonempty polynomial set $\mathbb{P}$.

Proof. Algorithm GenCharSet has the same structure as CharSet. While replacing each $\mathbb{B}_i$ by an arbitrary medial set $\mathbb{M}_i$ of $\mathbb{F}_i$, and letting each enlarged polynomial set $\mathbb{F}_{i+1}$ be $\mathbb{F}_i \cup \mathbb{R}_i \cup \mathbb{M}_i$, we should get a sketch similar to (2.2.3), but each $\mathbb{M}_i$ is no longer a subset of $\mathbb{F}_i$. Then, the termination of GenCharSet is guaranteed by Lemmas 2.2.1 and 2.2.4. From the formation of each $\mathbb{F}_i$ and the pseudo-remainder formula, the correctness is easily proved by an argument similar to the correctness proof of CharSet. □

By taking different medial sets, one may get different variants of algorithm CharSet. In particular, if the basic set is taken as a medial set, then GenCharSet is identical to CharSet. Now let CharSetN designate the algorithm obtained from CharSet by replacing $\mathbb{F} \cup \mathbb{R}$ in the last line with $\mathbb{C} \cup \mathbb{R}$. Then CharSetN computes a medial set of the input polynomial set. Replacing step C2.1 in GenCharSet by

C2.1. Compute $\mathbb{C} \leftarrow$ CharSetN($\mathbb{F}$),

one obtains immediately a modification of CharSet as mentioned at the beginning of this section.

If one intends to compute triangular sets only, the algorithm may have plenty of scope for variation. Various modifications of CharSet lead naturally to modifications of the characteristic-series algorithms, for which we omit the details. The reader may also refer to Chou (1988), Ko (1988), Chou and Gao (1990b), and other relevant work for variants, modifications, and extensions.

Let $[\mathbb{P}, \mathbb{Q}]$ be a polynomial system. From (2.2.7) one obtains the following zero decomposition

$$\operatorname{Zero}(\mathbb{P}/\mathbb{Q}) = \bigcup_{i=1}^{e} \operatorname{Zero}(\mathbb{C}_i/\mathbb{I}_i \cup \mathbb{Q}), \tag{2.2.8}$$

in which $\mathbb{C}_i$ is an ascending set and $\mathbb{I}_i = \operatorname{ini}(\mathbb{C}_i)$ for each i. In (2.2.8), one can delete the component $\operatorname{Zero}(\mathbb{C}_i/\mathbb{I}_i \cup \mathbb{Q})$ when $0 \in \operatorname{prem}(\mathbb{Q}, \mathbb{C}_i)$ for some i. So we may assume that $0 \notin \mathrm{prem}(\mathbb{Q}, \mathbb{C}_i)$ for any i. Moreover, one can replace $\mathbb{I}_i \cup \mathbb{Q}$ in (2.2.8) by $\mathbb{D}_i = \mathbb{I}_i \cup \mathrm{prem}(\mathbb{Q}, \mathbb{C}_i)$ for each i, so that

$$\mathrm{Zero}(\mathbb{P}/\mathbb{Q}) = \bigcup_{i=1}^{e} \mathrm{Zero}(\mathbb{C}_i/\mathbb{D}_i), \tag{2.2.9}$$

where each $[\mathbb{C}_i, \mathbb{D}_i]$ is clearly a fine triangular system.

Definition 2.2.7. A finite set or sequence Ψ of (fine) triangular systems $\mathfrak{T}_1, \ldots, \mathfrak{T}_e$ in $\boldsymbol{K}[\boldsymbol{x}]$ is called a *(fine) triangular series*. It is called a *(fine) triangular series* of a polynomial system $\mathfrak{P}$ in $\boldsymbol{K}[\boldsymbol{x}]$ if (2.1.8) holds.

A (fine) triangular series of $[\mathbb{P}, \emptyset]$ is also called a *(fine) triangular series* of the polynomial set $\mathbb{P}$.

Ψ is called a *characteristic series* of $\mathfrak{P} = [\mathbb{P}, \mathbb{Q}]$ if (2.1.8) holds with $\mathfrak{T}_i = [\mathbb{T}_i, \mathbb{U}_i]$ and $\mathrm{prem}(\mathbb{P}, \mathbb{T}_i) = \{0\}$ for every i.

When $\Psi = \emptyset$, it is understood that $\mathrm{Zero}(\mathfrak{P}) = \emptyset$.

Clearly, the set of fine triangular systems $[\mathbb{C}_1, \mathbb{D}_1], \ldots, [\mathbb{C}_e, \mathbb{D}_e]$ in (2.2.9) is a characteristic series of $[\mathbb{P}, \mathbb{Q}]$.

Remark 2.2.4. *Weak-medial sets* and *quasi-medial sets* may be similarly defined. The corresponding weak- or quasi-characteristic sets can be computed by the algorithm obtained from GenCharSet by replacing the medial set with a weak-medial or a quasi-medial set. One can also compute weak-characteristic series from polynomial sets or polynomial systems by devising similar algorithms.

Remark 2.2.5. A (weak-, quasi-) medial set computed by CharSetN from $\mathbb{P}$ is called a *(weak-, quasi-) N-characteristic set* of $\mathbb{P}$. For a (weak-, quasi-) N-characteristic set $\mathbb{C}$, the zero relations (2.2.5) and (2.2.6) do not hold any more; we only have

$$\mathrm{Zero}(\mathbb{P}) \subset \mathrm{Zero}(\mathbb{C}).$$

It is worth noting that (weak-, quasi-) N-characteristic sets are sometimes sufficient for applications such as solving systems of algebraic equations. If, in particular, $\mathbb{C}$ has only finitely many zeros, whether every zero of $\mathbb{C}$ is also a zero of $\mathbb{P}$ can be verified by evaluation.

Remark 2.2.6. To determine whether a (weak-, quasi-) N-characteristic set $\mathbb{C}$ is indeed a (weak-, quasi-) characteristic set, one has to follow algorithm GenCharSet to verify whether all the polynomials in the input set have pseudo-remainder 0 with respect to $\mathbb{C}$. Experiments show that in most cases the pseudo-remainders are 0, i.e., GenCharSet terminates after the first iteration of the while-loop. The verification of 0 pseudo-remainders often takes a great amount of computing time. There are some strategies which can be used to partially avoid the verification of 0 pseudo-remainders. This is done by examining the factor-relations of some initials and removed factors (see Wang 1992b).

Most of the algorithms presented in this book have been implemented by the author in Maple, a popular computer algebra system. In particular, a package that implements a number of characteristic-set-based algorithms has been publicly available with the Maple share library since early 1991. The current version of the package can be obtained via WWW as: http://www-leibniz.imag.fr/ATINF/Dongming.Wang/charsets-2.0.tar.Z. This book focuses on the development of theory and algorithms. Implementation issues will not be discussed, neither will any experimental timing statistics and comparison among the algorithms be provided. The reader may consult relevant research publications for more information. Nevertheless, a number of remarks are given as links for efficient implementation of the algorithms. In general, one can skip reading the remarks if only the theoretical aspect is of concern.

2.3 Seidenberg's algorithm refined

The goal of this section is to present a decomposition algorithm that splits polynomial systems whenever pseudo-division is performed. Using this algorithm, triangular series are computed instead of characteristic series. One advantage of this is that the verification of 0 remainders is completely avoided. We employ a pure top-down elimination from x_n to x_1 which is essentially due to Seidenberg (1956a, b). In comparison, the elimination in CharSet may be considered as performed simultaneously for all the variables.

As a triangular set, not necessarily fine, may not be well behaved, it is impossible to set up the whole theory for characteristic sets in the quasi sense. Characteristic-set computation in the standard or the weak sense often leads to rapid increase of polynomial size. For in this case, any polynomial or its initial has to be reduced with respect to the others in an ascending set. To control the increase of polynomial size and for other reasons, we use triangular system $[\mathbb{T}, \mathbb{U}]$, in which $\mathrm{prem}(I, \mathbb{T})$ for all $I \in \mathrm{ini}(\mathbb{T})$ are collected, together with other polynomials, as $\mathbb{U}$.

Moreover, computing a characteristic set of $\mathbb{P} \cup \{I\}$ as in CharSer may have to perform pseudo-divisions which have been done already in the way of computing the characteristic set $\mathbb{C}$ of $\mathbb{P}$. In other words, there may be repeated computation of pseudo-remainders which is unnecessary. To avoid such repetition and to keep a maximal amount of information for subsequent computation, we shall retain partially triangularized systems with the data structures of triplets and quadruplets.

Before describing the elimination algorithm, let us first prove the following simple lemma.

Lemma 2.3.1. Let T be a nonconstant polynomial with $\mathrm{ini}(T) = I$ and $[\mathbb{P}, \mathbb{Q}]$ a polynomial system in $\boldsymbol{K}[\boldsymbol{x}]$, and $\mathbb{R} = \mathrm{prem}(\mathbb{P}, T) \setminus \{0\}$. Then

$$\begin{aligned}\mathrm{Zero}(\mathbb{P} \cup \{T\}/\mathbb{Q}) = &\, \mathrm{Zero}(\mathbb{R} \cup \{T\}/\mathbb{Q} \cup \{I\}) \\ &\cup \mathrm{Zero}(\mathbb{P} \cup \{I, \mathrm{red}(T)\}/\mathbb{Q}).\end{aligned} \tag{2.3.1}$$

Proof. For every polynomial $P \in \mathbb{P}$, pseudo-dividing P by T in x_i leads to a pseudo-remainder formula of the form

$$I^q P = AT + R, \tag{2.3.2}$$

where $A, R \in \boldsymbol{K}[\boldsymbol{x}]$ and the integer $q > 0$. For any

$$\bar{\boldsymbol{x}} \in \mathrm{Zero}(\mathbb{P} \cup \{T\}/\mathbb{Q}),$$

we have

$$T(\bar{\boldsymbol{x}}) = 0 \text{ and } P(\bar{\boldsymbol{x}}) = 0, \quad \forall P \in \mathbb{P},$$

so $R(\bar{\boldsymbol{x}}) = 0$ for all $R \in \mathbb{R}$. Clearly, $Q(\bar{\boldsymbol{x}}) \neq 0$ for all $Q \in \mathbb{Q}$. If $I(\bar{\boldsymbol{x}}) \neq 0$, then

$$\bar{\boldsymbol{x}} \in \mathrm{Zero}(\mathbb{R} \cup \{T\}/\mathbb{Q} \cup \{I\}). \tag{2.3.3}$$

Otherwise, we have $I(\bar{\boldsymbol{x}}) = 0$ and thus $\mathrm{red}(T)(\bar{\boldsymbol{x}}) = 0$; therefore

$$\bar{\boldsymbol{x}} \in \mathrm{Zero}(\mathbb{P} \cup \{I, \mathrm{red}(T)\}/\mathbb{Q}). \tag{2.3.4}$$

This shows that the left-hand side is contained in the right-hand side of (2.3.1). To show the opposite, one sees that if $\bar{\boldsymbol{x}}$ satisfies (2.3.4), then $T(\bar{\boldsymbol{x}}) = 0$ and thus $\bar{\boldsymbol{x}} \in \mathrm{Zero}(\mathbb{P} \cup \{T\}/\mathbb{Q})$. Otherwise, let (2.3.3) hold. By (2.3.2) we have $P(\bar{\boldsymbol{x}}) = 0$ for all $P \in \mathbb{P}$, so $\bar{\boldsymbol{x}} \in \mathrm{Zero}(\mathbb{P} \cup \{T\}/\mathbb{Q})$ as well. □

For any integer $1 \leq i \leq n$ and polynomial set $\mathbb{P}$, the set of those polynomials in $\mathbb{P}$ which involve the variables $x_1, \ldots, x_i$ only is denoted by $\mathbb{P}^{(i)}$. Symbolically,

$$\mathbb{P}^{(i)} \triangleq \mathbb{P} \cap \boldsymbol{K}[x_1, \ldots, x_i].$$

Moreover, let

$$\mathbb{P}^{[i]} \triangleq \mathbb{P} \setminus \mathbb{P}^{(i)}, \quad \mathbb{P}^{\langle i \rangle} \triangleq \mathbb{P}^{(i)} \setminus \mathbb{P}^{(i-1)}.$$

For any polynomial system $\mathfrak{P} = [\mathbb{P}, \mathbb{Q}]$, define

$$\mathfrak{P}^{(i)} \triangleq [\mathbb{P}^{(i)}, \mathbb{Q}^{(i)}], \quad \mathfrak{P}^{\langle i \rangle} \triangleq [\mathbb{P}^{\langle i \rangle}, \mathbb{Q}^{\langle i \rangle}].$$

A polynomial set $\mathbb{P}$ is said to be of *level* i, denoted as $\mathrm{level}(\mathbb{P}) = i$, if $\mathbb{P} \subset \boldsymbol{K}[x_1, \ldots, x_i]$ and $\mathbb{P}^{\langle i \rangle} \neq \emptyset$, i.e., i is the smallest integer such that $\mathbb{P} \subset \boldsymbol{K}[x_1, \ldots, x_i]$. The level of $\mathbb{P}$ is also called the *level* of $\mathfrak{P}$.

Now we introduce a data structure called *triplet*, which will be used in the presentation of several algorithms.

Data structure. A *triplet* of level i $(1 \leq i \leq n)$ is a list $[\mathbb{P}, \mathbb{Q}, \mathbb{T}]$ of three elements, where

a. $[\mathbb{P}, \mathbb{Q}]$ is a polynomial system of level i in $\boldsymbol{K}[\boldsymbol{x}]$;
b. $\mathbb{T}$, if nonempty, is a triangular set in $\boldsymbol{K}[\boldsymbol{x}]$ with $\mathbb{T}^{(i)} = \emptyset$.

When speaking about a polynomial system $[\mathbb{P}, \mathbb{Q}]$, we are concerned with $\mathrm{Zero}(\mathbb{P}/\mathbb{Q})$. Trivially, $\mathbb{P}$ may be written as $\mathbb{P} = \mathbb{P}^{(i)} \cup \mathbb{P}^{[i]}$ for every i. It may happen that, for some i, $\mathbb{P}^{(i)}$ is of level i and $\mathbb{P}^{[i]}$ can be ordered as a triangular set $\mathbb{T}$. In this case, $[\mathbb{P}^{(i)}, \mathbb{Q}, \mathbb{T}]$ is a triplet, with which $\mathrm{Zero}(\mathbb{P}^{(i)} \cup \mathbb{T}/\mathbb{Q})$ is of concern.

Our elimination procedure will start with a triplet $[\mathbb{P}, \mathbb{Q}, \mathbb{T}]$ with $\mathbb{T} = \emptyset$. The variables x_i are eliminated and the obtained, triangularized polynomials are adjoined to $\mathbb{T}$ successively for $i = n, n-1, \ldots, 1$.

Let i be a positive integer and $[\mathbb{P}, \mathbb{Q}]$ a polynomial system of level i. Clearly, $\mathbb{F} = \mathbb{P}^{\langle i \rangle} \neq \emptyset$ and every polynomial in $\mathbb{F}$ has class i. We want to eliminate the variable x_i for the polynomials in $\mathbb{F}$, so that after the elimination only one polynomial has class i. For this purpose, let us take one polynomial T from $\mathbb{F}$ which has minimal degree in x_i and pseudo-divide all the polynomials in $\mathbb{F} \setminus \{T\}$ by T in x_i. Meanwhile, $\mathrm{ini}(T)$ is assumed to be nonzero and the case in which $\mathrm{ini}(T)$ happens to be 0 is considered disjunctively by replacing T with $\mathrm{ini}(T)$ and $\mathrm{red}(T)$. Then, we reset $\mathbb{F}$ to be $\{T\} \cup \mathrm{prem}(\mathbb{F}, T) \setminus \{0\}$ and repeat the above process. In this way, we shall finally get a single polynomial T in $\mathbb{F}$ which has class i and a set of other polynomial systems of level $\leq i$.

The procedure explained above is described in the following algorithmic form.

Algorithm Elim: $[T, \mathbb{F}, \mathbb{G}, \Delta] \leftarrow \mathrm{Elim}(\mathbb{P}, \mathbb{Q}, i)$. Given an integer $i > 0$ and a polynomial system $[\mathbb{P}, \mathbb{Q}]$ of level i in $\boldsymbol{K}[\boldsymbol{x}]$, this algorithm computes a polynomial T of class i, a polynomial system $[\mathbb{F}, \mathbb{G}]$ of level $\leq i-1$, and a set Δ of polynomial systems of level $\leq i$ such that

$$\mathrm{Zero}(\mathbb{P}/\mathbb{Q}) = \mathrm{Zero}(\mathbb{F} \cup \{T\}/\mathbb{G}) \cup \bigcup_{[\mathbb{P}^*, \mathbb{Q}^*] \in \Delta} \mathrm{Zero}(\mathbb{P}^*/\mathbb{Q}^*). \tag{2.3.5}$$

E1. Set $T \leftarrow 0$, $\mathbb{F} \leftarrow \mathbb{P}$, $\mathbb{G} \leftarrow \mathbb{Q}$, $\Delta \leftarrow \emptyset$.
E2. While $\mathbb{F}^{\langle i \rangle} \neq \{T\}$, do:
  E2.1. Let T be an element of $\mathbb{F}^{\langle i \rangle}$ with minimal degree in x_i.
  E2.2. Set

$$\Delta \leftarrow \Delta \cup \{[\mathbb{F} \setminus \{T\} \cup \{\mathrm{red}(T), \mathrm{ini}(T)\}, \mathbb{G}]\},$$
$$\mathbb{G} \leftarrow \mathbb{G} \cup \{\mathrm{ini}(T)\}.$$

  E2.3. Compute $\mathbb{F} \leftarrow \{T\} \cup \mathrm{prem}(\mathbb{F}, T) \setminus \{0\}$.
E3. Set $\mathbb{F} \leftarrow \mathbb{F} \setminus \{T\}$.

Proof. Since $\mathbb{P}$ is of level i, initially $\mathbb{F}^{\langle i \rangle}$ is neither empty nor equal to $\{T\} = \{0\}$. One sees clearly that every substep of E2 terminates. As in each iteration of this while-loop $\deg(T, x_i)$ decreases at least by 1, after a finite number of steps all the nonzero pseudo-remainders of the polynomials in $\mathbb{F}$ with respect to T will have class $<i$. Then, the set $\mathbb{F}^{\langle i \rangle}$ becomes $\{T\}$ and the while-loop terminates.

The zero relation (2.3.5) follows from repeated application of the relation (2.3.1) in Lemma 2.3.1. □

Note that step E2.2 can be skipped when $\mathrm{ini}(T)$ is a constant, and the pseudo-remainders need be computed in step E2.3 actually only for the polynomials in $\mathbb{F}^{[i-1]} \setminus \{T\}$.

Example 2.3.1. The following polynomial set

$$\mathbb{P} = \{x^{31} - x^6 - x - y, x^8 - z, x^{10} - t\},$$

popularized by L. Robbiano (according to C. Traverso and L. Donati), was considered in Wang (1993). Here and later on it will be used to illustrate several algorithms. One may observe that $\mathbb{P}$ is already a triangular set with respect to the variable ordering $x \prec y \prec z \prec t$. But, for our purpose, we order the variables as $t \prec z \prec y \prec x$.

To see how Elim works, consider the polynomial system $[\mathbb{P}, \emptyset]$ of level 4 as input. Initially, set

$$T \leftarrow 0, \ \mathbb{F} \leftarrow \mathbb{P}, \ \mathbb{G} \leftarrow \emptyset, \ \Delta \leftarrow \emptyset$$

in step E1.

Now come to the while-loop. First, take $T = x^8 - z$ from $\mathbb{F}^{[3]} = \mathbb{F}$ in step E2.1 which has minimal degree 8 in x and initial $I = 1$. Since I is a constant, we can skip step E2.2. Pseudo-dividing the two other polynomials in $\mathbb{F} = \mathbb{P}$ by T, one gets two nonzero pseudo-remainders

$$R_1 = z^3x^7 - x^6 - x - y, \quad R_2 = zx^2 - t,$$

where $\mathrm{lv}(R_1) = \mathrm{lv}(R_2) = x$. So in step E2.3, update $\mathbb{F} \leftarrow \{T, R_1, R_2\}$.

For the second loop, take $T = R_2$ from $\mathbb{F}^{[3]} = \mathbb{F}$ in step E2.1 which has minimal degree 2 in x and initial $I = z$. In step E2.2, set

$$\Delta \leftarrow \{[\{x^8 - z, R_1, z, -t\}, \emptyset]\}, \quad \mathbb{Q} \leftarrow \{z\}.$$

Similarly, pseudo-dividing the two other polynomials in $\mathbb{F}$ by $T = R_2$ yields the pseudo-remainders

$$R_3 = -z^5 + t^4, \quad R_4 = t^3z^3x - z^3x - z^3y - t^3$$

with $\mathrm{lv}(R_3) = z$ and $\mathrm{lv}(R_4) = x$. Then set $\mathbb{F} \leftarrow \{R_2, R_3, R_4\}$ in step E2.3.

For the third loop, set $T \leftarrow R_4$ in step E2.1, where $\deg(R_4, x) = 1 < \deg(R_2, x)$ and the initial $t^3z^3 - z^3$ of R_4 is simplified by $z \in \mathbb{Q}$ to $I = t^3 - 1$. In step E2.2 the polynomial system

$$[\{R_2, R_3, -z^3y - t^3, t^3 - 1\}, \{z\}]$$

is added to Δ and the polynomial $t^3 - 1$ to $\mathbb{Q}$. Pseudo-dividing R_2 by $T = R_4$, we have

$$R_5 = \mathrm{prem}(R_2, R_4) = z^6y^2 + 2t^3z^3y - t^7z^5 + 2t^4z^5 - tz^5 + t^6$$

with $\mathrm{lv}(R_5) = y$. Finally, set $\mathbb{F} \leftarrow \{R_4, R_3, R_5\}$ and the while-loop terminates.

The algorithm terminates after deleting T from $\mathbb{F}$ in step E3. The output consists of $T = R_4$, the polynomial system

$$[\mathbb{F}, \mathbb{G}] = [\{R_3, R_5\}, \{z, t^3 - 1\}]$$

and the set Δ of two other polynomial systems.

Now, let us explain how to decompose a polynomial system $[\mathbb{P}, \mathbb{Q}]$ into triangular systems by Elim as the main subalgorithm. This is done by performing an elimination top-down from x_n to x_1. More concretely, for each x_i, $i = n, \ldots, 1$, one proceeds as follows.

The iteration starts with $i = n$. If $\mathbb{P}^{\langle i\rangle} = \emptyset$, then go for next i. Otherwise, let $T \in \mathbb{P}^{\langle i\rangle}$ have minimal degree in x_i. Then

$$\mathbb{P} = 0, \mathbb{Q} \neq 0 \iff \begin{cases} \mathbb{P}^* = 0, I = 0, \mathrm{red}(T) = 0, & \mathbb{Q} \neq 0; \text{ or} \\ \mathrm{prem}(\mathbb{P}, T) = 0, T = 0, & \mathbb{Q} \neq 0, I \neq 0, \end{cases}$$

where

$$\mathbb{P}^* = \mathbb{P} \setminus \{T\}, \quad I = \mathrm{ini}(T).$$

Therefore we have

$$\begin{aligned} \mathrm{Zero}(\mathbb{P}/\mathbb{Q}) &= \mathrm{Zero}(\mathbb{P}^* \cup \{I, \mathrm{red}(T)\}/\mathbb{Q}) \cup \mathrm{Zero}(\mathrm{prem}(\mathbb{P}, T) \cup \{T\}/\mathbb{Q} \cup \{I\}) \\ &= \cdots \qquad\qquad\qquad\qquad \text{(repeat recursively)} \\ &= \bigcup_{i=1}^{e} \mathrm{Zero}(\mathbb{T}_i/\mathbb{U}_i). \end{aligned}$$

The above sketch is made precise in the following algorithm.

Algorithm TriSer: $\Psi \leftarrow \mathrm{TriSer}(\mathbb{P}, \mathbb{Q})$. Given a polynomial system $[\mathbb{P}, \mathbb{Q}]$ in $\boldsymbol{K}[\boldsymbol{x}]$, this algorithm computes a fine triangular series Ψ of $[\mathbb{P}, \mathbb{Q}]$.

T1. Set $\Psi \leftarrow \emptyset$, $\Phi \leftarrow \{[\mathbb{P}, \mathbb{Q}, \emptyset]\}$.

T2. While $\Phi \neq \emptyset$, do:

T2.1. Let $[\mathbb{F}, \mathbb{G}, \mathbb{T}']$ be an element of Φ and set $\Phi \leftarrow \Phi \setminus \{[\mathbb{F}, \mathbb{G}, \mathbb{T}']\}$.

T2.2. Compute $[\mathbb{T}, \mathbb{U}, \Omega] \leftarrow \mathrm{PriTriSys}(\mathbb{F}, \mathbb{G})$.

T2.3. Set $\Phi \leftarrow \Phi \cup \{[\mathbb{F}^*, \mathbb{G}^*, \mathbb{T}^* \cup \mathbb{T}']: [\mathbb{F}^*, \mathbb{G}^*, \mathbb{T}^*] \in \Omega\}$. If $\mathbb{T} \neq [1]$, then set $\Psi \leftarrow \Psi \cup \{[\mathbb{T} \cup \mathbb{T}', \mathbb{U}]\}$.

The subalgorithm PriTriSys is described as follows.

Algorithm PriTriSys: $[\mathbb{T}, \mathbb{U}, \Omega] \leftarrow \mathrm{PriTriSys}(\mathbb{P}, \mathbb{Q})$. Given a polynomial system $[\mathbb{P}, \mathbb{Q}]$ in $\boldsymbol{K}[\boldsymbol{x}]$, this algorithm computes a (fine triangular) system $[\mathbb{T}, \mathbb{U}]$ and a set Ω of triplets such that

$$\mathrm{Zero}(\mathbb{P}/\mathbb{Q}) = \mathrm{Zero}(\mathbb{T}/\mathbb{U}) \cup \bigcup_{[\mathbb{P}^*, \mathbb{Q}^*, \mathbb{T}^*] \in \Omega} \mathrm{Zero}(\mathbb{P}^* \cup \mathbb{T}^*/\mathbb{Q}^*).$$

P1. Set $\mathbb{T} \leftarrow \emptyset$, $\mathbb{F} \leftarrow \mathbb{P}$, $\mathbb{U} \leftarrow \mathbb{Q}$, $\Omega \leftarrow \emptyset$.

P2. For $i = \mathrm{level}(\mathbb{P}), \ldots, 1$ do:

P2.1. If $\mathbb{F} \cap \boldsymbol{K} \setminus \{0\} \neq \emptyset$, then set $\mathbb{T} \leftarrow [1]$, and the algorithm terminates. If $\mathrm{level}(\mathbb{F}) < i$, then go to P2 for next i.

P2.2. Compute $[T, \mathbb{F}, \mathbb{U}, \Delta] \leftarrow \mathrm{Elim}(\mathbb{F}, \mathbb{U}, i)$ and set

$$\Omega \leftarrow \Omega \cup \{\delta \cup [\mathbb{T}]: \ \delta \in \Delta\}.$$

P2.3. Compute $\mathbb{U} \leftarrow \mathrm{prem}(\mathbb{U}, T)$.

P2.4. If $0 \in \mathbb{U}$, then set $\mathbb{T} \leftarrow [1]$, and the algorithm terminates; else set $\mathbb{T} \leftarrow [T] \cup \mathbb{T}$.

In step T2 of TriSer, the set Φ of triplets increases and decreases, and meanwhile the triangular systems $[\mathbb{T}, \mathbb{U}]$ are produced. This procedure terminates when Φ becomes empty. Within the while-loop, for each triplet $[\mathbb{F}, \mathbb{G}, \mathbb{T}]$ of level l taken from Φ the variables are eliminated, successively from x_l to x_1, by the subalgorithm Elim.

As before, when $\mathrm{Zero}(\mathbb{P}/\mathbb{Q}) = \emptyset$ is detected in TriSer, we have $e = 0$ and $\Psi = \emptyset$.

Example 2.3.2. Let us recall Example 2.3.1 and illustrate TriSer with the input system $[\mathbb{P}, \emptyset]$. The sets Ψ and Φ are initially set to $\emptyset$ and $\{[\mathbb{P}, \emptyset, \emptyset]\}$, respectively.

Consider the while-loop. First, the only triplet in Φ is taken and deleted from Φ in step T2.1. We turn to PriTriSys in step T2.2; first iterate for $i = 4$. Call of Elim in step P2.2 yields the polynomial $T = R_4$, the polynomial system

$$[\mathbb{F}, \mathbb{G}] = [\{R_3, R_5\}, \{z, t^3 - 1\}]$$

and the set Δ as given in Example 2.3.1. Thus, two triplets are formed from the two polynomial systems of Δ and are added to Φ.

Since the two polynomials in $\mathbb{G}$ have leading variables $\prec x$, the execution of step P2.3 is trivial and does not update the value of any variable. In step P2.4, set $\mathbb{T} \leftarrow [R_4]$.

For $i = 3$ and 2, the polynomials R_5 and R_3 in $\mathbb{F}$ are chosen as T in step P2.2, respectively, and no elimination is necessary. As the pseudo-remainders of the two polynomials in $\mathbb{G}$ with respect to R_5 and R_3 are themselves, $\mathbb{G}$ is not updated in step P2.3. Therefore, we obtain the first triangular system $[\mathbb{T}_1, \mathbb{U}_1]$ with

$$\mathbb{T}_1 = [R_3, R_5, R_4], \quad \mathbb{U}_1 = \{z, t^3 - 1\},$$

which is added to Ψ in step T2.3.

Now there are two triplets in Φ which remain to be considered. For the first $[\{T, R_1, z, -t\}, \emptyset, \emptyset]$, the two polynomials T, R_1 have leading variable x, of which R_1 has lower degree 7 and initial $z^3 \rightsquigarrow z$. Here and elsewhere, $\rightsquigarrow$ stands for "simplified to." One may split the computation to two cases according to $z = 0$ and $z \neq 0$ by strictly following the described algorithm, which is somewhat complicated. Actually, we may simplify T and R_1 by $z = 0$ and $t = 0$ and make the resulting polynomials squarefree. Then, the second triangular set $\mathbb{T}_2 = [t, z, y, x]$ is obtained immediately, with $\mathbb{U}_2 = \emptyset$. For the other triplet

$$[\mathbb{F}, \mathbb{G}, \mathbb{T}] = [\{R_3, R_2, -z^3y - t^3, t^3 - 1\}, \{z\}, \emptyset],$$

the polynomials

$$R_2, \quad -z^3y - t^3, \quad R_3, \quad t^3 - 1$$

have leading variables x, y, z, t, respectively, and thus already constitute a triangular set. Hence, we get

$$\mathbb{T}_3 = [t^3 - 1, R_3, -z^3y - t^3, R_2], \quad \mathbb{U}_3 = \{z\}.$$

Proof of TriSer. Termination. We only need to prove that the while-loop terminates. For any triplet ψ taken from Φ in step T2.1 of TriSer, let $\mathbb{F}$ be the first component of ψ and $\mathbb{P}^*$ the first component of some polynomial system in Δ produced by

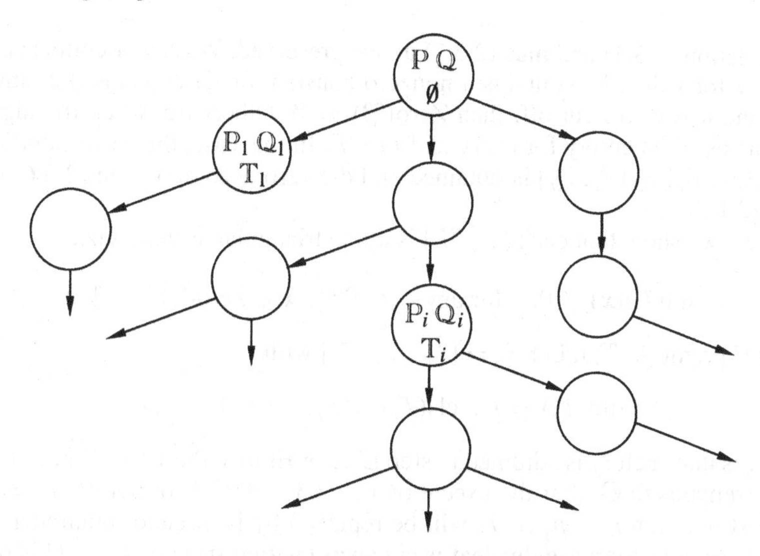

Fig. 3. Multibranch tree $\mathcal{T}$

Elim from ψ. Then, from the replacement of T by its initial and reductum in step E2.2 of Elim one sees clearly that either

$$\operatorname{level}(\mathbb{P}^*) < \operatorname{level}(\mathbb{F}) \quad \text{or} \quad \operatorname{level}(\mathbb{P}^*) = \operatorname{level}(\mathbb{F}) = l.$$

In the latter case, the minimal degree in x_l of the polynomials in $\mathbb{P}^{*\langle l \rangle}$ is smaller than that of the polynomials in $\mathbb{F}^{\langle l \rangle}$. Since both level and degree are positive integers, any steadily decreasing sequence of levels or minimal degrees is finite. Therefore, the while-loop can only have finitely many iterations. This proves the termination of TriSer.

Correctness. Let us view algorithm TriSer as for computing a multibranch tree $\mathcal{T}$ starting from its root with which the triplet $[\mathbb{P}, \mathbb{Q}, \emptyset]$ is associated (see Fig. 3).

Set $\mathfrak{P} = [\mathbb{P}, \mathbb{Q}]$. With each node or leaf i of $\mathcal{T}$, a triplet $[\mathbb{P}_i, \mathbb{Q}_i, \mathbb{T}_i]$ is associated such that after the execution of every step[1] of TriSer the zero relation

$$\operatorname{Zero}(\mathfrak{P}) = \bigcup_{i \text{ over all leaves of } \mathcal{T}} \operatorname{Zero}(\mathbb{P}_i \cup \mathbb{T}_i/\mathbb{Q}_i) \tag{2.3.6}$$

is preserved. This is because the relation (2.3.1) implies that

$$\operatorname{Zero}(\mathbb{P} \cup \mathbb{T}/\mathbb{Q}) = \operatorname{Zero}(\mathbb{F} \cup \{T\} \cup \mathbb{T}/\mathbb{G}) \cup \bigcup_{[\mathbb{P}^*, \mathbb{Q}^*] \in \Delta} \operatorname{Zero}(\mathbb{P}^* \cup \mathbb{T}/\mathbb{Q}^*) \tag{2.3.7}$$

for any $\mathbb{T}$, and because $\operatorname{Zero}(\mathbb{F} \cup \{T\} \cup \mathbb{T}/\mathbb{G})$ remains unchanged when step P2.3 is executed. The branches are generated clearly by the subalgorithm Elim with the

1 For steps P2.2 and P2.3, the polynomial T is taken into account of the triplet in process. Namely, $\mathbb{P}_i$ corresponds to $\mathbb{F} \cup \{T\}$.

zero relation (2.3.1) and thus (2.3.7) above preserved. We can of course cut those leaves i for which $\mathbb{P}_i$ contains a nonzero constant or $\mathbb{Q}_i$ contains 0 at any time. If all the leaves are cut off, then $\text{Zero}(\mathfrak{P}) = \emptyset$. Otherwise, when the algorithm terminates, $\mathbb{P}_i$ is empty for every leaf i of $\mathcal{T}$. In this case, the corresponding pair $\mathfrak{T}_i = [\mathbb{T}_i, \mathbb{U}_i] = [\mathbb{T}_i, \mathbb{Q}_i]$ is obtained and the zero decomposition (2.3.6) has the form (2.1.8).

Next we show that each $[\mathbb{T}_i, \mathbb{U}_i]$ is a fine triangular system, viz.,

$$\text{ini}(T)(\bar{\boldsymbol{x}}) \neq 0, \quad \text{for any } T \in \mathbb{T}_i^{\langle p \rangle}, \ \bar{\boldsymbol{x}} \in \text{Zero}(\mathbb{T}_i^{(p-1)}/\mathbb{U}_i),$$

and $0 \notin \text{prem}(\mathbb{U}_i, \mathbb{T}_i)$. Let $\mathbb{T}_i = [T_1, \ldots, T_r]$ with

$$\text{ini}(T_j) = I_j, \ \text{cls}(T_j) = p_j, \quad j = 1, \ldots, r.$$

One sees that each I_j is adjoined in step E2.2 of Elim to the set $\mathbb{G}$. Since $\text{cls}(I_j) < p_j$, I_j remains in $\mathbb{G}$ after the execution of P2.3 and P2.4 for iteration $i = p_j$. In the next iteration $i = p_{j-1}$, I_j will be replaced by its pseudo-remainder (which is nonzero, for otherwise this leaf is cut away) with respect to T_{j-1}. This pseudo-remainder will further be replaced by its nonzero pseudo-remainder with respect to T_{j-2} in the iteration $i = p_{j-2}$, and so on. Therefore,

$$\text{prem}(I_j, \mathbb{T}_i^{\{j-1\}}) = \text{prem}(I_j, [T_1, \ldots, T_{\{j-1\}}])$$

is contained in $\mathbb{U}_i$ for all j. From the pseudo-remainder formula (2.1.2), one knows that any zero of I_j which is also a zero of $\mathbb{T}_i^{\{j-1\}}$ must be a zero of $\text{prem}(I_j, \mathbb{T}_i^{\{j-1\}}) \in \mathbb{U}_i$. Hence, $I_j(\bar{\boldsymbol{x}}) \neq 0$ for every j and $\bar{\boldsymbol{x}} \in \text{Zero}(\mathbb{T}_i^{\{j-1\}}/\mathbb{U}_i)$.

Since all the polynomials in $\mathbb{U}_i$ are actually the nonzero pseudo-remainders of some polynomials with respect to $\mathbb{T}_i$, one sees that $0 \notin \text{prem}(\mathbb{U}_i, \mathbb{T}_i)$ for every i. Therefore, each $[\mathbb{T}_i, \mathbb{U}_i]$ is a fine triangular system and the proof is complete. □

Algorithm TriSer implements the strategies of top-down elimination and splitting mentioned at the beginning of this section. It is structurally simple and practically effective. Note that the second component of a triangular system computed by TriSer may contain numerous polynomials, which increases the solution size of the problem. Fortunately, this drawback will disappear when the computed fine triangular systems are made regular, simple, or irreducible (see Theorems 3.4.6, 4.3.11, and 5.1.11).

By TriSer the decomposition tree as in Fig. 3 is computed depth-first. When the basic ideas of the algorithm are understood, one can design the corresponding breadth-first algorithm without essential difficulty.

Definition 2.3.1. Any (fine) triangular system computed by algorithm PriTriSys from a polynomial system $\mathfrak{P}$ in $\boldsymbol{K}[\boldsymbol{x}]$ is called a *(fine) principal triangular system* of $\mathfrak{P}$.

Proposition 2.3.2. Let $\mathbb{P} \subset \boldsymbol{K}[\boldsymbol{x}]$ and $[\mathbb{T}, \mathbb{U}]$ be a principal triangular system of $[\mathbb{P}, \emptyset]$. Then $\mathbb{T}$ is a quasi-medial set of $\mathbb{P}$.

Proof. It is clear that $\mathbb{T} \subset \text{Ideal}(\mathbb{P})$ and $\mathbb{T}$ is a quasi-ascending set. So we only need to prove that $\mathbb{T}$ has rank not higher than that of any quasi-basic set $\mathbb{B}$ of $\mathbb{P}$, i.e., $\mathbb{T} \precsim \mathbb{B}$. For this purpose, let

$$\mathbb{B} = [B_1, \ldots, B_s], \quad \mathbb{T} = [T_1, \ldots, T_r]$$

and $p_i = \text{cls}(B_i)$. Since $B_1 \in \mathbb{P}$ and $\text{cls}(B_1) = p_1$, $\mathbb{P}^{\langle p_1 \rangle} \neq \emptyset$ and thus $\mathbb{T}$ contains an element of class p_1. This implies that $\text{cls}(T_1) \leq \text{cls}(B_1)$. If $\text{cls}(T_1) < \text{cls}(B_1)$, then $\mathbb{T} \prec \mathbb{B}$ and the proposition is already proved. Otherwise, $\text{cls}(T_1) = \text{cls}(B_1)$. From the elimination for each i, one knows that $\text{ldeg}(T_1) \leq \text{ldeg}(B_1)$. Hence either $T_1 \prec B_1$ or $T_1 \sim B_1$. In the former case, the proposition is proved. Suppose otherwise the latter happens.

Similarly, $\mathbb{T}$ should contain a polynomial of class p_2 and thus $\text{cls}(T_2) \leq \text{cls}(B_2)$, etc. Using the same argument, one knows that either there is a $j \leq \min(r, s)$ such that

$$T_1 \sim B_1, \ldots, T_{j-1} \sim B_{j-1}, \quad \text{while} \quad T_j \prec B_j,$$

or

$$s = r, \quad \text{and} \quad T_1 \sim B_1, \ldots, T_r \sim B_r.$$

In any case, $\mathbb{T} \precsim \mathbb{B}$ and the proposition is proved. □

Remark 2.3.1. It appears that algorithm TriSer may produce a large number of branches. Nevertheless, the branch problem here is actually not more serious than that in CharSer. This is so partially because for many of the branches produced the corresponding polynomial systems have no zeros. In this situation, with more polynomials in the second component of a polynomial system, higher possibility is created to discard the system. Some analysis shows that the number of involved pseudo-divisions for the triangularization process in TriSer is similar to that in CharSer. Due to the advantages explained before, the computation for every individual branch in TriSer is less expensive. However, at the implementation level heuristic detection of redundant components is always necessary and profitable.

2.4 Subresultant-based algorithm

The decomposition algorithm TriSerS presented in this section has the same functionality and employs the same strategies of splitting and top-down elimination as TriSer. For the difference: TriSerS is based on computing subresultant chains. Let us recall the theory of subresultants and the relations between PRS and subresultant chains reviewed in Sect. 1.3. It has been widely recognized that forming subresultant chains is one of the most efficient ways to compute PRS. In our case, the process allows in particular to decompose any polynomial system into *simple systems* (see Sect. 3.3). First we demonstrate how the computation of subresultant chains is incorporated into TriSerS as the core operation.

The subresultant chain of two polynomials has the well-known block structure as shown in Theorem 1.3.4 and Fig. 1 which has been extensively studied, for example, in Collins (1967), Brown and Traub (1971), Loos (1983), and Mishra (1993). For our purpose, it is sufficient to use the existing results without entering into details of the theory of subresultants. As before, let $\boldsymbol{R}$ be a commutative ring with identity and $\boldsymbol{K}$ a field of characteristic 0. For the decomposition algorithms based on subresultant chains, the following lemma is of particular importance.

Lemma 2.4.1. Let $S_{\mu+1}$ and S_μ be two polynomials in $\boldsymbol{R}[x]$ with $\deg(S_{\mu+1}, x) \geq \deg(S_\mu, x) > 0$ and let $S_{\mu+1}, S_\mu, \dots, S_0$ be the subresultant chain of $S_{\mu+1}$ and S_μ with respect to x, with PSC chain $R_{\mu+1}, R_\mu, \dots, R_0$. Then for any $1 \leq i \leq \mu$,

$$S_i \neq 0, S_{i-1} = \cdots = S_0 = 0 \Longleftrightarrow R_i \neq 0, R_{i-1} = \cdots = R_0 = 0.$$

Proof. Corollary 7.7.9 in Mishra (1993, p. 262). □

Recall the SRS $S_{d_2}, \dots, S_{d_r}$ of $S_{\mu+1}$ and S_μ with respect to x_k in Definition 1.3.4. We rename these regular subresultants $H_2, \dots, H_r$ and set $P_1 = S_{\mu+1}$, $P_2 = S_\mu$. Clearly, $H_2 \backsim P_2$. Let $\boldsymbol{x}^{\{i\}}$ stand for $x_1, \dots, x_i$ or $(x_1, \dots, x_i)$, and similarly for $\bar{\boldsymbol{x}}^{\{i\}}$, etc.

Lemma 2.4.2. Let P_1 and P_2 be two polynomials in $\boldsymbol{K}[\boldsymbol{x}^{\{k\}}]$ with $\deg(P_1, x_k) \geq \deg(P_2, x_k) > 0$, $H_2, \dots, H_r$ be the SRS of P_1 and P_2 with respect to x_k, $I = \mathrm{lc}(P_2, x_k)$, and $I_i = \mathrm{lc}(H_i, x_k)$ for $i = 2, \dots, r$. Then

a. for any $2 \leq i \leq r$ and $\bar{\boldsymbol{x}}^{\{k-1\}} \in \mathrm{Zero}(\{I_{i+1}, \dots, I_r\}/II_i)$,

$$\gcd(P_1(\bar{\boldsymbol{x}}^{\{k-1\}}, x_k), P_2(\bar{\boldsymbol{x}}^{\{k-1\}}, x_k)) = H_i(\bar{\boldsymbol{x}}^{\{k-1\}}, x_k);$$

b. $$\mathrm{Zero}(\{P_1, P_2\}/I) = \bigcup_{i=2}^{r} \mathrm{Zero}(\{H_i, I_{i+1}, \dots, I_r\}/II_i). \tag{2.4.1}$$

Proof. a. Let $\mathfrak{S}$: $S_{\mu+1}, S_\mu, \dots, S_0$ be the subresultant chain of $P_1 = S_{\mu+1}$ and $P_2 = S_\mu$ with respect to x_k, with PSC chain $R_{\mu+1}, R_\mu, \dots, R_0$ and block indices $d_1, d_2, \dots, d_r$. Then, $H_i = S_{d_i}$ and $I_i = R_{d_i}$ for $2 \leq i \leq r$.

By Definition 1.3.4, for any $0 \leq j \leq \mu$ and $j \notin \{d_2, \dots, d_r\}$, S_j is defective, so R_j is identically zero. Let

$$\bar{\boldsymbol{x}}^{\{k-1\}} \in \mathrm{Zero}(\{I_{i+1}, \dots, I_r\}/II_i).$$

Then $R_j(\bar{\boldsymbol{x}}^{\{k-1\}}) = 0$ for $0 \leq j \leq d_i - 1$. Set

$$\left.\begin{aligned} \bar{S}_j &= S_j(\bar{\boldsymbol{x}}^{\{k-1\}}, x_k), & 0 \leq j \leq \mu + 1, \\ \bar{P}_i &= P_i(\bar{\boldsymbol{x}}^{\{k-1\}}, x_k), & i = 1, 2, \\ \bar{H}_i &= H_i(\bar{\boldsymbol{x}}^{\{k-1\}}, x_k), & 2 \leq i \leq r. \end{aligned}\right\} \tag{2.4.2}$$

By Lemma 2.4.1, $\bar{S}_{d_i-1} = \cdots = \bar{S}_0 = 0$ and $\bar{H}_i = \bar{S}_{d_i}$ is a nonzero polynomial in x_k. Note that the specialization of $\boldsymbol{x}^{\{k-1\}}$ to $\bar{\boldsymbol{x}}^{\{k-1\}}$ induces a homomorphism that maps the coefficients of P_1 and P_2 in x_k to numbers in some extension field of $\boldsymbol{K}$. By Proposition 1.3.5, each $\bar{S}_j$ may differ from the jth subresultant of $\bar{P}_1$ and $\bar{P}_2$ with respect to x_k at most by a factor of some power of $I(\bar{\boldsymbol{x}}^{\{k-1\}}) \neq 0$. According to Theorem 1.3.4 about the block structure of subresultant chains, there exists an integer d, $d_i \leq d \leq \mu$, such that $\bar{S}_d \backsim \bar{S}_{d_i}$. It follows from Theorem 1.3.6 that $\bar{S}_d$ is similar to the last polynomial in the subresultant PRS of $\bar{P}_1$ and $\bar{P}_2$ with respect to x_k. Therefore,

$$\gcd(\bar{P}_1, \bar{P}_2) = \bar{S}_d \backsim \bar{S}_{d_i} = \bar{H}_i.$$

b. For any $\bar{\boldsymbol{x}}^{\{k-1\}} \in \mathrm{Zero}(\emptyset/I)$, there must be an i ($2 \le i \le r$) such that

$$I_i(\bar{\boldsymbol{x}}^{\{k-1\}}) \neq 0, \quad I_{i+1}(\bar{\boldsymbol{x}}^{\{k-1\}}) = \cdots = I_r(\bar{\boldsymbol{x}}^{\{k-1\}}) = 0.$$

Thus, according to part a, $\bar{H}_i = \gcd(\bar{P}_1, \bar{P}_2)$, where $\bar{H}_i$ and $\bar{P}_1$, $\bar{P}_2$ are as in (2.4.2). The zero relation follows immediately. □

Lemma 2.4.2 a may be simply stated as: $\gcd(P_1, P_2, x_k) = H_i$ when $I_{i+1} = 0, \ldots, I_r = 0$ and $II_i \neq 0$ for any $2 \le i \le r$. Here $\gcd(P_1, P_2, x_k)$ means the GCD of P_1 and P_2, considered as univariate polynomials in x_k. A similar wording will be used for squarefreeness in later chapters.

Now, we show how to decompose a polynomial system $[\mathbb{P}, \mathbb{Q}]$ in $\boldsymbol{K}[\boldsymbol{x}]$ into triangular systems by using subresultant chains. Again, let us perform a top-down elimination for x_k, $k = n, \ldots, 1$.

We start with $k = n$. If, trivially, $\mathbb{P}^{\langle k \rangle} = \emptyset$, then proceed for next k. Consider the simple case $|\mathbb{P}^{\langle k \rangle}| = 1$ and let $P \in \mathbb{P}^{\langle k \rangle}$ with $I = \mathrm{ini}(P)$. Then

$$\mathbb{P} = 0, \ \mathbb{Q} \neq 0 \iff \begin{cases} \mathbb{P} = 0, \ \mathbb{Q} \neq 0, \ I \neq 0; \ \text{or} \\ \mathbb{P} \setminus \{P\} = 0, \ I = 0, \ \mathrm{red}(P) = 0, \ \mathbb{Q} \neq 0. \end{cases}$$

Here two subsystems are produced. For the first, we have obtained a single polynomial P in x_k whose initial is assumed to be nonzero, so the process can continue for next k. For the second, the minimal degree in x_k of the polynomials of class k has decreased. So we can assume that the subsystem may be dealt with by induction.

Now come to the more general case $|\mathbb{P}^{\langle k \rangle}| > 1$. Let $P_1, P_2 \in \mathbb{P}^{\langle k \rangle}$ with P_2 having minimal degree in x_k and compute the SRS $H_2, \ldots, H_r$ of P_1 and P_2 with respect to x_k. Let $I = \mathrm{lc}(P_2, x_k)$ and $I_i = \mathrm{lc}(H_i, x_k)$ for $2 \le i \le r$ as in Lemma 2.4.2. Then

$$\mathbb{P} = 0, \ \mathbb{Q} \neq 0 \iff \begin{cases} \mathbb{P}_2 = 0, \ I = 0, \ \mathrm{red}(P_2) = 0, \quad \mathbb{Q} \neq 0; \ \text{or} \\ \begin{bmatrix} \mathbb{P}_{12} = 0, H_i = 0, & \mathbb{Q} \neq 0, I \neq 0, \\ I_{i+1} = 0, \ldots, I_r = 0 & I_i \neq 0 \end{bmatrix} \\ \qquad\qquad\qquad\qquad\qquad \text{for some } 2 \le i \le r, \end{cases}$$

where

$$\mathbb{P}_2 = \mathbb{P} \setminus \{P_2\}, \quad \mathbb{P}_{12} = \mathbb{P} \setminus \{P_1, P_2\}.$$

It follows that

$$\begin{aligned} \mathrm{Zero}(\mathbb{P}/\mathbb{Q}) &= \mathrm{Zero}(\mathbb{P}_2 \cup \{I, \mathrm{red}(P_2)\}/\mathbb{Q}) \\ &\quad \cup \bigcup_{i=2}^{r} \mathrm{Zero}(\mathbb{P}_{12} \cup \{H_i, I_{i+1}, \ldots, I_r\}/\mathbb{Q} \cup \{I, I_i\}) \\ &= \cdots \qquad\qquad \text{(repeat recursively)} \\ &= \bigcup_{i=1}^{e} \mathrm{Zero}(\mathbb{T}_i/\mathbb{U}_i). \end{aligned}$$

What has been explained above can be formalized as the following algorithm.

Algorithm TriSerS: $\Psi \leftarrow \mathrm{TriSerS}(\mathbb{P}, \mathbb{Q})$. Given a polynomial system $[\mathbb{P}, \mathbb{Q}]$ in $\boldsymbol{K}[\boldsymbol{x}]$, this algorithm computes a fine triangular series Ψ of $[\mathbb{P}, \mathbb{Q}]$.

T1. Set $\Phi \leftarrow \{[\mathbb{P}, \mathbb{Q}, n]\}$, $\Psi \leftarrow \emptyset$.

T2. While $\Phi \neq \emptyset$, do:

T2.1. Let $[\mathbb{T}, \mathbb{U}, l]$ be an element of Φ and set $\Phi \leftarrow \Phi \setminus \{[\mathbb{T}, \mathbb{U}, l]\}$.

T2.2. For $k = l, \ldots, 1$ do:

T2.2.1. If $\mathbb{T}^{\langle k \rangle} = \emptyset$, then go to T2.2.3; else repeat:

T2.2.1.1. Let P_2 be an element of $\mathbb{T}^{\langle k \rangle}$ with minimal degree in x_k and set

$$\Phi \leftarrow \Phi \cup \{[\mathbb{T} \setminus \{P_2\} \cup \{\mathrm{ini}(P_2), \mathrm{red}(P_2)\}, \mathbb{U}, k]\},$$
$$\mathbb{U} \leftarrow \mathbb{U} \cup \{\mathrm{ini}(P_2)\}.$$

If $|\mathbb{T}^{\langle k \rangle}| = 1$, then go to T2.2.2. Otherwise, let P_1 be an element of $\mathbb{T}^{\langle k \rangle} \setminus \{P_2\}$.

T2.2.1.2. Compute the SRS $H_2, \ldots, H_r$ of P_1 and P_2 with respect to x_k and set $I_i \leftarrow \mathrm{lc}(H_i, x_k)$ for $2 \leq i \leq r$. If $\mathrm{cls}(H_r) < k$, then set $\bar{r} \leftarrow r - 1$; else set $\bar{r} \leftarrow r$.

T2.2.1.3. Set

$$\Phi \leftarrow \Phi \cup \{[\mathbb{T} \setminus \{P_1, P_2\} \cup \{H_i, I_{i+1}, \ldots, I_r\}, \mathbb{U} \cup \{I_i\}, k]: 2 \leq i \leq \bar{r} - 1\},$$
$$\mathbb{T} \leftarrow \mathbb{T} \setminus \{P_1, P_2\} \cup \{H_r, H_{\bar{r}}\},$$
$$\mathbb{U} \leftarrow \mathbb{U} \cup \{I_{\bar{r}}\}.$$

T2.2.2. Compute $\mathbb{U} \leftarrow \mathrm{prem}(\mathbb{U}, P_2)$.

T2.2.3. If $\mathbb{T} \cap \boldsymbol{K} \setminus \{0\} \neq \emptyset$ or $0 \in \mathbb{U}$, then go to T2.

T2.3. Set $\Psi \leftarrow \Psi \cup \{[\mathbb{T}, \mathbb{U}]\}$, with $\mathbb{T}$ ordered as a triangular set.

Proof. The algorithm adopts a top-down elimination from x_n to x_1. For each x_k, a single polynomial P_2 of class k is first produced from $\mathbb{T}^{\langle k \rangle}$ so long as $\mathbb{T}^{\langle k \rangle} \neq \emptyset$ (step T2.2.1); this polynomial is then used to reduce the polynomials in $\mathbb{U}$ (step T2.2.2). There are two kinds of splitting in the algorithm. One is performed in step T2.2.1.1 according as the initial of the considered polynomial vanishes or not: either it is assumed to be nonvanishing or the polynomial is replaced by the initial and the reductum. The other kind of splitting is performed for SRS elimination in step T2.2.1.3 according to Lemma 2.4.2. At each time of splitting, one produced system (corresponding to the case $i = r$ or $r - 1$ in Lemma 2.4.2b) is taken to update the current system $[\mathbb{T}, \mathbb{U}]$ and the others are added to Φ. As in any case of splitting a polynomial system $\mathfrak{P}$ into subsystems $\mathfrak{P}_i$ the zero relation $\mathrm{Zero}(\mathfrak{P}) = \bigcup_i \mathrm{Zero}(\mathfrak{P}_i)$ is preserved, the decomposition (2.1.8) is obtained eventually. In view of steps T2.2.2 and T2.2.3, each computed triangular system as $\mathfrak{T}_i$ in (2.1.8) is fine.

The termination of the algorithm is guaranteed because in each case of splitting, new polynomial systems are generated from the current system in two ways: either replacing one polynomial by another with a lower degree in the common leading variable, or replacing two polynomials by one of the same class k. For the latter, some polynomials of class smaller than k may be added. Step T2.2.1 terminates obviously, as in each repetition two polynomials $P_1, P_2 \in \mathbb{T}^{\langle k \rangle}$ are replaced by one $H_{\bar{r}}$ of class k and sometimes plus a polynomial H_r of class $<k$ (see T2.2.1.3). □

The polynomial set in the following example, considered initially by M. Bronstein, can be found in Wu (1987), Gao and Chou (1992), and Wang (1998).

Example 2.4.1. Let $\mathbb{P} = \{P_1, P_2, P_3\}$ with

$$P_1 = x^2 + y^2 + z^2 - r^2, \quad P_2 = xy + z^2 - 1,$$
$$P_3 = xyz - x^2 - y^2 - z + 1$$

and $r \prec z \prec x \prec y$.

First assume that $\mathrm{ini}(P_2) = x \neq 0$ and compute the subresultant chain of P_3, P_2 and of P_1, P_2 with respect to y. We obtain P_3, P_2, F and P_1, P_2, G with

$$F = -x^4 - z^3x^2 + x^2 - z^4 + 2z^2 - 1,$$
$$G = x^4 + z^2x^2 - r^2x^2 + z^4 - 2z^2 + 1.$$

Thus, P_2, F and P_2, G are the SRS of P_3, P_2 and P_1, P_2 respectively. It follows that

$$\gcd(P_3, P_2, y) = \gcd(P_1, P_2, y) = P_2$$

when $F = G = 0$ and $x \neq 0$. From the subresultant chain of F and G calculated in Example 1.3.2, one sees that the SRS of F and G with respect to x is

$$G, \quad H^2x^2, \quad (z^4 - 2z^2 + 1)^2H^4,$$

where $H = z^3 - z^2 + r^2 - 1$. Hence,

$$\gcd(F, G, x) = \begin{cases} G & \text{when } H = 0, \\ x^2 & \text{when } z^4 - 2z^2 + 1 = 0,\ H \neq 0. \end{cases}$$

Since x is assumed to be nonvanishing, the latter case is discarded. Therefore, we get a fine triangular system $[\mathbb{T}_1, \mathbb{U}_1]$ with

$$\mathbb{T}_1 = [H, G, P_2], \quad \mathbb{U}_1 = \{x\}.$$

For the case $x = 0$, a new polynomial set is generated by replacing P_2 with $\mathrm{ini}(P_2) = x$ and $\mathrm{red}(P_2) = z^2 - 1$. Following the same procedure, one can obtain from this polynomial set the second triangular system $[\mathbb{T}_2, \emptyset]$ with

$$\mathbb{T}_2 = [r^4 - 4r^2 + 3, z + r^2 - 2, x, y^2 - r^2 + 1].$$

It follows that

$$\mathrm{Zero}(\mathbb{P}) = \mathrm{Zero}(\mathbb{T}_1/x) \cup \mathrm{Zero}(\mathbb{T}_2).$$

Example 2.4.2. By using algorithm TriSerS the polynomial set $\mathbb{P}$ in Example 2.3.1 can be decomposed into the following reduced triangular systems

$$\mathfrak{T}_1 = [[-z^5 + t^4, T_2, T_3], \{t(t^3 - 1), z\}],$$
$$\mathfrak{T}_2 = [[t, z, y, x], \emptyset]$$
$$\mathfrak{T}_3 = [[t(t^3 - 1), -z^5 + t, tzy^2 + 2z^3y + 1, zx^2 - t], \{z\}],$$

where

$$T_2 = -tzy^2 - 2z^3y + t^8 - 2t^5 - t^3 + t^2,$$
$$T_3 = t^4x - tx - ty - z^2,$$

such that $\mathrm{Zero}(\mathbb{P}) = \bigcup_{i=1}^{3} \mathrm{Zero}(\mathfrak{T}_i)$.

For comparing the triangular set in $\mathfrak{T}_1$ with $\mathbb{T}_1 = [R_3, R_5, R_4]$ in Example 2.3.2, we note that $t^3T_2 = \mathrm{prem}(R_5, R_3, z)$ and $-t^3T_3 = \mathrm{prem}(z^2R_4, R_3, z)$.

Example 2.4.3. Let $\mathbb{P} = \{P_1, P_2, P_3\}$ with

$$P_1 = z(x^2 + y^2 - c) + 1, \quad P_2 = y(x^2 + z^2 - c) + 1,$$
$$P_3 = x(y^2 + z^2 - c) + 1.$$

This set of polynomials, originating from a paper by V. W. Noonburg, has been considered by Gao and Chou (1992) and Wang (1998). Under the variable ordering $c \prec z \prec y \prec x$, $\mathbb{P}$ can be decomposed by algorithm TriSerS into 7 fine triangular systems $[\mathbb{T}_1, \mathbb{U}_1], \ldots, [\mathbb{T}_7, \mathbb{U}_7]$ such that

$$\mathrm{Zero}(\mathbb{P}) = \bigcup_{i=1}^{7} \mathrm{Zero}(\mathbb{T}_i/\mathbb{U}_i),$$

where

$$\mathbb{T}_1 = [2cz^4 - 2z^3 - c^2z^2 - 2cz - 1, (cz + 1)y + cz^2 - z, 2z^2x + cz + 1],$$
$$\mathbb{T}_2 = [2z^4 - 3cz^2 + z + c^2, zy - z^2 + c, x - z],$$
$$\mathbb{T}_3 = [z^3 - cz - 1, (z^2 - c)y^2 + y - cz^2 + z + c^2, yx - z^2 + c],$$
$$\mathbb{T}_4 = [2z^4 - 3cz^2 + z + c^2, (2z^3 - 2cz + 2)y - cz^2 - z + c^2, P_3],$$
$$\mathbb{T}_5 = [2z^3 - cz + 1, y - z, 2z^2x - cx + 1],$$
$$\mathbb{T}_6 = [c, 2z^3 + 1, y - z, 2z^2x + 1],$$
$$\mathbb{T}_7 = [4c^3 - 27, 9z + 2c^2, 6cy^2 - 9y - 4c^2, 3yx + 2c];$$
$$\mathbb{U}_1 = \{c, z, cz + 1\},$$
$$\mathbb{U}_2 = \{z, z^2 - c, 2z^2 - c\},$$
$$\mathbb{U}_3 = \{z^2 - c, y\},$$
$$\mathbb{U}_4 = \{z^2 - c, z^3 - cz + 1, z^3 - cz - 1\},$$
$$\mathbb{U}_5 = \{z, 2z^2 - c\},$$
$$\mathbb{U}_6 = \{z\},$$
$$\mathbb{U}_7 = \{c, y\}.$$

In computing these triangular systems, some intermediate polynomials were factorized over **Q**. See Remark 2.4.2.

Two slightly different data structures are adopted for algorithms TriSer and TriSerS. We do so mainly to follow our early idea on the algorithm design and to show the two possibilities. It is possible to use the data structure of one algorithm for the other.

Remark 2.4.1. For the implementation of TriSer and TriSerS, some details have to be taken into account for the sake of efficiency. For example, a polynomial system $[\mathbb{P}, \mathbb{Q}]$ is readily found to have no zero whenever $\mathbb{P}$ contains a nonzero constant or $0 \in \mathbb{Q}$. Any factor of a polynomial in $\mathbb{P}$, when it occurs as a factor in some polynomial in $\mathbb{Q}$, may be removed, and so may any such factor of other polynomials in $\mathbb{Q}$. Heuristic reduction and simplification of some polynomials by the others should be adopted. The usual GCD and squarefree decomposition may be used in combination with the conditional GCD and squarefree computation. Here is a more technical trick: for any $[\mathbb{P}, \mathbb{Q}]$, when $|\mathbb{P}^{\langle 1 \rangle}| \geq 2$, $\text{Zero}(\mathbb{P}/\mathbb{Q})$ is likely to be empty and the emptiness may be tested first by computing the GCD of the polynomials in $\mathbb{P}^{\langle 1 \rangle}$.

Remark 2.4.2. To reduce cost for computing triangular series by CharSer, TriSer, or TriSerS, polynomial systems may be split by heuristically factorizing some intermediate polynomials at appropriate stage. If some polynomial in a polynomial set $\mathbb{P}$ can be factorized, for instance, into two polynomials and thus $[\mathbb{P}, \mathbb{Q}]$ can be split into two polynomial systems, say, $[\mathbb{P}', \mathbb{Q}]$ and $[\mathbb{P}'', \mathbb{Q}]$, such that

$$\text{Zero}(\mathbb{P}/\mathbb{Q}) = \text{Zero}(\mathbb{P}'/\mathbb{Q}) \cup \text{Zero}(\mathbb{P}''/\mathbb{Q}),$$

then one may proceed to decompose $[\mathbb{P}', \mathbb{Q}]$ and $[\mathbb{P}'', \mathbb{Q}]$, respectively, instead of $[\mathbb{P}, \mathbb{Q}]$. Polynomial factorization is expensive in general, but making proper use of it may improve the efficiency of the decomposition algorithms. This issue will be treated in more detail in Chap. 4.

As we have seen in the previous sections, the procedures for computing decomposition (2.1.8) with fine triangular systems are not complex. However, a fine triangular system may have "undesired behavior," so much more sophisticated algorithms will be developed in the following chapters for computing various kinds of triangular systems that have better behavior.

3 Projection and simple systems

The fine triangular systems computed by algorithms CharSer, TriSer, and TriSerS are not necessarily *perfect*. In other words, those triangular systems which have no zero are not necessarily detected. This issue is to be treated in this and the following chapters. To get some primitive idea, let us look at the following example.

Example 3.1. Consider the fine triangular set $\mathbb{T} = [T_1, T_2, T_3]$ with

$$T_1 = x^2 + u,$$
$$T_2 = y^2 + 2xy - u,$$
$$T_3 = (x + y)z + 1$$

and $u \prec x \prec y \prec z$. Now $I = \text{ini}(T_3) = x + y$. We want to verify whether $\text{Zero}(\mathbb{T}) = \emptyset$. For this, there are four different techniques available.

Factorization. To understand the "undesired behavior" of $\mathbb{T}$, let us observe that T_2 factors as

$$T_2 \doteq (y + x)^2 = I^2$$

over $\mathbf{Q}(u, x)$ with minimal polynomial T_1 for x. It is then obvious that $\mathbb{T}$ has no zero.

Projection. Instead of algebraic factorization, we calculate

$$\text{prem}(I^2, T_2) = x^2 + u = T_1,$$

where $\deg(T_2, y) = 2$ is taken for the exponent of I. Thus the same conclusion is reached.

Squarefree decomposition. As another way, let us form

$$\text{prem}(T_2, \partial T_2/\partial y) = -4(x^2 + u) = -4T_1.$$

This says that T_2 is the square of some polynomial T when $T_1 = 0$. T can be easily determined to be $I = y + x$. Therefore, one can conclude that $\mathbb{T}$ has no zero.

GCD computation. Finally, we compute

$$\text{prem}(T_2, I_2) = -(x^2 + u) = -T_1.$$

It follows that I is the GCD of T_2 and I when $T_1 = 0$. So $\text{Zero}(\mathbb{T}) = \emptyset$ is verified as well.

Our aim in what follows is to develop the above techniques into systematic algorithms. This is done first by incorporating *projection* into some algorithms. In

Sects. 3.3 and 5.1, we shall consider the problem by means of other devices, for which the concepts of simple systems and regular systems will play a role. The perfectness of triangular systems may also be guaranteed when one arrives at an irreducible decomposition, the central theme of Chap. 4.

3.1 Projection

Let a polynomial system $[\mathbb{P}, \mathbb{Q}]$ in $\boldsymbol{K}[x_1, \ldots, x_n]$ be given. We want to eliminate the variables $x_n, \ldots, x_{k+1}$ $(0 \leq k < n)$ and to obtain finitely many other polynomial systems $[\mathbb{P}_1, \mathbb{Q}_1], \ldots, [\mathbb{P}_e, \mathbb{Q}_e]$ in $\boldsymbol{K}[x_1, \ldots, x_k]$ such that

$$\operatorname{Zero}(\mathbb{P}/\mathbb{Q}) \neq \emptyset \iff \bigcup_{i=1}^{e} \operatorname{Zero}(\mathbb{P}_i/\mathbb{Q}_i) \neq \emptyset.$$

When $k = 0$, $\operatorname{Zero}(\mathbb{P}/\mathbb{Q}) \neq \emptyset$ if and only if there exists an i such that $\mathbb{P}_i \setminus \{0\} = \emptyset$ and $0 \notin \mathbb{Q}_i$. It is also expected that for any

$$(\bar{x}_1, \ldots, \bar{x}_k) \in \bigcup_{i=1}^{e} \operatorname{Zero}(\mathbb{P}_i/\mathbb{Q}_i)$$

one can find $\bar{x}_{k+1}, \ldots, \bar{x}_n$ in some extension field $\tilde{\boldsymbol{K}}$ of $\boldsymbol{K}$ such that $(\bar{x}_1, \ldots, \bar{x}_n) \in \operatorname{Zero}(\mathbb{P}/\mathbb{Q})$. An elimination procedure meeting these two requirements only is relatively simple. However, the algorithms to be presented in Sect. 3.2 are somewhat involved mainly because we also want to establish the zero relationship between the given system and the eliminated (triangular) systems.

Basic lemmas

Recall the notations $\mathbb{P}^{(i)}$, $\mathbb{P}^{[i]}$ and $\mathbb{P}^{\langle i \rangle}$ introduced in Sect. 2.3. We continue writing $\boldsymbol{x}^{\{i\}}$ for $x_1, \ldots, x_i$ or $(x_1, \ldots, x_i)$ with $\boldsymbol{x} = \boldsymbol{x}^{\{n\}}$, and similarly $\bar{\boldsymbol{x}}^{\{i\}}$ for $\bar{x}_1, \ldots, \bar{x}_i$ or $(\bar{x}_1, \ldots, \bar{x}_i)$, etc. Unless stated otherwise, $\tilde{\boldsymbol{K}}$ always denotes some extension field of $\boldsymbol{K}$.

For any $\bar{x}_1, \ldots, \bar{x}_i \in \tilde{\boldsymbol{K}}$, the set of polynomials obtained from $\mathbb{P}$ by substituting $\bar{x}_1, \ldots, \bar{x}_i$ for $x_1, \ldots, x_i$ is denoted by $\mathbb{P}^{\langle \bar{x}, i \rangle}$. Symbolically,

$$\mathbb{P}^{\langle \bar{x}, i \rangle} \triangleq \mathbb{P}|_{\boldsymbol{x}^{\{i\}} = \bar{\boldsymbol{x}}^{\{i\}}} = \mathbb{P}|_{x_1 = \bar{x}_1, \ldots, x_i = \bar{x}_i}.$$

For any polynomial system $\mathfrak{P} = [\mathbb{P}, \mathbb{Q}]$, we have

$$\mathfrak{P}^{\langle \bar{x}, i \rangle} \triangleq [\mathbb{P}^{\langle \bar{x}, i \rangle}, \mathbb{Q}^{\langle \bar{x}, i \rangle}].$$

Definition 3.1.1. For any polynomial system $\mathfrak{P}$ in $\boldsymbol{K}[\boldsymbol{x}]$ and $1 \leq i \leq n-1$, the *projection* of $\operatorname{Zero}(\mathfrak{P})$ onto $\boldsymbol{x}^{\{i\}}$ is defined to be

$$\operatorname{Proj}_{\boldsymbol{x}^{\{i\}}} \operatorname{Zero}(\mathfrak{P}) \triangleq \{\bar{\boldsymbol{x}}^{\{i\}} \in \tilde{\boldsymbol{K}}^i : \exists \bar{x}_{i+1}, \ldots, \bar{x}_n \in \tilde{\boldsymbol{K}} \text{ such that } \bar{\boldsymbol{x}} \in \operatorname{Zero}(\mathfrak{P})\}.$$

Moreover, we define

$$\mathrm{Proj}_{\boldsymbol{x}}\mathrm{Zero}(\mathfrak{P}) \triangleq \mathrm{Zero}(\mathfrak{P})$$

for the extreme case $i = n$, and

$$\mathrm{ProjZero}(\mathfrak{P}) \triangleq \begin{cases} \emptyset & \text{if } \mathrm{Zero}(\mathfrak{P}) = \emptyset, \\ \{0\} & \text{otherwise,} \end{cases}$$

for the extreme case $i = 0$.

It is easy to see that

$$\mathrm{Proj}_{\boldsymbol{x}^{\{i\}}}\mathrm{Zero}(\mathfrak{P}) \neq \emptyset \iff \mathrm{Zero}(\mathfrak{P}) \neq \emptyset.$$

And, for i elements $\bar{x}_1, \ldots, \bar{x}_i \in \tilde{\boldsymbol{K}}$,

$$\bar{\boldsymbol{x}}^{\{i\}} \in \mathrm{Proj}_{\boldsymbol{x}^{\{i\}}}\mathrm{Zero}(\mathfrak{P}) \iff \mathrm{Zero}(\mathfrak{P}^{\langle \bar{x}, i\rangle}) \neq \emptyset.$$

For any polynomial system $\mathfrak{P} = [\mathbb{P}, \mathbb{Q}]$, if $\mathbb{P}^{[i]} = \mathbb{Q}^{[i]} = \emptyset$, then obviously $\mathrm{Proj}_{\boldsymbol{x}^{\{i\}}}\mathrm{Zero}(\mathfrak{P}) = \mathrm{Zero}(\mathfrak{P})$.

Lemma 3.1.1. Let $[\mathbb{P}, \mathbb{Q}]$ be a polynomial system of level $\leq i$ in $\boldsymbol{K}[\boldsymbol{x}]$. Suppose that $\mathbb{Q}^{[i]} \neq \emptyset$ and let $H_1, \ldots, H_h$ be all the polynomials in $\mathbb{Q}^{[i]}$. Denote by $H_{l1}, \ldots, H_{lm_l}$ all the nonzero coefficients of the terms in H_l with respect to those variables which are $\succ x_i$. Then

$$\mathrm{Proj}_{\boldsymbol{x}^{\{i\}}}\mathrm{Zero}(\mathbb{P}/\mathbb{Q}) = \bigcup_{1 \leq j_1 \leq m_1, \ldots, 1 \leq j_h \leq m_h} \mathrm{Zero}(\mathbb{P}/\mathbb{Q}_{j_1 \ldots j_h}), \tag{3.1.1}$$

$$\mathrm{Zero}(\mathbb{P}/\mathbb{Q}) = \bigcup_{1 \leq j_1 \leq m_1, \ldots, 1 \leq j_h \leq m_h} \mathrm{Zero}(\mathbb{P}/\mathbb{Q}'_{j_1 \ldots j_h}), \tag{3.1.2}$$

where

$$\mathbb{Q}_{j_1 \ldots j_h} = \mathbb{Q}^{(i)} \cup \{H_{1j_1}, \ldots, H_{hj_h}\}, \quad \mathbb{Q}'_{j_1 \ldots j_h} = \mathbb{Q} \cup \{H_{1j_1}, \ldots, H_{hj_h}\}.$$

Proof. We first prove (3.1.1). For any $\bar{\boldsymbol{x}}^{\{i\}} \in \mathrm{Proj}_{\boldsymbol{x}^{\{i\}}}\mathrm{Zero}(\mathbb{P}/\mathbb{Q})$, by definition there exist $\bar{x}_{i+1}, \ldots, \bar{x}_n \in \tilde{\boldsymbol{K}}$ such that $\bar{\boldsymbol{x}} \in \mathrm{Zero}(\mathbb{P}/\mathbb{Q})$. Clearly, $H_l(\bar{\boldsymbol{x}}) \neq 0$ and thus

$$H_{l1}(\bar{\boldsymbol{x}}^{\{i\}}), \ldots, H_{lm_l}(\bar{\boldsymbol{x}}^{\{i\}})$$

cannot be all 0 for each l; let j'_l be any integer such that $H_{lj'_l}(\bar{\boldsymbol{x}}^{\{i\}}) \neq 0$. Then

$$\bar{\boldsymbol{x}}^{\{i\}} \in \mathrm{Zero}(\mathbb{P}/\mathbb{Q}_{j'_1 \ldots j'_h}). \tag{3.1.3}$$

In the other direction, if $\bar{\boldsymbol{x}}^{\{i\}}$ belongs to the right-hand side of (3.1.1), then there must be some indices $j'_1, \ldots, j'_h$ such that (3.1.3) holds. Therefore,

$$H_l(\bar{\boldsymbol{x}}^{\{i\}}, x_{i+1}, \ldots, x_n) \not\equiv 0$$

for all l, so there are $\bar{x}_{i+1}, \ldots, \bar{x}_n \in \tilde{\boldsymbol{K}}$ such that $H_1 \cdots H_h(\bar{\boldsymbol{x}}) \neq 0$. This implies that $H_l(\bar{\boldsymbol{x}}) \neq 0$ for each l. Hence, $\bar{\boldsymbol{x}} \in \mathrm{Zero}(\mathbb{P}/\mathbb{Q})$ and thus $\bar{\boldsymbol{x}}^{\{i\}} \in \mathrm{Proj}_{\boldsymbol{x}^{\{i\}}}\mathrm{Zero}(\mathbb{P}/\mathbb{Q})$.

To show (3.1.2), one first sees that the right-hand side is obviously contained in the left-hand side. This is simply because

$$\mathrm{Zero}(\mathbb{P}/\mathbb{Q}'_{j_1 \ldots j_h}) \subset \mathrm{Zero}(\mathbb{P}/\mathbb{Q})$$

for each set of $j_1, \ldots, j_h$. On the other hand, for any $\bar{\boldsymbol{x}} \in \mathrm{Zero}(\mathbb{P}/\mathbb{Q})$ let j'_l be any integer such that $H_{lj'_l}(\bar{\boldsymbol{x}}^{\{i\}}) \neq 0$ for each l as before. Then

$$\bar{\boldsymbol{x}} \in \mathrm{Zero}(\mathbb{P}/\mathbb{Q}'_{j'_1 \ldots j'_h})$$

and thus $\bar{\boldsymbol{x}}$ belongs to the right-hand side of (3.1.2). □

Remark 3.1.1. The zero relations (3.1.1) and (3.1.2) in Lemma 3.1.1 can be complicated by replacing $\mathbb{P}$ on the right-hand side with $\mathbb{P} \cup \mathbb{H}_{j_1 \ldots j_h}$, where

$$\mathbb{H}_{j_1 \ldots j_h} = \{H_{lj}\colon\ 0 \leq j \leq j_l - 1, 1 \leq l \leq h\} \setminus \{0\}$$

and $H_{l0} = 0$ for $l = 1, \ldots, h$. This is considered of practical interest because the more polynomials in the system the easier the elimination may be, in particular, when the system has no zero. This modification of the zero relations would lead the subalgorithm ProjA described in Sect. 3.2 to a more complicated version.

Lemma 3.1.2. Let T be a polynomial in $\boldsymbol{K}[\boldsymbol{x}]$ with

$$\mathrm{cls}(T) = i > 0, \quad \mathrm{ini}(T) = I, \quad \mathrm{ldeg}(T) = d,$$

and $[\mathbb{P}, \mathbb{Q}]$ a polynomial system of level $l \leq i - 1$ with level$(\mathbb{Q}) \leq i$.

a. If $\mathbb{Q}^{\langle i \rangle} = \emptyset$, then for any $l \leq j \leq i - 1$

$$\mathrm{Proj}_{\boldsymbol{x}^{\{j\}}}\mathrm{Zero}(\mathbb{P} \cup \{T\}/\mathbb{Q} \cup \{I\}) = \mathrm{Proj}_{\boldsymbol{x}^{\{j\}}}\mathrm{Zero}(\mathbb{P}/\mathbb{Q} \cup \{I\}). \tag{3.1.4}$$

b. Suppose that $\mathbb{Q}^{\langle i \rangle} \neq \emptyset$ and let $H_1, \ldots, H_h$ be all the polynomials in $\mathbb{Q}^{\langle i \rangle}$. Set

$$R = \mathrm{prem}((H_1 \cdots H_h)^d, T), \quad \mathbb{Q}' = \mathbb{Q}^{\langle i-1 \rangle} \cup \{I, R\}.$$

Then, for any $l \leq j \leq i - 1$

$$\mathrm{Proj}_{\boldsymbol{x}^{\{j\}}}\mathrm{Zero}(\mathbb{P} \cup \{T\}/\mathbb{Q} \cup \{I\}) = \mathrm{Proj}_{\boldsymbol{x}^{\{j\}}}\mathrm{Zero}(\mathbb{P}/\mathbb{Q}'), \tag{3.1.5}$$

$$\mathrm{Zero}(\mathbb{P} \cup \{T\}/\mathbb{Q} \cup \{I\}) = \mathrm{Zero}(\mathbb{P} \cup \{T\}/\mathbb{Q}'). \tag{3.1.6}$$

Proof. a. In this case, all the polynomials in $\mathbb{Q}$ have class $<i$, i.e., $\mathbb{Q} \subset \boldsymbol{K}[\boldsymbol{x}^{\{i-1\}}]$. The left-hand side is obviously contained in the right-hand side of (3.1.4). For the other direction, consider any $l \leq j \leq i - 1$ and

$$\bar{\boldsymbol{x}}^{\{j\}} \in \mathrm{Proj}_{\boldsymbol{x}^{\{j\}}}\mathrm{Zero}(\mathbb{P}/\mathbb{Q} \cup \{I\}).$$

By definition there exist $\bar{x}_{j+1}, \ldots, \bar{x}_{i-1} \in \tilde{\boldsymbol{K}}$ such that $\bar{\boldsymbol{x}}^{\{i-1\}} \in \mathrm{Zero}(\mathbb{P}/\mathbb{Q} \cup \{I\})$. According to the fundamental theorem of algebra, $T(\bar{\boldsymbol{x}}^{\{i-1\}}, x_i)$ has a zero $\bar{x}_i \in \tilde{\boldsymbol{K}}$ for x_i. Thus, $\bar{\boldsymbol{x}}^{\{i\}}$ belongs to the left-hand side of (3.1.4).

b. To prove (3.1.5), first consider any

$$\bar{\boldsymbol{x}}^{\{j\}} \in \mathrm{Proj}_{\boldsymbol{x}^{\{j\}}}\mathrm{Zero}(\mathbb{P} \cup \{T\}/\mathbb{Q} \cup \{I\}). \tag{3.1.7}$$

Then there exist $\bar{x}_{j+1}, \ldots, \bar{x}_i \in \tilde{\boldsymbol{K}}$ such that

$$T(\bar{\boldsymbol{x}}^{\{i\}}) = 0, \;\; I(\bar{\boldsymbol{x}}^{\{i-1\}}) \neq 0, \;\; H_1 \cdots H_h(\bar{\boldsymbol{x}}^{\{i\}}) \neq 0.$$

By the pseudo-remainder formula

$$I^s(H_1 \cdots H_h)^d = AT + R \tag{3.1.8}$$

for some integer $s \geq 0$, we have $R(\bar{\boldsymbol{x}}^{\{i\}}) \neq 0$. Therefore, $\bar{\boldsymbol{x}}^{\{i\}} \in \mathrm{Zero}(\mathbb{P}/\mathbb{Q}')$, which implies that

$$\bar{\boldsymbol{x}}^{\{j\}} \in \mathrm{Proj}_{\boldsymbol{x}^{\{j\}}}\mathrm{Zero}(\mathbb{P}/\mathbb{Q}'). \tag{3.1.9}$$

Now let (3.1.9) hold; then there exist $\bar{x}_{j+1}, \ldots, \bar{x}_i \in \tilde{\boldsymbol{K}}$ such that $\bar{\boldsymbol{x}}^{\{i\}} \in \mathrm{Zero}(\mathbb{P}/\mathbb{Q}')$. Note that, while $T, H_1, \ldots, H_h$ are regarded as polynomials in $\boldsymbol{K}(\boldsymbol{x}^{\{i-1\}})[x_i]$, T contains a factor not occurring in any of $H_1, \ldots, H_h$ if and only if $R \neq 0$. Since $R(\bar{\boldsymbol{x}}^{\{i\}}) \neq 0$, $T(\bar{\boldsymbol{x}}^{\{i-1\}}, x_i)$ must contain a factor, say, T', which is not a factor of any $H_l(\bar{\boldsymbol{x}}^{\{i-1\}}, x_i)$, $1 \leq l \leq h$. Hence, there must be an $\tilde{x}_i$ in some algebraic-extension field of $\boldsymbol{K}(\bar{\boldsymbol{x}}^{\{i-1\}})$ and thus of $\boldsymbol{K}$ such that

$$T(\bar{\boldsymbol{x}}^{\{i-1\}}, \tilde{x}_i) = 0 \;\text{ while }\; H_1 \cdots H_h(\bar{\boldsymbol{x}}^{\{i-1\}}, \tilde{x}_i) \neq 0$$

(actually, any zero of T' does). Therefore,

$$(\bar{\boldsymbol{x}}^{\{i-1\}}, \tilde{x}_i) \in \mathrm{Zero}(\mathbb{P} \cup \{T\}/\mathbb{Q} \cup \{I\}),$$

so (3.1.7) holds. This completes the proof of (3.1.5).

Finally, from the formula (3.1.8) it is easy to see that under the condition $I \neq 0$, $H_1 \cdots H_h \neq 0$ if and only if $R \neq 0$. Hence (3.1.6) holds true. □

Projection for triangular systems

Definition 3.1.2. A triangular system $\mathfrak{T}$ in $\boldsymbol{K}[\boldsymbol{x}]$ is said to be *perfect* over $\tilde{\boldsymbol{K}}$ ($\supset \boldsymbol{K}$) if $\tilde{\boldsymbol{K}}$-Zero$(\mathfrak{T}) \neq \emptyset$.

A triangular set $\mathbb{T} \subset \boldsymbol{K}[\boldsymbol{x}]$ is said to be *perfect* over $\tilde{\boldsymbol{K}}$ if $[\mathbb{T}, \mathrm{ini}(\mathbb{T})]$ is perfect over $\tilde{\boldsymbol{K}}$.

A triangular set or system in $\boldsymbol{K}[\boldsymbol{x}]$ is said to be *perfect* (without reference to any specific field) if it is perfect over some suitable extension of $\boldsymbol{K}$.

Consider a fine triangular system $[\mathbb{T}, \mathbb{U}]$ with

$$\mathbb{T} = [T_1, \ldots, T_r].$$

Let $\operatorname{cls}(T_i) = p_i$ for each i; clearly, $0 < p_1 < \cdots < p_r \leq n$. In general, for each i and any

$$\bar{\boldsymbol{x}}^{\{p_i\}} \in \operatorname{Zero}(\mathbb{T}^{\{i\}}/\mathbb{U}^{\langle p_i \rangle})$$

the existence of $\bar{x}_{p_i+1}, \ldots, \bar{x}_n \in \tilde{\boldsymbol{K}}$ such that $\bar{\boldsymbol{x}} \in \operatorname{Zero}(\mathbb{T}/\mathbb{U})$ is not guaranteed. In other words,

$$[\mathbb{T}^{[p_i]\langle \bar{x}, p_i \rangle}, \mathbb{U}^{[p_i]\langle \bar{x}, p_i \rangle}]$$

is not necessarily perfect. We explain how to deal with this situation by means of projection demonstrated in Lemmas 3.1.1 and 3.1.2. Here, *projection* is meant to carry out the task in either of the following two cases. It is considered first with respect to T_r.

Case A. If $p_r = n$, this case is skipped. If $p_r < n$ and $\mathbb{U}^{[p_r]} = \emptyset$, then proceed with case B below. Suppose, otherwise, that $p_r < n$ and $\mathbb{U}^{[p_r]} \neq \emptyset$. Let $H_1, \ldots, H_h$ be all the polynomials in $\mathbb{U}^{[p_r]}$ and denote by $H_{l1}, \ldots, H_{lm_l}$ all the nonzero coefficients of the terms in H_l with respect to those variables which are $\succ x_{p_r}$ for each l. Then, by Lemma 3.1.1

$$\operatorname{Zero}(\mathbb{T}/\mathbb{U}) = \bigcup_{1 \leq j_1 \leq m_1, \ldots, 1 \leq j_h \leq m_h} \operatorname{Zero}(\mathbb{T}/\mathbb{U}_{j_1 \ldots j_h}), \qquad (3.1.10)$$

where $\mathbb{U}_{j_1 \ldots j_h} = \mathbb{U} \cup \{H_{1j_1}, \ldots, H_{hj_h}\}$. To simplify notations, let

$$\mathcal{J} = \{j_1 \ldots j_h \colon 1 \leq j_1 \leq m_1, \ldots, 1 \leq j_h \leq m_h\};$$

i.e., $\mathcal{J}$ is the set of indices of $\mathbb{U}_{j_1 \ldots j_h}$. Then, for any $\bar{\boldsymbol{x}}^{\{p_r\}} \in \operatorname{Zero}(\mathbb{T}/\mathbb{U}^{\langle p_r \rangle})$, there exist $\bar{x}_{p_r+1}, \ldots, \bar{x}_n \in \tilde{\boldsymbol{K}}$ such that $H_1 \cdots H_h(\bar{\boldsymbol{x}}) \neq 0$ if and only if

$$H_{1j_1} \cdots H_{hj_h}(\bar{\boldsymbol{x}}^{\{p_r\}}) \neq 0 \quad \text{for some } j_1 \ldots j_h \in \mathcal{J}.$$

Or equivalently, we have

$$\operatorname{Proj}_{\boldsymbol{x}^{\{p_r\}}} \operatorname{Zero}(\mathbb{T}/\mathbb{U}) = \bigcup_{j \in \mathcal{J}} \operatorname{Zero}(\mathbb{T}/\mathbb{U}_j^{\langle p_r \rangle}).$$

Case B. Consider each triangular system $[\mathbb{T}, \mathbb{U}_j]$, $j \in \mathcal{J}$, and note that $\operatorname{Zero}(\mathbb{T}/\mathbb{U}_j \cup \operatorname{ini}(\mathbb{T})) = \operatorname{Zero}(\mathbb{T}/\mathbb{U}_j)$. If $\mathbb{U}_j^{\langle p_r \rangle} = \emptyset$, then

$$\operatorname{Proj}_{\boldsymbol{x}^{\{p_r-1\}}} \operatorname{Zero}(\mathbb{T}/\mathbb{U}_j) = \operatorname{Zero}(\mathbb{T}^{\{r-1\}}/\mathbb{U}_j^{\langle p_r - 1 \rangle})$$

according to Lemma 3.1.2 a. In this case, proceed next for T_{r-1}.

Otherwise, let $K_1, \ldots, K_k$ be all the polynomials in $\mathbb{U}_j^{\langle p_r \rangle}$. Compute

$$R = \operatorname{prem}((K_1 \cdots K_k)^{\operatorname{ldeg}(T_r)}, T_r), \quad \mathbb{U}'_j = \mathbb{U}_j \setminus \mathbb{U}_j^{\langle p_r \rangle} \cup \{R\}.$$

If $R = 0$, then $\operatorname{Zero}(\mathbb{T}/\mathbb{U}_j) = \emptyset$ and the triangular system $[\mathbb{T}, \mathbb{U}_j]$ is removed. In the case $R \neq 0$, application of Lemma 3.1.2 b yields

$$\begin{aligned}&\operatorname{Proj}_{\boldsymbol{x}^{\{p_r-1\}}}\operatorname{Zero}(\mathbb{T}/\mathbb{U}_j) = \operatorname{Proj}_{\boldsymbol{x}^{\{p_r-1\}}}\operatorname{Zero}(\mathbb{T}^{\{r-1\}}/\mathbb{U}_j'^{(p_r)}),\\ &\operatorname{Zero}(\mathbb{T}/\mathbb{U}_j) = \operatorname{Zero}(\mathbb{T}/\mathbb{U}_j').\end{aligned} \tag{3.1.11}$$

Combining (3.1.10) and (3.1.11) results in

$$\operatorname{Zero}(\mathbb{T}/\mathbb{U}) = \bigcup_{j\in\mathcal{J}} \operatorname{Zero}(\mathbb{T}/\mathbb{U}_j').$$

Meanwhile, we have

$$\operatorname{Proj}_{\boldsymbol{x}^{\{p_{r-1}\}}}\operatorname{Zero}(\mathbb{T}/\mathbb{U}) = \bigcup_{j\in\mathcal{J}} \operatorname{Proj}_{\boldsymbol{x}^{\{p_{r-1}\}}}\operatorname{Zero}(\mathbb{T}^{\{r-1\}}/\mathbb{U}_j'^{(p_r)}).$$

The above projection cases A and B can be repeated for each triangular system $[\mathbb{T}^{\{r-1\}}, \mathbb{U}_j'^{(p_r)}]$ with respect to T_{r-1}, and so forth. In this way, either all the split triangular systems are removed and thus $\operatorname{Zero}(\mathbb{T}/\mathbb{U}) = \emptyset$, or a finite sequence of polynomial sets $\mathbb{U}_1^*, \dots, \mathbb{U}_s^*$ is finally obtained such that

$$\operatorname{Zero}(\mathbb{T}/\mathbb{U}) = \bigcup_{i=1}^{s} \operatorname{Zero}(\mathbb{T}/\mathbb{U}_i^*). \tag{3.1.12}$$

In particular, when projection is needed only for $x_n, \dots, x_{k+1}$, let t be such that $p_t < k+1 \leq p_{t+1}$. Then, the projection is performed first for both cases A and B with respect to $T_r, \dots, T_{t+1}$, and finally for case A with $p = k$ in addition. Then

$$\operatorname{Proj}_{\boldsymbol{x}^{\{k\}}}\operatorname{Zero}(\mathbb{T}/\mathbb{U}) = \bigcup_{i=1}^{s} \operatorname{Zero}(\mathbb{T}^{(k)}/\mathbb{U}_i^{*(k)}).$$

Definition 3.1.3. Let $\mathfrak{T} = [\mathbb{T}, \mathbb{U}]$ be a fine triangular system in $\boldsymbol{K}[\boldsymbol{x}]$ and k a nonnegative integer. $\mathfrak{T}$ is said to possess

- the *projection property* of dimension k if

$$\operatorname{Zero}(\mathfrak{T}^{(i)}) \subset \operatorname{Proj}_{\boldsymbol{x}^{\{i\}}}\operatorname{Zero}(\mathfrak{T}) \tag{3.1.13}$$

  holds for $i = k$ and all $i \in \{\operatorname{cls}(T) : T \in \mathbb{T}, \operatorname{cls}(T) > k\}$;
- the *strong projection property* of dimension k if (3.1.13) holds for all $k \leq i < n$.

  When the dimension is not mentioned, it is meant that $k = 0$.

Lemmas 3.1.1 and 3.1.2 ensure that the above-computed triangular systems $[\mathbb{T}, \mathbb{U}_i^*]$, $1 \leq i \leq s$, all possess the projection property of dimension k.

We do not describe the above projection procedure for triangular systems as a formal algorithm because it is a special case of algorithm TriSerP in Sect. 3.2.

Case A here is so designed that projection is performed once for all the variables $x_n, \ldots, x_{p_r+1}$. This is mainly for some practical consideration. Of course, one can modify the procedure in order to project for one variable each time (see Remark 3.2.1).

For an arbitrary polynomial system $\mathfrak{P}$, using CharSer, TriSer, or TriSerS one can compute a fine triangular series Ψ of $\mathfrak{P}$. If $\Psi = \emptyset$, then $\mathrm{Zero}(\mathfrak{P}) = \emptyset$. Otherwise, for each $\mathfrak{T} = [\mathbb{T}, \mathbb{U}] \in \Psi$ one can project for $x_n, \ldots, x_{k+1}$ to determine the polynomial sets corresponding to $\mathbb{U}_i^*$ in (3.1.12). When $\mathrm{Zero}(\mathfrak{T}) = \emptyset$, it will be detected in the way of projection. Thus, either $\mathrm{Zero}(\mathfrak{T}) = \emptyset$ is detected for all $\mathfrak{T} \in \Psi$, or a zero decomposition of the form

$$\mathrm{Zero}(\mathfrak{P}) = \bigcup_{i=1}^{e} \mathrm{Zero}(\mathfrak{T}_i)$$

is finally reached, such that

$$\mathrm{Proj}_{\boldsymbol{x}^{\{k\}}} \mathrm{Zero}(\mathfrak{P}) = \bigcup_{i=1}^{e} \mathrm{Zero}(\mathfrak{T}_i^{(k)})$$

and each $\mathfrak{T}_i$ is a fine triangular system possessing the projection property of dimension k. In fact, for any $\bar{\boldsymbol{x}}^{\{k\}} \in \mathrm{Zero}(\mathfrak{T}_i^{(k)})$ the zeros of $\mathfrak{T}_i^{[k]\langle \bar{x},k\rangle}$ for $x_{k+1}, \ldots, x_n$ can be successively determined from the triangular system. As a consequence,

$$\mathrm{Zero}(\mathfrak{P}^{\langle \bar{x},k\rangle}) \neq \emptyset.$$

Therefore, the requirements we have specified at the beginning of this section are all satisfied. In particular, when $k = 0$, $\mathrm{Zero}(\mathfrak{P}) = \emptyset$ if and only if $e = 0$.

Example 3.1.1. Consider the triangular set $\mathbb{T}_1 = [T_1, T_2, T_3]$ with

$$\begin{aligned} T_1 &= z^3 - z^2 + r^2 - 1, \\ T_2 &= x^4 + z^2x^2 - r^2x^2 + z^4 - 2z^2 + 1, \\ T_3 &= xy + z^2 - 1, \end{aligned}$$

which have been computed in Example 2.4.1. We want to project $[\mathbb{T}_1, \{x\}]$ with $k = 0$. No projection is needed with respect to T_3. To project with respect to T_2, compute

$$R = \mathrm{prem}(x^4, T_2) = R_1x^2 + R_2,$$

where $R_1 = -z^2 + r^2$ and $R_2 = -z^4 + 2z^2 - 1$. Thus, $[\mathbb{T}_1, \{x\}]$ is split to

$$[\mathbb{T}_1, \{R_1, R\}], \quad [\mathbb{T}_1, \{R_2, R\}].$$

For projection with respect to T_1, we need to compute

$$\begin{aligned} R_1^* &= \mathrm{prem}(R_1^3, T_1) \\ &= (-3r^4 + 5r^2 - 3)z^2 - (3r^4 - 4r^2 + 1)z + r^6 - 4r^4 + 6r^2 - 2, \\ R_2^* &= \mathrm{prem}(R_2^3, T_1) \\ &= (-8r^2 + 4r^6 - 6r^4 + 11)z^2 - (12r^4 - 29r^2 + 17)z \\ &\quad - r^8 - 4r^6 + 16r^4 - 11r^2 - 1. \end{aligned}$$

Replacing R_1 and R_2 in the two triangular systems by R_1^* and R_2^*, respectively, we obtain

$$\mathfrak{T}_1 = [\mathbb{T}_1, \{R_1^*, R\}], \quad \mathfrak{T}_2 = [\mathbb{T}_1, \{R_2^*, R\}].$$

As all the coefficients of R_i^* with respect to r and z are constants, no further splitting is needed for each $\mathfrak{T}_i$. Therefore,

$$\mathrm{Zero}(\mathbb{T}_1/x) = \mathrm{Zero}(\mathfrak{T}_1) \cup \mathrm{Zero}(\mathfrak{T}_2)$$

and each $\mathfrak{T}_i$ possesses the projection property. In particular, for any $(\bar{r}, \bar{z}) \in \mathrm{Zero}(T_1/R_i^*)$,

$$\mathrm{Zero}([\bar{T}_2, \bar{T}_3]/x) \neq \emptyset,$$

where $\bar{T}_i = T_i|_{r=\bar{r}, z=\bar{z}}$, for $i = 1, 2, 3$. Nevertheless, the original $[\mathbb{T}_1, \{x\}]$ does not satisfy this property. This can be seen easily by taking $\bar{r} = \bar{z} = 1$; then

$$\bar{T}_1 = R_1^*|_{r=\bar{r}, z=\bar{z}} = R_2^*|_{r=\bar{r}, z=\bar{z}} = 0, \;\; \bar{T}_2 = x^3, \;\; \bar{T}_3 = xy.$$

It follows that $(1, 1) \in \mathrm{Zero}(T_1)$ and $(1, 1) \notin \mathrm{Zero}(T_1/R_i^*)$. Now,

$$\mathrm{Zero}([\bar{T}_2, \bar{T}_3]/x) = \emptyset.$$

Finally, we note that projection of $\mathfrak{T}_3 = [\mathbb{T}_2, \emptyset]$ in Example 2.4.1 does not modify the triangular system. Therefore, the polynomial set $\mathbb{P}$ given there can be decomposed into three triangular systems $\mathfrak{T}_1, \mathfrak{T}_2, \mathfrak{T}_3$ such that

$$\mathrm{Zero}(\mathbb{P}) = \bigcup_{i=1}^{3} \mathrm{Zero}(\mathfrak{T}_i)$$

and each $\mathfrak{T}_i$ possesses the projection property.

Refer to Remark 3.1.1 and $\mathbb{H}_{j_1 \ldots j_h}$ defined therein. If the modification indicated there is incorporated into the above projection process for $[\mathbb{T}, \mathbb{U}]$, then in the corresponding places $\mathbb{T}$ should be replaced by $\mathbb{T} \cup \mathbb{H}_j$, $j \in \mathcal{J}$. In this case, one obtains the projection method of Wu (1990). Usually, $\mathbb{T} \cup \mathbb{H}_j$ is no more a triangular set, so its triangular series has to be further computed. For this reason, $\mathbb{H}_j$ was also abandoned by Gao and Chou (1992).

The projection case B is clearly expensive when $\mathbb{U}_j^{\langle p_r \rangle} \neq \emptyset$. For the pseudo-remainder

$$\mathrm{prem}\Big(\prod_{K \in \mathbb{U}_j^{\langle p_r \rangle}} K^{\mathrm{ldeg}(T_r)}, T_r\Big)$$

is difficult to compute. This projection process can be considerably improved by eliminating polynomials from $\mathbb{U}_j^{\langle p_r \rangle}$ via GCD computation and normalization. See the concepts and computation of *regular systems* and *normal triangular sets* in Sects. 5.1 and 5.2.

We shall show in Sect. 3.2 how the projection process explained above can be effectively embedded into algorithm TriSer, so that one does not need to compute a triangular series before projection.

3.2 Zero decomposition with projection

Refer to the data structure of triplet introduced in Sect. 2.3. Quadruplet is defined now to help understand the algorithms presented in this section.

Data structure. A *quadruplet* of level i ($1 \le i \le n$) is a list $[\mathbb{P}, \mathbb{Q}, \mathbb{T}, \mathbb{U}]$ of four elements such that $[\mathbb{P}, \mathbb{Q}, \mathbb{T}]$ is a triplet of level i, $\mathrm{level}(\mathbb{Q}) = q \le p$, and $\mathbb{U}$ is a polynomial set in $\boldsymbol{K}[\boldsymbol{x}]$ with $\mathbb{U}^{(q)} = \emptyset$, where

$$p = \begin{cases} \mathrm{cls}(\mathrm{op}(1, \mathbb{T})) & \text{if } \mathbb{T} \ne \emptyset, \\ n & \text{otherwise.} \end{cases} \tag{3.2.1}$$

For any polynomial system $[\mathbb{P}, \mathbb{Q}]$, one may write $\mathbb{P}$ and $\mathbb{Q}$ as

$$\mathbb{P} = \mathbb{P}^{(i)} \cup \mathbb{P}^{[i]}, \quad \mathbb{Q} = \mathbb{Q}^{(q)} \cup \mathbb{Q}^{[q]}$$

for some i and q such that $\mathrm{level}(\mathbb{P}^{(i)}) = i$, $\mathbb{P}^{[i]}$ can be ordered as a triangular set $\mathbb{T}$, and $q = \mathrm{level}(\mathbb{Q}^{(q)}) \le p$, where p is defined in (3.2.1). Let $\mathbb{U} = \mathbb{Q}^{[q]}$. Then, $[\mathbb{P}^{(i)}, \mathbb{Q}^{(q)}, \mathbb{T}, \mathbb{U}]$ is a quadruplet, with which $\mathrm{Zero}\,(\mathbb{P}^{(i)} \cup \mathbb{T}/\mathbb{Q}^{(q)} \cup \mathbb{U})$ is of concern.

The subalgorithm ProjA below implements Lemma 3.1.1. The polynomial system $[\mathbb{P}, \mathbb{Q}]$ is split by projection into finitely many subsystems, of which one is separated as $[\mathbb{P}, \mathbb{Q}', \mathbb{T}, \mathbb{U}']$ (in step P2.4) and the others are put into Θ. Those polynomials corresponding to $H_1, \dots, H_h$ in Lemma 3.1.1 are moved from $\mathbb{Q}$ to $\mathbb{U}$, forming the output sets $\mathbb{Q}'$ and $\mathbb{U}'$ (in step P1).

Algorithm ProjA: $[\mathbb{Q}', \mathbb{U}', \Theta] \leftarrow \mathrm{ProjA}(\mathbb{P}, \mathbb{Q}, \mathbb{T}, \mathbb{U}, i)$. Given an integer $i > 0$ and a quadruplet $[\mathbb{P}, \mathbb{Q}, \mathbb{T}, \mathbb{U}]$ of level $\le i$, this algorithm computes a polynomial set $\mathbb{Q}'$ of level $\le i$, a polynomial set $\mathbb{U}' = \mathbb{U} \cup \mathbb{Q}^{[i]}$, and a set Θ of quadruplets of level $\le i$ such that

$$\mathrm{Proj}_{\boldsymbol{x}^{\{i\}}} \mathrm{Zero}(\mathbb{P}/\mathbb{Q}) = \mathrm{Zero}(\mathbb{P}/\mathbb{Q}') \cup \bigcup_{[\mathbb{P},\mathbb{Q}^*,\mathbb{T},\mathbb{U}'] \in \Theta} \mathrm{Zero}(\mathbb{P}/\mathbb{Q}^*), \tag{3.2.2}$$

$$\mathrm{Zero}(\mathbb{P}/\mathbb{Q}) = \mathrm{Zero}(\mathbb{P}/\mathbb{Q}' \cup \mathbb{Q}^{[i]}) \cup \bigcup_{[\mathbb{P},\mathbb{Q}^*,\mathbb{T},\mathbb{U}'] \in \Theta} \mathrm{Zero}(\mathbb{P}/\mathbb{Q}^* \cup \mathbb{Q}^{[i]}), \tag{3.2.3}$$

where $\mathrm{level}(\mathbb{Q}^*) \le i$.

P1. Set $\mathbb{Q}' \leftarrow \mathbb{Q}^{(i)}$, $\mathbb{U}' \leftarrow \mathbb{U} \cup \mathbb{Q}^{[i]}$, $\Theta \leftarrow \emptyset$.

P2. If $\mathbb{Q}^{[i]} \ne \emptyset$, then do:

P2.1. Let $H_1, \dots, H_h$ be all the polynomials in $\mathbb{Q}^{[i]}$.

P2.2. For $l = 1, \dots, h$ do:

P2.2.1. Compute

$$V_l \leftarrow \{x_j \colon \deg(H_l, x_j) > 0, i < j \le n\}.$$

P2.2.2. Let $\mathcal{H}_l$ be the set of all the nonzero coefficients of H_l with respect to V_l. If $\mathcal{H}_l \cap \boldsymbol{K} \ne \emptyset$, then set $m_l \leftarrow 1$, $H_{l1} \leftarrow 1$; else let $H_{l1}, \dots, H_{lm_l}$ be all the polynomials in $\mathcal{H}_l$.

P2.3. Form

$$\Theta \leftarrow \{[\mathbb{P}, \mathbb{Q}' \cup \{H_{1j_1}, \dots, H_{hj_h}\}, \mathbb{T}, \mathbb{U}']: \\ 1 \le j_1 \le m_1, \dots, 1 \le j_h \le m_h\}.$$

P2.4. Set

$$\mathbb{Q}' \leftarrow \mathbb{Q}' \cup \{H_{11}, \dots, H_{h1}\}, \quad \Theta \leftarrow \Theta \setminus \{[\mathbb{P}, \mathbb{Q}', \mathbb{T}, \mathbb{U}']\}.$$

Proof. No recursive loop is involved in this algorithm, so the termination is obvious.

To see (3.2.2) and (3.2.3), we first note that in step P2.2.2, if $\mathcal{H}_l \cap \boldsymbol{K} \neq \emptyset$, then H_l has at least one coefficient which is a nonzero constant. In this case, for any $\bar{\boldsymbol{x}}^{\{i\}} \in \tilde{\boldsymbol{K}}^i$ there always exist $\bar{x}_{i+1}, \dots, \bar{x}_n \in \tilde{\boldsymbol{K}}$ such that $H_l(\bar{\boldsymbol{x}}) \neq 0$, so one does not need to consider the coefficients of H_l with respect to V_l. In other words, H_l is not needed. This is treated by simply taking $m_l = 1$ and $H_{l1} = 1$.

Except for this minor modification, $[\mathbb{P}, \mathbb{Q}']$ here corresponds to the subsystem in Lemma 3.1.1 for the indices $j_1 = 1, \dots, j_h = 1$, while the $[\mathbb{P}, \mathbb{Q}^*]$'s put into Θ correspond to the subsystems in Lemma 3.1.1 for all the other indices. Therefore, (3.2.2) and (3.2.3) are actually an alternative form of (3.1.1) and (3.1.2) in Lemma 3.1.1. □

Now, we are ready to present the elimination algorithm with projection. This algorithm is modified from TriSer by: (i) replacing the reduction step P2.3 in PriTriSys with step T2.2.4 below for the projection case B in which there are polynomials of class i but no polynomial of class $>i$ to be "projected"; (ii) inserting two steps T2.2.3 and T2.3 for the projection case A in which there are polynomials of classes $>i$ to be "projected."

Algorithm TriSerP: $\Psi \leftarrow \mathrm{TriSerP}(\mathbb{P}, \mathbb{Q}, k)$. Given a polynomial system $[\mathbb{P}, \mathbb{Q}]$ in $\boldsymbol{K}[\boldsymbol{x}]$ and an integer k $(0 \le k < n)$, this algorithm computes either an empty set Ψ, that means, $\mathrm{Zero}(\mathbb{P}/\mathbb{Q}) = \emptyset$, or a finite nonempty set

$$\Psi = \{[\mathbb{P}_1, \mathbb{Q}_1, \mathbb{T}_1, \mathbb{U}_1], \dots, [\mathbb{P}_e, \mathbb{Q}_e, \mathbb{T}_e, \mathbb{U}_e]\},$$

where each $[\mathbb{P}_i, \mathbb{Q}_i, \mathbb{T}_i, \mathbb{U}_i]$ is a quadruplet of level $\le k$ with $\mathrm{level}(\mathbb{Q}_i) \le k$, such that

a. $$\mathrm{Zero}(\mathbb{P}/\mathbb{Q}) = \bigcup_{i=1}^{e} \mathrm{Zero}(\mathbb{P}_i \cup \mathbb{T}_i/\mathbb{Q}_i \cup \mathbb{U}_i); \tag{3.2.4}$$

b. $$\mathrm{Proj}_{\boldsymbol{x}^{\{k\}}} \mathrm{Zero}(\mathbb{P}/\mathbb{Q}) = \bigcup_{i=1}^{e} \mathrm{Zero}(\mathbb{P}_i/\mathbb{Q}_i); \tag{3.2.5}$$

c. for any $1 \le i \le e$ and

$$j \in \{k\} \cup \{\mathrm{cls}(T): \ T \in \mathbb{T}_i\}, \\ (\bar{x}_1, \dots, \bar{x}_j) \in \mathrm{Zero}(\mathbb{P}_i \cup \mathbb{T}_i^{(j)}/\mathbb{Q}_i \cup \mathbb{U}_i^{(j)}),$$

$[\mathbb{T}_i^{[j]\langle\bar{x},j\rangle}, \mathbb{U}_i^{[j]\langle\bar{x},j\rangle}]$ is a perfect triangular system, and so is $[\mathbb{T}_i, \mathbb{U}_i]$.

T1. Set $\Psi \leftarrow \emptyset$, $\Phi \leftarrow \{[\mathbb{P}, \mathbb{Q}, \emptyset, \emptyset]\}$.

T2. While $\Phi \neq \emptyset$, do:

T2.1. Let $[\mathbb{F}, \mathbb{G}, \mathbb{T}, \mathbb{U}]$ be an element of Φ and set

$$\Phi \leftarrow \Phi \setminus \{[\mathbb{F}, \mathbb{G}, \mathbb{T}, \mathbb{U}]\}, \quad l \leftarrow \mathrm{level}(\mathbb{F}).$$

T2.2. For $\imath = l, \ldots, k+1$ do:

T2.2.1. If $\mathbb{F} \cap \boldsymbol{K} \setminus \{0\} \neq \emptyset$, then go to T2. If $\mathrm{level}(\mathbb{F}) < \imath$, then go to T2.2 for next $\imath$.

T2.2.2. Compute $[T, \mathbb{F}, \mathbb{G}, \Delta] \leftarrow \mathrm{Elim}(\mathbb{F}, \mathbb{G}, \imath)$ and set

$$\Phi \leftarrow \Phi \cup \{\delta \cup [\mathbb{T}, \mathbb{U}]\colon\ \delta \in \Delta\}.$$

T2.2.3. Compute $[\mathbb{G}, \mathbb{U}, \Theta] \leftarrow \mathrm{ProjA}(\mathbb{F} \cup \{T\}, \mathbb{G}, \mathbb{T}, \mathbb{U}, \imath)$ and set $\Phi \leftarrow \Phi \cup \Theta$.

T2.2.4. If $\mathbb{G}^{[\imath-1]} \neq \emptyset$, then compute

$$\mathbb{G} \leftarrow \mathbb{G}^{(\imath-1)} \cup \left\{\mathrm{prem}\left(\prod_{G \in \mathbb{G}^{[\imath-1]}} G^{\mathrm{ldeg}(T)}, T\right)\right\}.$$

T2.2.5. If $0 \in \mathbb{G}$, then go to T2; else set $\mathbb{T} \leftarrow [T] \cup \mathbb{T}$.

T2.3. Compute $[\mathbb{G}, \mathbb{U}, \Theta] \leftarrow \mathrm{ProjA}(\mathbb{F}, \mathbb{G}, \mathbb{T}, \mathbb{U}, k)$ and set $\Phi \leftarrow \Phi \cup \Theta$.

T2.4. Set $\Psi \leftarrow \Psi \cup \{[\mathbb{F}, \mathbb{G}, \mathbb{T}, \mathbb{U}]\}$.

We may assume that $\mathbb{P}_i \cap \boldsymbol{K} \setminus \{0\} = \emptyset$ and $0 \notin \mathbb{Q}_i$ for each $\psi_i = [\mathbb{P}_i, \mathbb{Q}_i, \mathbb{T}_i, \mathbb{U}_i] \in \Psi$. For, otherwise, $\mathrm{Zero}(\mathbb{P}_i \cup \mathbb{T}_i/\mathbb{Q}_i \cup \mathbb{U}_i) = \emptyset$ and ψ_i can be simply deleted from Ψ. If $k = 0$, then $\mathrm{Zero}(\mathbb{P}/\mathbb{Q}) \neq \emptyset$ if and only if $e \geq 1$. Hence, when $k = 0$ and $e \geq 1$, $\mathbb{P}_i \setminus \{0\} = \emptyset$ and $[\mathbb{T}_i, \mathbb{U}_i]$ possesses the projection property for all $1 \leq i \leq e$.

Example 3.2.1. See Example 2.3.2. Let $k = 0$ and perform the elimination with projection. For $z \in \mathbb{U}_1$, we need to compute in step T2.2.4 the pseudo-remainder of z^5, instead of that of z, with respect to R_3. It is $-t^4 \rightsquigarrow t$, so $\mathbb{U}_1$ is replaced by $\{t, t^3 - 1\}$. Similarly, for $z \in \mathbb{U}_3$ we need to compute the pseudo-remainder of z^5 with respect to R_3, which is $-t^4 \rightsquigarrow t$, and then the pseudo-remainder of t^3 with respect to $t^3 - 1$, which is the constant 1. Hence, $\mathbb{U}_3$ is simplified to $\emptyset$. The projection steps T2.2.3 and T2.3 are trivially executed for this example.

Proof of TriSerP. Termination. Define, for any polynomial system $[\mathbb{P}, \mathbb{Q}]$, a triple

$$\mathrm{Index}(\mathbb{P}/\mathbb{Q}) \triangleq \langle d, l, p\rangle,$$

where

$$\begin{aligned} d &= \min\{\deg(P, x_l)\colon\ P \in \mathbb{P}^{\langle l\rangle}\},\\ l &= \mathrm{level}(\mathbb{P}),\\ p &= \max(l, \mathrm{level}(\mathbb{Q})). \end{aligned}$$

We order two triples as $\langle d_1, l_1, p_1\rangle \prec \langle d_2, l_2, p_2\rangle$ if

$$p_1 < p_2; \text{ or}$$
$$p_1 = p_2 \text{ while } l_1 < l_2; \text{ or}$$
$$p_1 = p_2, l_1 = l_2 \text{ while } d_1 < d_2.$$

For a quadruplet ψ taken from Ψ in step T2.1 of TriSerP, let $\mathbb{F}$, $\mathbb{G}$ be the first two components of ψ and $\mathbb{P}^*$, $\mathbb{Q}^*$ the two components of some polynomial system in Δ produced by Elim or the first two components of some quadruplet in Θ produced by ProjA from ψ. Then we always have

$$\text{Index}(\mathbb{P}^*/\mathbb{Q}^*) \prec \text{Index}(\mathbb{F}/\mathbb{G}).$$

Since each component of the triple Index($\mathbb{P}/\mathbb{Q}$) is a positive integer, any steadily decreasing sequence of such index triples is finite. Therefore, the while-loop of TriSerP has only finitely many iterations. The termination is proved.

Correctness. This is to show that the computed Ψ satisfies the properties a, b, and c in the specification of TriSerP.

a. Similar to TriSer, algorithm TriSerP can also be viewed as for computing a multibranch tree $\mathcal{T}$. With the root of $\mathcal{T}$, the quadruplet $[\mathbb{P}, \mathbb{Q}, \emptyset, \emptyset]$ is associated, and with each node or leaf i, a quadruplet $[\mathbb{P}_i, \mathbb{Q}_i, \mathbb{T}_i, \mathbb{U}_i]$ is associated such that after the execution of every step of TriSerP the zero relation (2.3.6), when $\mathbb{Q}_i$ on the right-hand side is replaced by $\mathbb{Q}_i \cup \mathbb{U}_i$, is preserved. To see this, one only needs to note that in the present case, the branches are generated also by the subalgorithm ProjA with the zero relation (3.2.3) preserved, while (3.2.3) implies that

$$\text{Zero}(\mathbb{P} \cup \mathbb{T}/\mathbb{Q} \cup \mathbb{U}) = \text{Zero}(\mathbb{P} \cup \mathbb{T}/\mathbb{G} \cup \mathbb{U}) \cup \bigcup_{[\mathbb{P},\mathbb{Q}^*,\mathbb{T},\mathbb{U}]\in\Theta} \text{Zero}(\mathbb{P} \cup \mathbb{T}/\mathbb{Q}^* \cup \mathbb{U}'),$$

where $\mathbb{U}' = \mathbb{U} \cup \mathbb{Q}^{[i]}$. Zero($\mathbb{F} \cup \{T\} \cup \mathbb{T}/\mathbb{G} \cup \mathbb{U}$) also remains unchanged when step T2.2.4 is executed.

Cutting those leaves i of $\mathcal{T}$ for which $\mathbb{P}_i$ contains a nonzero constant or $0 \in \mathbb{Q}_i$ and assuming that not all the leaves are cut off, we obtain the zero decomposition (3.2.4). From the correctness proof of TriSer, one sees clearly that $[\mathbb{T}_i, \mathbb{U}_i]$ here is also a triangular system.

b. First let $\bar{\boldsymbol{x}}^{\{k\}} \in \tilde{\boldsymbol{K}}^k$ belong to the right-hand side of (3.2.5); then there is an i such that $\bar{\boldsymbol{x}}^{\{k\}} \in \text{Zero}(\mathbb{P}_i/\mathbb{Q}_i)$. By property c to be proved, there exist $\bar{x}_{k+1}, \ldots, \bar{x}_n \in \tilde{\boldsymbol{K}}$ such that

$$(\bar{x}_{k+1}, \ldots, \bar{x}_n) \in \text{Zero}(\mathbb{T}_i^{\langle\bar{x},k\rangle}/\mathbb{U}_i^{\langle\bar{x},k\rangle}).$$

Hence

$$\bar{\boldsymbol{x}} \in \text{Zero}(\mathbb{P}_i \cup \mathbb{T}_i/\mathbb{Q}_i \cup \mathbb{U}_i). \tag{3.2.6}$$

By (3.2.4), $\bar{\boldsymbol{x}} \in \text{Zero}(\mathbb{P}/\mathbb{Q})$. It follows that

$$\bar{\boldsymbol{x}}^{\{k\}} \in \text{Proj}_{\boldsymbol{x}^{\{k\}}}\text{Zero}(\mathbb{P}/\mathbb{Q}). \tag{3.2.7}$$

Now suppose that (3.2.7) holds, so there exist $\bar{x}_{k+1}, \ldots, \bar{x}_n \in \tilde{\boldsymbol{K}}$ such that $\bar{\boldsymbol{x}} \in \mathrm{Zero}(\mathbb{P}/\mathbb{Q})$. By (3.2.4), there must be an i such that (3.2.6) holds. In particular, we have

$$\bar{\boldsymbol{x}}^{\{k\}} \in \mathrm{Zero}(\mathbb{P}_i/\mathbb{Q}_i) \subset \bigcup_{i=1}^{e} \mathrm{Zero}(\mathbb{P}_i/\mathbb{Q}_i).$$

Thus, (3.2.5) is proved.

c. Let $\mathbb{F}$, $\mathbb{G}$, $\mathbb{T}$, $\mathbb{U}$, and T be as in TriSerP. We first show two assertions:

1. if step T2.2.3 is executed for some ι, then after the execution, for any $(\bar{x}_1, \ldots, \bar{x}_\iota) \in \mathrm{Zero}(\mathbb{F} \cup \{T\}/\mathbb{G})$,

$$\mathrm{Zero}(\mathbb{T}^{\langle \bar{x}, \iota \rangle}/\mathbb{U}^{\langle \bar{x}, \iota \rangle}) \neq \emptyset; \tag{3.2.8}$$

2. if step T2.2.4 is executed for some ι, then after the execution, for any j, $\mathrm{level}(\mathbb{F}) \leq j \leq \iota - 1$, and $(\bar{x}_1, \ldots, \bar{x}_j) \in \mathrm{Proj}_{\boldsymbol{x}^{\{j\}}}\mathrm{Zero}(\mathbb{F}/\mathbb{G})$,

$$\mathrm{Zero}([T] \cup \mathbb{T}^{\langle \bar{x}, j \rangle}/\mathbb{U}^{\langle \bar{x}, j \rangle}) \neq \emptyset. \tag{3.2.9}$$

If $0 \in \mathbb{G}$, then $\mathrm{Zero}(\mathbb{F}/\mathbb{G}) = \emptyset$. In this case, the property is trivial and need not be considered.

To avoid confusion of notations, the quadruplet $[\mathbb{F}, \mathbb{G}, \mathbb{T}, \mathbb{U}]$ in what follows will always be referred to before the execution of the step under discussion, and the corresponding components after the execution, if updated, will be referred to with a superscript asterisk. The proof proceeds by induction on $|\mathbb{T}|$.

Case i. $\mathbb{T} = \emptyset$.

A. Let ψ and ψ^* be the quadruplets corresponding to $[\mathbb{F}, \mathbb{G}, \mathbb{T}, \mathbb{U}]$ before and after the execution of step T2.2.3 in TriSerP, respectively. Then $\psi = [\mathbb{F}, \mathbb{G}, \emptyset, \emptyset]$ and $\psi^* = [\mathbb{F}, \mathbb{G}^*, \emptyset, \mathbb{U}^*]$, where $\mathbb{U}^* = \mathbb{G}^{[\iota]}$. Let $\bar{\boldsymbol{x}}^{\{\iota\}} \in \mathrm{Zero}(\mathbb{F} \cup \{T\}/\mathbb{G}^*)$. By (3.2.2), there exist $\bar{x}_{\iota+1}, \ldots, \bar{x}_n \in \tilde{\boldsymbol{K}}$ such that

$$\bar{\boldsymbol{x}} \in \mathrm{Zero}(\mathbb{F} \cup \{T\}/\mathbb{G}).$$

Since $\mathbb{U}^* \subset \mathbb{G}$, $U(\bar{\boldsymbol{x}}) \neq 0$ for any $U \in \mathbb{U}^*$. Hence, $\bar{\boldsymbol{x}} \in \mathrm{Zero}(\emptyset/\mathbb{U}^*)$ and (3.2.8) holds.

B. For step T2.2.4, we have $\psi = [\mathbb{F}, \mathbb{G}, \emptyset, \mathbb{U}]$ and $\psi^* = [\mathbb{F}, \mathbb{G}^*, \emptyset, \mathbb{U}]$, where

$$\mathbb{G}^* = \begin{cases} \mathbb{G}^{(\iota-1)} \cup \{\mathrm{prem}(\prod_{G \in \mathbb{G}^{[\iota-1]}} G^{\mathrm{ldeg}(T)}, T)\} & \text{if } \mathbb{G}^{[\iota-1]} \neq \emptyset, \\ \mathbb{G} & \text{otherwise.} \end{cases}$$

In both cases, for any $\mathrm{level}(\mathbb{F}) \leq j \leq \iota - 1$ and $\bar{\boldsymbol{x}}^{\{j\}} \in \mathrm{Proj}_{\boldsymbol{x}^{\{j\}}}\mathrm{Zero}(\mathbb{F}/\mathbb{G}^*)$, by (3.1.4) and (3.1.5), and noting that $\mathrm{Zero}(T/\mathbb{G} \cup \{\mathrm{ini}(T)\}) = \mathrm{Zero}(T/\mathbb{G})$, there exist $\bar{x}_{j+1}, \ldots, \bar{x}_\iota \in \tilde{\boldsymbol{K}}$ such that

$$\bar{\boldsymbol{x}}^{\{\iota\}} \in \mathrm{Zero}(\mathbb{F} \cup \{T\}/\mathbb{G}). \tag{3.2.10}$$

Now for (3.2.10), by A above there exist $\bar{x}_{\iota+1}, \ldots, \bar{x}_n \in \tilde{\boldsymbol{K}}$ such that $\bar{\boldsymbol{x}} \in \mathrm{Zero}(\emptyset/\mathbb{U})$. Therefore, $\bar{\boldsymbol{x}} \in \mathrm{Zero}([T]/\mathbb{U})$ and (3.2.9) holds.

Case ii. $\mathbb{T} \neq \emptyset$.

By induction we suppose that the property in assertion 2 is satisfied after the execution of step T2.2.4 for $\iota = p$, where $p = \mathrm{cls}(\mathrm{op}(1, \mathbb{T}))$. Observe that steps T2.2.5 and T2.2.1 are trivial, the execution of step T2.2.2 does not update $\mathbb{T}$ and $\mathbb{U}$, and for this step any zero of $[\mathbb{F}^* \cup \{T\}, \mathbb{G}^*]$ is also a zero of $[\mathbb{F}, \mathbb{G}]$ by (2.3.5). Hence, we have the following assertion 2′ which corresponds to 2 for $j = \mathrm{level}(\mathbb{F})$:

2′. if step T2.2.2 is executed for some ι, then after the execution, for any $(\bar{x}_1, \ldots, \bar{x}_\iota) \in \mathrm{Proj}_{\boldsymbol{x}^{\{\iota\}}}\mathrm{Zero}(\mathbb{F} \cup \{T\}/\mathbb{G})$,

$$\mathrm{Zero}(\mathbb{T}^{\langle \bar{x}, \iota\rangle}/\mathbb{U}^{\langle \bar{x}, \iota\rangle}) \neq \emptyset.$$

A. For step T2.2.3, we have $\psi = [\mathbb{F}, \mathbb{G}, \mathbb{T}, \mathbb{U}]$ and $\psi^* = [\mathbb{F}, \mathbb{G}^*, \mathbb{T}, \mathbb{U}^*]$, where $\mathbb{U}^* = \mathbb{U} \cup \mathbb{G}^{[\iota]}$. For any $\bar{\boldsymbol{x}}^{\{\iota\}} \in \mathrm{Zero}(\mathbb{F} \cup \{T\}/\mathbb{G}^*)$, according to (3.2.2) there exist $\bar{x}_{\iota+1}, \ldots, \bar{x}_p \in \tilde{\boldsymbol{K}}$ such that

$$\bar{\boldsymbol{x}}^{\{p\}} \in \mathrm{Zero}(\mathbb{F} \cup \{T\}/\mathbb{G}).$$

Therefore, by 2′ there exist $\bar{x}_{p+1}, \ldots, \bar{x}_n \in \tilde{\boldsymbol{K}}$ such that $\bar{\boldsymbol{x}} \in \mathrm{Zero}(\mathbb{T}/\mathbb{U})$. Since $\mathbb{U}^{*(p)} = \mathbb{G}^{[\iota]} \subset \mathbb{G}$,

$$\bar{\boldsymbol{x}} \in \mathrm{Zero}(\mathbb{T}/\mathbb{U}^{*(p)} \cup \mathbb{U}) = \mathrm{Zero}(\mathbb{T}/\mathbb{U}^*),$$

so (3.2.8) holds.

B. Similar to B in case i, for any $\mathrm{level}(\mathbb{F}) \leq j \leq \iota - 1$ and $\bar{\boldsymbol{x}}^{\{j\}} \in \mathrm{Zero}(\mathbb{F}/\mathbb{G}^*)$, by (3.1.4) and (3.1.5), and noting that $\mathrm{Zero}(T/\mathbb{G} \cup \{\mathrm{ini}(T)\}) = \mathrm{Zero}(T/\mathbb{G})$, there exist $\bar{x}_{j+1}, \ldots, \bar{x}_\iota \in \tilde{\boldsymbol{K}}$ such that

$$\bar{\boldsymbol{x}}^{\{\iota\}} \in \mathrm{Zero}(\mathbb{F} \cup \{T\}/\mathbb{G}).$$

By A in the present case, there exist $\bar{x}_{\iota+1}, \ldots, \bar{x}_n \in \tilde{\boldsymbol{K}}$ such that

$$\bar{\boldsymbol{x}} \in \mathrm{Zero}(\mathbb{T}/\mathbb{U}).$$

Hence, $\bar{\boldsymbol{x}} \in \mathrm{Zero}([T] \cup \mathbb{T}/\mathbb{U})$ and (3.2.9) holds as well. By now the two assertions 1 and 2 have been proved.

Next, we show that after the execution of step T2.3, (3.2.8) holds for any $\bar{\boldsymbol{x}}^{\{\iota\}} \in \mathrm{Zero}(\mathbb{F}/\mathbb{G})$.

If $\mathbb{T} = \emptyset$, then step T2.2 is trivially executed and the execution of step T2.3 is the same as that of step T2.2.3 for $\iota = k$ in A of case i, noting that the polynomial T does not play any special role in ProjA. Therefore, for any $\bar{\boldsymbol{x}}^{\{k\}} \in \mathrm{Zero}(\mathbb{F}/\mathbb{G}^*)$, there are $\bar{x}_{k+1}, \ldots, \bar{x}_n \in \tilde{\boldsymbol{K}}$ such that $\bar{\boldsymbol{x}}$ is not a zero of any polynomial in $\mathbb{U}^* \subset \mathbb{G}$. Hence, $\bar{\boldsymbol{x}} \in \mathrm{Zero}(\emptyset/\mathbb{U}^*)$ and (3.2.8) holds.

If $\mathbb{T} \neq \emptyset$, then step T2.2.4 must have been executed before, say for $\iota = p > k$, where $p = \mathrm{cls}(\mathrm{op}(1, \mathbb{T}))$. Now the execution of step T2.3 is the same as that of step T2.2.3 for $\iota = k$ in A of case ii. Therefore, for any $\bar{\boldsymbol{x}}^{\{k\}} \in \mathrm{Zero}(\mathbb{F}/\mathbb{G}^*)$, there exist $\bar{x}_{k+1}, \ldots, \bar{x}_n \in \tilde{\boldsymbol{K}}$ such that $\bar{\boldsymbol{x}} \in \mathrm{Zero}(\mathbb{T}/\mathbb{U}^*)$, so (3.2.8) holds as well.

Clearly, the final $[\mathbb{F}, \mathbb{G}, \mathbb{T}, \mathbb{U}]$ is some $\psi_i = [\mathbb{P}_i, \mathbb{Q}_i, \mathbb{T}_i, \mathbb{U}_i] \in \Psi$ in the specification of TriSerP. In the way of computing ψ_i, step T2.2.3 must have been executed for all $\iota \in \{\mathrm{cls}(T): T \in \mathbb{T}_i\}$ and step T2.3 for $\iota = k$. From the splitting process and the zero relations that are preserved between the original and the split systems, we know that any $[\mathbb{P}_i \cup \mathbb{T}_i^{(j)}, \mathbb{Q}_i \cup \mathbb{U}_i^{(j)}]$ is produced from some corresponding $[\mathbb{F} \cup \{T\}, \mathbb{G}]$ as in the assertion 1 for $\iota = j$ such that any

$$(\bar{x}_1, \ldots, \bar{x}_j) \in \mathrm{Zero}(\mathbb{P}_i \cup \mathbb{T}_i^{(j)}/\mathbb{Q}_i \cup \mathbb{U}_i^{(j)})$$

is also a zero of $[\mathbb{F} \cup \{T\}, \mathbb{G}]$. Therefore, it follows from 1 that

$$\mathrm{Zero}(\mathbb{T}_i^{[j]\langle\bar{x},j\rangle}/\mathbb{U}_i^{[j]\langle\bar{x},j\rangle}) \neq \emptyset.$$

In other words, $[\mathbb{T}_i^{[j]\langle\bar{x},j\rangle}, \mathbb{U}_i^{[j]\langle\bar{x},j\rangle}]$ is perfect for any $j \in \{k\} \cup \{\mathrm{cls}(T): T \in \mathbb{T}_i\}$. Since

$$\mathrm{Zero}(\mathbb{T}_i^{\langle\bar{x},k\rangle}/\mathbb{U}_i^{\langle\bar{x},k\rangle}) \neq \emptyset \Longrightarrow \mathrm{Zero}(\mathbb{T}_i/\mathbb{U}_i) \neq \emptyset,$$

by definition the triangular system $[\mathbb{T}_i, \mathbb{U}_i]$ is also perfect.

This completes the correctness proof of TriSerP. □

Remark 3.2.1. The second "if-condition" in step T2.2.1 of TriSerP may be modified so that projection step T2.2.3 is also executed when $\mathrm{level}(\mathbb{F}) < \iota$. Then, ProjA is called for every ι and V_l in step P2.2.1 contains x_ι only for each call. This may simplify the presentation and proof slightly. In this case, properties b and c in the specification may be modified accordingly:

b′. for any $k \leq j < n$, $\mathrm{Proj}_{\boldsymbol{x}^{\{j\}}} = \bigcup_{i=1}^{e} \mathrm{Zero}(\mathbb{P}_i \cup \mathbb{T}_i^{(j)}/\mathbb{Q}_i \cup \mathbb{U}^{(j)})$;

c′. for any $1 \leq i \leq e$ and $k \leq j < n$, $\bar{\boldsymbol{x}}^{\{j\}} \in \mathrm{Zero}(\mathbb{P}_i \cup \mathbb{T}_i^{(j)}/\mathbb{Q}_i \cup \mathbb{U}^{(j)})$, $[\mathbb{T}_i^{[j]\langle\bar{x},j\rangle}, \mathbb{U}_i^{[j]\langle\bar{x},j\rangle}]$ is a perfect triangular system, and thus so is $[\mathbb{T}_i^{[j]}, \mathbb{U}_i^{[j]}]$.

If $k = 0$, then each $[\mathbb{T}_i, \mathbb{U}_i]$ possesses the strong projection property. However, if splitting also occurs when $\mathrm{level}(\mathbb{F}) < \iota \neq k$, there is a critical drawback: Elim in step T2.2.2 may be called repeatedly for the same $\mathbb{F}$.

Remark 3.2.2. The projection step T2.2.4 can be modified by using a more complicated procedure as follows. Instead of forming

$$\mathrm{prem}\left(\prod_{G \in \mathbb{G}^{[\iota-1]}} G^{\mathrm{ldeg}(T)}, T\right),$$

after squarefreeing T one computes the GCD of T and each polynomial $G \in \mathbb{G}^{[\iota-1]}$ with respect to x_ι, say, by pseudo-division, and deletes it as a factor from T and G. After the deletion of all such common divisors, the GCD of T and every polynomial in $\mathbb{G}^{[\iota-1]}$ should be 1. Then, $\mathrm{Zero}(T/\mathbb{G}^{[\iota-1]}) \neq \emptyset$ if and only if T is of positive degree in x_ι (see Seidenberg 1956a). Along with computing the GCDs, the system is split into finitely many other systems so that the necessary zero relations are preserved. This technique will be reflected in algorithm SimSer. In fact, another projection algorithm can be derived from SimSer.

Algorithm TriSerP provides a quantifier elimination procedure and thus a decision procedure for the existential theory of algebraically closed fields. As a corollary of this algorithm, we have the following projection theorem.

Theorem 3.2.1 (Projection theorem of elimination theory – affine case). Let $\{\mathbb{F}_i(\boldsymbol{x}, \boldsymbol{y}) : 1 \leq i \leq s\}$ be a set of finite conjunctions of polynomial equations and inequations over $\boldsymbol{K}$ in the variables

$$\boldsymbol{x} = (x_1, \ldots, x_n), \quad \boldsymbol{y} = (y_1, \ldots, y_m).$$

Then there is a finite set of $\mathbb{G}_j(\boldsymbol{x})$ of which each one is a finite conjunction of polynomial equations and inequations over $\boldsymbol{K}$ having the following property: for every point $\bar{\boldsymbol{x}} = (\bar{x}_1, \ldots, \bar{x}_n)$ of the affine space $\mathbf{V}^n$ over some extension field $\tilde{\boldsymbol{K}}$ of $\boldsymbol{K}$ there is a point $\bar{\boldsymbol{y}} = (\bar{y}_1, \ldots, \bar{y}_m)$ of the affine space $\mathbf{W}^m$ over some algebraic-extension field of $\tilde{\boldsymbol{K}}$ such that $(\bar{\boldsymbol{x}}, \bar{\boldsymbol{y}})$ satisfies at least one of the $\mathbb{F}_i(\boldsymbol{x}, \boldsymbol{y})$ if and only if $\bar{\boldsymbol{x}}$ satisfies one of the $\mathbb{G}_j(\boldsymbol{x})$.

One proof of this theorem, contained in the classical decision method of A. Tarski, was clarified by Jacobson (1974, sect. 5.4, pp. 305 f). Another proof appeared in Seidenberg (1956a, b). A recent proof was given by Wu (1990).

For every polynomial system $[\mathbb{P}_i, \mathbb{Q}_i]$ in (3.2.4), one can further compute its triangular series by algorithm CharSer, TriSer, or TriSerS. The corresponding zero decompositions may be merged with (3.2.4). As a consequence, there is an algorithm which computes, for any polynomial system $[\mathbb{P}, \mathbb{Q}]$ and integer $0 \leq k < n$, a set Ψ which is either empty, that means, $\text{Zero}(\mathbb{P}/\mathbb{Q}) = \emptyset$, or of the form

$$\{[\mathbb{P}_1, \mathbb{Q}_1, \mathbb{T}_1, \mathbb{U}_1], \ldots, [\mathbb{P}_e, \mathbb{Q}_e, \mathbb{T}_e, \mathbb{U}_e]\}$$

such that a, b, and c in the specification of TriSerP are all satisfied and moreover each $[\mathbb{P}_i \cup \mathbb{T}_i, \mathbb{Q}_i \cup \mathbb{U}_i]$ is a (fine) triangular system possessing the projection property of dimension k, where $\mathbb{P}_i$ is ordered as triangular set. In this case, we call $n - k$ the *dimension* of projection and say that the elimination is performed with *full* projection if the dimension is n, and *without* projection if the dimension is 0.

Example 3.2.2. Let $\mathbb{P} = \{P_1, \ldots, P_4\}$ with

$$\begin{aligned}
P_1 &= (x - u)^2 + (y - v)^2 - 1, \\
P_2 &= v^2 - u^3, \\
P_3 &= 2v(x - u) + 3u^2(y - v), \\
P_4 &= (3wu^2 - 1)(2wv - 1).
\end{aligned}$$

This set of polynomials was communicated by P. Vermeer from the Department of Computer Science, Purdue University in April 1990. It has been used as a test example in Wang (1993).

Under the variable ordering $x \prec y \prec u \prec v \prec w$, $\mathbb{P}$ can be decomposed by TriSerP with projection for w, v, u into five fine triangular systems $\mathfrak{T}_i = [\mathbb{T}_i, \mathbb{U}_i]$

such that the zero decomposition (2.1.8) holds with $\mathbb{Q} = \emptyset$ and $e = 5$, and each $\mathfrak{T}_i$ possesses the (strong) projection property of dimension 2. Listed below are the triangular sets $\mathbb{T}_i$ and the corresponding $\mathbb{U}_i$ which will be used in Example 7.4.2.

$$\begin{aligned}
\mathbb{T}_1 &= [T_{11}, T_{12}, P_3, P_4],\\
\mathbb{T}_2 &= [T_{21}, T_{22}, T_{23}, P_3, P_4],\\
\mathbb{T}_3 &= [T_{31}, T_{32}, T_{33}, P_3, P_4],\\
\mathbb{T}_4 &= [T_{41}, y, 12xu + 2u - 9x^2 - 2x + 9, v^2 + u^2 - 2xu + x^2 - 1, P_4],\\
\mathbb{T}_5 &= [x, 729y^4 - 956y^2 - 529, u(85u - 81y^2 + 72), u(3uv + 2v - 3uy), P_4],
\end{aligned}$$

where

$$\begin{aligned}
T_{11} &= 729y^6 - (1458x^3 - 729x^2 + 4158x + 1685)y^4 + (729x^6 - 1458x^5\\
&\quad - 2619x^4 - 4892x^3 - 297x^2 + 5814x + 427)y^2 + 729x^8 + 216x^7\\
&\quad - 2900x^6 - 2376x^5 + 3870x^4 + 4072x^3 - 1188x^2 - 1656x + 529,\\
T_{12} &= [2187y^4 - 6(729x^3 + 162x^2 + 2079x + 478)y^2 + 2187x^6 - 1944x^5\\
&\quad - 10125x^4 - 4800x^3 + 2501x^2 + 4968x - 1587]u + 4x^2 T_{32},\\
T_{21} &= 243x^2 + 36x + 85,\\
T_{22} &= 10460353203y^6 - 6377292(8523x + 4535)y^4 + 648(155380149x\\
&\quad + 61648)y^2 - 16(2250218592x - 1609630283),\\
T_{23} &= (81y^2 + 162x^3 - 36x^2 - 154x - 72)u + 72x^3 - 4x^2,\\
T_{31} &= (81x^2 + 18x + 28)(729x^4 + 972x^3 - 1026x^2 + 1684x + 765),\\
T_{32} &= 27(18x - 1)y^2 + 243x^4 + 756x^3 - 270x^2 + 124x + 279,\\
T_{33} &= -T_{21}u^2 + T_{23},\\
T_{41} &= 27x^4 + 4x^3 - 54x^2 - 36x + 23,
\end{aligned}$$

and

$$\begin{aligned}
\mathbb{U}_1 &= \{x, y, T_{21}, \mathrm{ini}(T_{12}), T_{32}, 729(2187x^6 - 1134x^5 - 7326x^4 + 4144x^3\\
&\quad + 2015x^2 - 6498x - 2268)y^4 - 2(1594323x^9 + 2007666x^8\\
&\quad + 2591595x^7 + 6800112x^6 - 12642075x^5 + 2179818x^4\\
&\quad + 4872429x^3 - 12546172x^2 - 7821216x - 1084104)y^2\\
&\quad + 1594323x^{12} + 590490x^{11} - 12328119x^{10} - 6466230x^9\\
&\quad + 22602402x^8 + 8733636x^7 - 22926870x^6 + 11418356x^5\\
&\quad + 35613711x^4 + 1579842x^3 - 13321235x^2 - 318366x + 1199772\},\\
\mathbb{U}_2 &= \{x, y, 4194x - 935, -6561y^2 + 16344x + 4132, 1162261467xy^4\\
&\quad - 26244(35676x - 79985)y^2 - 40(61438590x + 29843347)\},\\
\mathbb{U}_3 &= \{x, y, 18x - 1, T_{21}, 8474827586184x^5 - 6240413571255x^4\\
&\quad + 7521969157884x^3 + 2321430215166x^2 + 3035377934972x\\
&\quad + 1281758320845, U\},
\end{aligned}$$

$\mathbb{U}_4 = \{9x^2 + 2x - 9, 6x + 1, x^3 + 54x^2 + 27x - 52\}$,
$\mathbb{U}_5 = \{y, 5653y^2 - 2116, U\}$.

The polynomial U in $\mathbb{U}_3$ and $\mathbb{U}_5$ is somewhat too large to be produced here. It is irreducible of degrees 15, 10, 1 in x, y, u, respectively, and consists of 91 terms.

A triangular series of $\mathbb{P}$ can also be computed easily by TriSer or TriSerS with respect to the same variable ordering. One may obtain with TriSer five fine triangular systems in which the triangular sets are the same as the above $\mathbb{T}_i$, and with TriSerS four fine triangular systems in which some of the triangular sets are slightly different from the corresponding $\mathbb{T}_i$ above.

Applications of projection include solving parametric algebraic systems, automatic derivation of locus equations, implicitization of parametric objects, and determining existence conditions of singularities, which will be discussed in Sects. 7.1, 7.3 and 7.4.

3.3 Decomposition into simple systems

In this section, we introduce the concept of simple systems, which possess nice properties other than those of perfect triangular systems. We extend algorithm TriSerS to compute such simple systems. For any polynomial system $\mathfrak{P} = [\mathbb{P}, \mathbb{Q}]$, define

$$\breve{\mathfrak{P}} = \mathbb{P} \cup \mathbb{Q}.$$

Recall the notations $\boldsymbol{x}^{\{i\}} \triangleq (x_1, \ldots, x_i)$ and $\bar{\boldsymbol{x}}^{\{i\}} \triangleq (\bar{x}_1, \ldots, \bar{x}_i)$, etc.

For any $P \in \boldsymbol{K}[\boldsymbol{x}^{\{k\}}]$ and $\bar{\boldsymbol{x}}^{\{k-1\}}$ in some extension field $\tilde{\boldsymbol{K}}$ of $\boldsymbol{K}$, the polynomial $P(\bar{\boldsymbol{x}}^{\{k-1\}}, x_k)$ is said to be *squarefree* (with respect to x_k) if

$$\gcd(P(\bar{\boldsymbol{x}}^{\{k-1\}}, x_k), \frac{\partial P}{\partial x_k}(\bar{\boldsymbol{x}}^{\{k-1\}}, x_k)) \in \tilde{\boldsymbol{K}}.$$

For example, $x_2^2 - x_1$ is squarefree with respect to x_2 for $x_1 = 1$, but not for $x_1 = 0$.

Definition 3.3.1. A pair $\mathfrak{S} = [\mathbb{T}, \tilde{\mathbb{T}}]$ in which $\mathbb{T}$ and $\tilde{\mathbb{T}}$ are either triangular sets in $\boldsymbol{K}[\boldsymbol{x}]$ or the empty set is called a *simple system* if

a. $\mathbb{T} \cap \tilde{\mathbb{T}} = \emptyset$ and $\breve{\mathfrak{S}}$ can be reordered as a triangular set;
b. for every $P \in \breve{\mathfrak{S}}$ of class p and any $\bar{\boldsymbol{x}}^{\{p-1\}} \in \text{Zero}(\mathfrak{S}^{(p-1)})$,

$$\text{ini}(P)(\bar{\boldsymbol{x}}^{\{p-1\}}) \neq 0 \ \text{ and } \ P(\bar{\boldsymbol{x}}^{\{p-1\}}, x_p) \text{ is squarefree.}$$

A triangular set $\mathbb{T} \subset \boldsymbol{K}[\boldsymbol{x}]$ is said to be *simple* or called a *simple set* if $[\mathbb{T}, \emptyset]$ is a simple system, or there exists another triangular set $\tilde{\mathbb{T}}$ such that $[\mathbb{T}, \tilde{\mathbb{T}}]$ is a simple system.

While talking about a triangular system $\mathfrak{T}$, we sometimes say that $\mathfrak{T}$ is *simple*. Naturally, this means that $\mathfrak{T}$ is a simple system. The concept of simple systems is due to Thomas (1937, chap. VI). What he called a simple system is a reduced primitive simple system in our definition.

Example 3.3.1. Let $\mathbb{P} = \{P_1, P_2, P_3\}$ with

$$P_1 = x_2^2 - x_1,$$
$$P_2 = x_2x_3^3 - 2x_1x_3^2 + x_3^2 + x_1x_2x_3 - 2x_2x_3 + x_1,$$
$$P_3 = x_2x_3x_4 + x_4 + x_1x_3 + x_2$$

and $x_1 \prec \cdots \prec x_4$. The polynomials P_1, P_2, P_3 are all irreducible over $\mathbf{Q}$. One sees that

$\text{ini}(P_1) = 1$, $I_2 = \text{ini}(P_2) = x_2$ and $I_3 = \text{ini}(P_3) = x_2x_3 + 1$,
$\mathbb{T} = [P_1, P_2, P_3]$ is a triangular set,
$\mathfrak{T} = [\mathbb{T}, \{I_2, I_3\}]$ is a fine and reduced triangular system.

However, $\mathfrak{T}$ is not a simple system. First, $\text{cls}(I_3) = \text{cls}(P_2)$ and $\text{cls}(I_2) = \text{cls}(P_1)$, so condition a is violated. Second, one may verify that P_2 has a factorization $P_2 \doteq (x_2x_3 + 1)(x_3 - x_2)^2$ over $\mathbf{Q}(x_1, x_2)$ with x_2 having minimal polynomial P_1. Thus, P_2 is not squarefree with respect to x_3 for any $(x_1, x_2) \in \text{Zero}(P_1/I_2)$.

Example 3.3.2. The polynomials and triangular systems are as in Example 2.4.1. $[\mathbb{T}_2, \emptyset]$ is not a simple system because $y^2 - r^2 + 1$ is not squarefree with respect to y when $r = \pm 1 \in \text{Zero}(T)$, where $T = r^4 - 4r^2 + 3$. Since $\text{lv}(G) = x \in \mathbb{U}_1$ and thus $\mathbb{T}_1 \cup \mathbb{U}_1$ cannot be ordered as a triangular set, $[\mathbb{T}_1, \mathbb{U}_1]$ is not a simple system either.

As further illustration, consider $\mathfrak{T} = [\mathbb{T}_1, \{T\}]$, which is a triangular system. This can be verified as follows: $\text{ini}(P_2) = x = 0$ and $H = G = 0$ only if $z = \pm 1$ and $r = \pm 1$ or $r^2 = 3$. This is possible only if $T = 0$. Hence, if $H = G = 0$ and $T \neq 0$, then $x \neq 0$. For $\mathfrak{T}$, condition a is satisfied. However, neither is $\mathfrak{T}$ a simple system because H is not squarefree with respect to z, for example, when $27r^2 - 31 = 0$ (noting that $27r^2 - 31$ and T are relatively prime).

Definition 3.3.2. A triangular system $\mathfrak{T}$ in $\boldsymbol{K}[\boldsymbol{x}]$ is said to be *primitive* if every $P \in \check{\mathfrak{T}}$ is primitive with respect to its leading variable.

Lemma 3.3.1. Let $[\mathbb{T}, \tilde{\mathbb{T}}]$ be a simple system in $\boldsymbol{K}[\boldsymbol{x}]$ and

$$\mathbb{T}^* = [\text{pp}(T, \text{lv}(T)) \colon\ T \in \mathbb{T}],\quad \tilde{\mathbb{T}}^* = [\text{pp}(T, \text{lv}(T)) \colon\ T \in \tilde{\mathbb{T}}].$$

Then $[\mathbb{T}^*, \tilde{\mathbb{T}}^*]$ is a primitive simple system such that

$$\text{Zero}(\mathbb{T}^*/\tilde{\mathbb{T}}^*) = \text{Zero}(\mathbb{T}/\tilde{\mathbb{T}}).$$

Proof. Note that the primitive part of any polynomial has the same class as the polynomial itself, so $\mathbb{T}^*$, $\tilde{\mathbb{T}}^*$ and $\mathbb{T}^* \cup \tilde{\mathbb{T}}^*$ can all be ordered as triangular sets. Hence, we only need to see that for any $T \in \mathbb{T} \cup \tilde{\mathbb{T}}$ of class p and

$$\bar{\boldsymbol{x}}^{\{p-1\}} \in \text{Zero}(\mathbb{T}^{(p-1)}/\tilde{\mathbb{T}}^{(p-1)}),$$

$\text{cont}(T, x_p)(\bar{\boldsymbol{x}}^{\{p-1\}}) \neq 0$ and thus $\text{cont}(T, x_p)$ can be removed from T. This is obvious because $\text{cont}(T, x_p)$ is a divisor of $\text{ini}(T)$, while $\text{ini}(T)(\bar{\boldsymbol{x}}^{\{p-1\}}) \neq 0$ by definition. □

In view of this lemma, we shall feel free to make simple systems primitive, in particular for example calculations.

Lemma 3.3.2. Let P_1 and P_2 be two polynomials in $\boldsymbol{K}[\boldsymbol{x}^{\{k\}}]$ with $\deg(P_1, x_k) \geq \deg(P_2, x_k) > 0$, $H_2, \ldots, H_r$ be the SRS of P_1 and P_2 with respect to x_k and

$$I = \mathrm{lc}(P_2, x_k), \quad I_i = \mathrm{lc}(H_i, x_k), \; 2 \leq i \leq r.$$

Let $\mathbb{P}, \mathbb{Q} \subset \boldsymbol{K}[\boldsymbol{x}^{\{k-1\}}]$ be two polynomial sets and assume that

$$I(\bar{\boldsymbol{x}}^{\{k-1\}}) \neq 0 \text{ and } P_2(\bar{\boldsymbol{x}}^{\{k-1\}}, x_k) \text{ is squarefree}$$

for any $\bar{\boldsymbol{x}}^{\{k-1\}} \in \mathrm{Zero}(\mathbb{P}/\mathbb{Q})$. Then

$$\mathrm{Zero}(\mathbb{P} \cup \{P_2\}/\mathbb{Q} \cup \{P_1\}) = \bigcup_{i=2}^{r} \mathrm{Zero}(\mathbb{P} \cup \mathbb{P}_i/\mathbb{Q} \cup \{I_i\}), \tag{3.3.1}$$

where $\mathbb{P}_i = \{\mathrm{pquo}(P_2, H_i, x_k), I_{i+1}, \ldots, I_r\}$ for each i.

Proof. For any $\bar{\boldsymbol{x}}^{\{k-1\}} \in \mathrm{Zero}(\mathbb{P}/\mathbb{Q})$, there must be an i $(2 \leq i \leq r)$ such that $I_i(\bar{\boldsymbol{x}}^{\{k-1\}}) \neq 0$, and $I_{i+1}(\bar{\boldsymbol{x}}^{\{k-1\}}) = \cdots = I_r(\bar{\boldsymbol{x}}^{\{k-1\}}) = 0$. According to Lemma 2.4.2 a,

$$H_i(\bar{\boldsymbol{x}}^{\{k-1\}}, x_k) = \gcd(P_1(\bar{\boldsymbol{x}}^{\{k-1\}}, x_k), P_2(\bar{\boldsymbol{x}}^{\{k-1\}}, x_k)).$$

The zero relation (3.3.1) is established. □

Observe that on the right-hand side of (3.3.1), P_1 does not appear and the only polynomial of class k is $\mathrm{pquo}(P_2, H_i, x_k)$ for each i. In this sense, the polynomial P_1 is eliminated by means of splitting. The purpose of splitting in the following lemma is to make an arbitrary polynomial squarefree.

Lemma 3.3.3. Let P be a polynomial in $\boldsymbol{K}[\boldsymbol{x}^{\{k\}}]$ with $\deg(P, x_k) > 1$ and $I = \mathrm{lc}(P, x_k)$, $H_2, \ldots, H_r$ be the SRS of P and its derivative $\partial P/\partial x_k$ with respect to x_k, and

$$H_2^* = H_2, \quad H_i^* = H_i/I, \; 3 \leq i \leq r;$$
$$I_i = \mathrm{lc}(H_i^*, x_k), \; 2 \leq i \leq r.$$

Then

$$\mathrm{Zero}(P/I) = \bigcup_{i=2}^{r} \mathrm{Zero}(\{Q_i, I_{i+1}, \ldots, I_r\}/I I_i), \tag{3.3.2}$$

$$\mathrm{Zero}(\emptyset/PI) = \bigcup_{i=2}^{r} \mathrm{Zero}(\{I_{i+1}, \ldots, I_r\}/Q_i I I_i), \tag{3.3.3}$$

where $Q_i = \mathrm{pquo}(P, H_i^*, x_k)$ for each i. Moreover, $Q_i(\bar{\boldsymbol{x}}^{\{k-1\}}, x_k)$ is squarefree for any $2 \leq i \leq r$ and $\bar{\boldsymbol{x}}^{\{k-1\}} \in \mathrm{Zero}(\{I_{i+1}, \ldots, I_r\}/I I_i)$.

Proof. Obviously, $\mathrm{lc}(\partial P/\partial x_k, x_k) = \deg(P, x_k)I$. It is also easy to see from the definition of subresultants that I divides H_i for $3 \le i \le r$. As a fundamental fact in algebra, we know that for any $2 \le i \le r$ and $\bar{\boldsymbol{x}}^{\{k-1\}} \in \mathrm{Zero}(\{I_{i+1}, \ldots, I_r\}/II_i)$,

$$P(\bar{\boldsymbol{x}}^{\{k-1\}}, x_k)/\gcd(P(\bar{\boldsymbol{x}}^{\{k-1\}}, x_k), \frac{\partial P}{\partial x_k}(\bar{\boldsymbol{x}}^{\{k-1\}}, x_k))$$

is squarefree and has the same set of zeros as $P(\bar{\boldsymbol{x}}^{\{k-1\}}, x_k)$ for x_k. The squarefreeness of $Q_i(\bar{\boldsymbol{x}}^{\{k-1\}}, x_k)$ and the zero relations (3.3.2) and (3.3.3) follow from this fact and Lemma 2.4.2 a. □

Definition 3.3.3. A finite set or sequence of simple systems $\mathfrak{S}_1, \ldots, \mathfrak{S}_e$ in $\boldsymbol{K}[\boldsymbol{x}]$ is called a *simple series*. It is called a *simple series* of a polynomial system $\mathfrak{P}$ if the following zero decomposition holds:

$$\mathrm{Zero}(\mathfrak{P}) = \bigcup_{i=1}^{e} \mathrm{Zero}(\mathfrak{S}_i). \tag{3.3.4}$$

A simple series of $[\mathbb{P}, \emptyset]$ is also called a *simple series* of the polynomial set $\mathbb{P}$.

The algorithm below is devised to compute a simple series of any given polynomial system. It employs an elimination process again top-down from x_n to x_1 with splitting, modified from algorithm TriSerS. For each x_k (in the for-loop S2.2), there are four major steps:

S2.2.1 producing from $\mathbb{T}^{\langle k\rangle} \neq \emptyset$ a single polynomial P_2 of class k;
S2.2.2 making P_2 squarefree with respect to x_k;
S2.2.3 eliminating the polynomials from $\tilde{\mathbb{T}}^{\langle k\rangle} \neq \emptyset$ by P_2;
S2.2.4 producing a single polynomial P_1 squarefree with respect to x_k from $\tilde{\mathbb{T}}^{\langle k\rangle} \neq \emptyset$.

There are three kinds of splitting performed:

i. in steps S2.2.1.1 and S2.2.4.1 according as the initial of the considered polynomial vanishes or not (either the initial is assumed to be nonvanishing or the polynomial is replaced by its initial and reductum);
ii. in steps S2.2.1.3 and S2.2.3.2 according to Lemmas 2.4.2 b and 3.3.2 for basic elimination;
iii. in steps S2.2.2.2 and S2.2.4.3 according to Lemma 3.3.3 for squarefreeness.

Algorithm SimSer: $\Psi \leftarrow \mathrm{SimSer}(\mathbb{P}, \mathbb{Q})$. Given a polynomial system $[\mathbb{P}, \mathbb{Q}]$ in $\boldsymbol{K}[\boldsymbol{x}]$, this algorithm computes a simple series Ψ of $[\mathbb{P}, \mathbb{Q}]$.

S1. Set $\Phi \leftarrow \{[\mathbb{P}, \mathbb{Q}, n]\}$, $\Psi \leftarrow \emptyset$.
S2. While $\Phi \neq \emptyset$, do:
  S2.1. Let $[\mathbb{T}, \tilde{\mathbb{T}}, l]$ be an element of Φ and set $\Phi \leftarrow \Phi \setminus \{[\mathbb{T}, \tilde{\mathbb{T}}, l]\}$.
  S2.2. For $k = l, \ldots, 1$ do:
    S2.2.1. While $\mathbb{T}^{\langle k\rangle} \neq \emptyset$, do:

S2.2.1.1. Let P_2 be an element of $\mathbb{T}^{\langle k\rangle}$ with minimal degree in x_k and set

$$\Phi \leftarrow \Phi \cup \{[\mathbb{T} \setminus \{P_2\} \cup \{\mathrm{ini}(P_2), \mathrm{red}(P_2)\}, \tilde{\mathbb{T}}, k]\},$$
$$\tilde{\mathbb{T}} \leftarrow \tilde{\mathbb{T}} \cup \{\mathrm{ini}(P_2)\}.$$

If $|\mathbb{T}^{\langle k\rangle}| = 1$, then go to S2.2.2; else take a polynomial P_1 from $\mathbb{T}^{\langle k\rangle} \setminus \{P_2\}$.

S2.2.1.2. Compute the SRS $H_2, \ldots, H_r$ of P_1 and P_2 with respect to x_k and set $I_i \leftarrow \mathrm{lc}(H_i, x_k)$ for $2 \le i \le r$. If $\mathrm{cls}(H_r) < k$, then set $\bar{r} \leftarrow r - 1$; else set $\bar{r} \leftarrow r$.

S2.2.1.3. Set

$$\Phi \leftarrow \Phi \cup \{[\mathbb{T} \setminus \{P_1, P_2\} \cup \{H_i, I_{i+1}, \ldots, I_r\}, \tilde{\mathbb{T}} \cup \{I_i\}, k]\colon\ 2 \le i \le \bar{r} - 1\},$$
$$\mathbb{T} \leftarrow \mathbb{T} \setminus \{P_1, P_2\} \cup \{H_r, H_{\bar{r}}\},$$
$$\tilde{\mathbb{T}} \leftarrow \tilde{\mathbb{T}} \cup \{I_{\bar{r}}\}.$$

S2.2.2. If $\mathbb{T}^{\langle k\rangle} = \emptyset$, then go to S2.2.4. If $\deg(P_2, x_k) = 1$, then go to S2.2.3; else:

S2.2.2.1. Compute the SRS $H_2, \ldots, H_r$ of P_2 and its derivative $\partial P_2/\partial x_k$ with respect to x_k and set

$$H_2^* \leftarrow H_2, \quad H_i^* \leftarrow H_i/\mathrm{ini}(P_2), \quad i = 3, \ldots, r,$$
$$I_i \leftarrow \mathrm{lc}(H_i^*, x_k), \quad i = 2, \ldots, r.$$

If $\tilde{\mathbb{T}}^{\langle k\rangle} = \emptyset$, then set $\bar{k} \leftarrow k - 1$; else set $\bar{k} \leftarrow k$.

S2.2.2.2. Set

$$\Phi \leftarrow \Phi \cup \{[\mathbb{T} \setminus \{P_2\} \cup \{\mathrm{pquo}(P_2, H_i^*, x_k), I_{i+1}, \ldots, I_r\}, \tilde{\mathbb{T}} \cup \{I_i\}, \bar{k}]\colon\ 2 \le i \le r - 1\},$$
$$\mathbb{T} \leftarrow \mathbb{T} \setminus \{P_2\} \cup \{\mathrm{pquo}(P_2, H_r^*, x_k)\},$$
$$\tilde{\mathbb{T}} \leftarrow \tilde{\mathbb{T}} \cup \{I_r\},$$
$$P_2 \leftarrow \mathrm{pquo}(P_2, H_r^*, x_k).$$

S2.2.3. While $\tilde{\mathbb{T}}^{\langle k\rangle} \neq \emptyset$ and $\mathrm{cls}(P_2) = k$, do:

S2.2.3.1. Let P_1 be a polynomial in $\tilde{\mathbb{T}}^{\langle k\rangle}$, compute the SRS $H_2, \ldots, H_r$ of P_1 and P_2 if $\deg(P_1, x_k) \ge \deg(P_2, x_k)$, or of P_2 and P_1 otherwise, with respect to x_k and set $I_i \leftarrow \mathrm{lc}(H_i, x_k)$ for $2 \le i \le r$.

S2.2.3.2. Set

$$\Phi \leftarrow \Phi \cup \{[\mathbb{T} \setminus \{P_2\} \cup \{\mathrm{pquo}(P_2, H_i, x_k), I_{i+1}, \ldots, I_r\}, \tilde{\mathbb{T}} \setminus \{P_1\} \cup \{I_i\}, k]\colon\ 2 \le i \le r - 1\},$$
$$\mathbb{T} \leftarrow \mathbb{T} \setminus \{P_2\} \cup \{\mathrm{pquo}(P_2, H_r, x_k)\},$$
$$\tilde{\mathbb{T}} \leftarrow \tilde{\mathbb{T}} \setminus \{P_1\} \cup \{I_r\},$$
$$P_2 \leftarrow \mathrm{pquo}(P_2, H_r, x_k).$$

S2.2.4. If $\tilde{\mathbb{T}}^{\langle k \rangle} \neq \emptyset$, then:

S2.2.4.1. Set

$$P_1 \leftarrow \prod_{P \in \tilde{\mathbb{T}}^{\langle k \rangle}} P,$$
$$\Phi \leftarrow \Phi \cup \{[\mathbb{T} \cup \{\mathrm{ini}(P_1)\}, \tilde{\mathbb{T}} \setminus \tilde{\mathbb{T}}^{\langle k \rangle} \cup \{\mathrm{red}(P_1)\}, k]\},$$
$$\tilde{\mathbb{T}} \leftarrow \tilde{\mathbb{T}} \cup \{\mathrm{ini}(P_1)\}.$$

If $\deg(P_1, x_k) = 1$, then go to S2.2.5.

S2.2.4.2. Compute the SRS $H_2, \ldots, H_r$ of P_1 and its derivative $\partial P_1/\partial x_k$ with respect to x_k and set

$$H_2^* \leftarrow H_2, \quad H_i^* \leftarrow H_i/\mathrm{ini}(P_1), \quad i = 3, \ldots, r,$$
$$I_i \leftarrow \mathrm{lc}(H_i^*, x_k), \quad i = 2, \ldots, r.$$

S2.2.4.3. Set

$$\Phi \leftarrow \Phi \cup \{[\mathbb{T} \cup \{I_{i+1}, \ldots, I_r\}, \tilde{\mathbb{T}} \setminus \tilde{\mathbb{T}}^{\langle k \rangle} \cup \{\mathrm{pquo}(P_1, H_i^*, x_k), I_i\}, k-1]: 2 \leq i \leq r-1\},$$
$$\tilde{\mathbb{T}} \leftarrow \tilde{\mathbb{T}} \setminus \tilde{\mathbb{T}}^{\langle k \rangle} \cup \{\mathrm{pquo}(P_1, H_r^*, x_k), I_r\}.$$

S2.2.5. Set $\mathbb{T} \leftarrow \mathbb{T} \setminus \{0\}$, $\tilde{\mathbb{T}} \leftarrow \tilde{\mathbb{T}} \setminus (\boldsymbol{K} \setminus \{0\})$. If $\mathbb{T} \cap \boldsymbol{K} \neq \emptyset$ or $0 \in \tilde{\mathbb{T}}$, then go to S2.

S2.3. Set $\Psi \leftarrow \Psi \cup \{[\mathbb{T}, \tilde{\mathbb{T}}]\}$, with $\mathbb{T}$ and $\tilde{\mathbb{T}}$ ordered as triangular sets when they are nonempty.

Proof. Correctness. Let us first note that the interchange of P_1 and P_2 in step S2.2.3.1 when $\deg(P_1, x_k) < \deg(P_2, x_k)$ does not cause any problem. To see this, we claim that Lemma 2.4.2a is still valid when I is set to $\mathrm{lc}(P_1, x_k)$ instead of $\mathrm{lc}(P_2, x_k)$. The leading coefficient I needs to be considered as shown in the proof because the subresultants may differ by a factor of some power of I when the coefficients of P_1 and P_2 with respect to x_k are specialized. According to Proposition 1.3.5, it does not matter which leading coefficient of P_1 and P_2 is taken as I and assumed to be nonvanishing. Therefore, (3.3.1) in Lemma 3.3.2 still holds when $\deg(P_1, x_k) < \deg(P_2, x_k)$ and $H_2, \ldots, H_r$ is the SRS of P_2 and P_1 with respect to x_k (while I remains unchanged). [It may happen that

$$I_2(\bar{\boldsymbol{x}}^{\{k-1\}}) = \cdots = I_r(\bar{\boldsymbol{x}}^{\{k-1\}}) = 0$$

for some $\bar{\boldsymbol{x}}^{\{k-1\}} \in \mathrm{Zero}(\emptyset/I)$ (cf. the proof of Lemma 3.3.2). In this case, $P_1(\bar{\boldsymbol{x}}^{\{k-1\}}, x_k) \equiv 0$, so $\mathrm{Zero}(P_2/P_1 I) = \emptyset$. Hence, the case need not be considered.]

Next we see that in each case of splitting in SimSer, one split system is taken to update the current system $[\mathbb{T}, \tilde{\mathbb{T}}]$; this system corresponds to that for $i = r$ in (2.4.1) and (3.3.1)–(3.3.3), with one exception: for $i = r - 1$ in (2.4.1) when $\deg(H_r, x_k) = 0$. The other split systems are added to Φ. By (2.4.1) and (3.3.1)–(3.3.3) and the evident zero relation for the first kind of splitting, an associated zero decomposition of the form $\mathrm{Zero}(\mathbb{P}/\mathbb{Q}) = \bigcup_\alpha \mathrm{Zero}(\mathbb{P}_\alpha/\mathbb{Q}_\alpha)$ holds all the time, where the union ranges over all the split systems. Thus the decomposition (3.3.4) with $\mathfrak{P} = [\mathbb{P}, \mathbb{Q}]$ should be obtained eventually. The computed pairs of ordered polynomial sets in Ψ are simple systems by definition.

Termination. One first notes that steps S2.2.1 and S2.2.3 terminate obviously because in each loop of S2.2.1 two polynomials $P_1, P_2 \in \mathbb{T}^{\langle k \rangle}$ are replaced by one $H_{\bar{r}}$ of class k (see S2.2.1.3), and in each loop of S2.2.3 one polynomial $P_1 \in \tilde{\mathbb{T}}^{\langle k \rangle}$ is deleted (see S2.2.3.2). In any case of splitting, the split polynomial systems are obtained from the current system either by replacing one or two polynomials with another having lower degree in their common leading variable x_k (as in most of the cases), or by replacing two or more polynomials with a single one of the same class k (as in S2.2.1.3 when $\bar{r} = 2$ and in S2.2.4.3 when $|\tilde{\mathbb{T}}^{\langle k \rangle}| > 1$), sometimes having polynomials of classes $< k$ added as well. Hence, the while-loop S2 has only finitely many iterations. □

Remark 3.3.1. Steps S2.2.2.1 and S2.2.2.2 in SimSer can be skipped when P_2 is any of the pquo(P_2, H_i^*, x_k) produced in S2.2.2.2 or the pquo(P_2, H_i, x_k) produced in S2.2.3.2 previously, because in this case P_2 is known to be conditionally squarefree with respect to x_k.

The strategies mentioned in Remark 2.4.1 should also be implemented to avoid unnecessary computations for TriSerP and SimSer. Some further reduction may sometimes simplify simple systems and make the result more canonical. For example, one can require that simple systems are made primitive and reduced. This issue will be addressed in Sect. 5.2, though the settlement does not contribute much to the theoretical development and practical application of the method.

One motivation for computing simple systems comes from the work of Thomas (1937). The functionality and some individual steps of SimSer are similar to those of Thomas' method. However, the algorithm here is described differently in terms of structure and elementary operations.

Example 3.3.3. Let $\mathbb{P}$, P_i, $\mathfrak{T}$ be as in Example 3.3.1 and

$$\mathbb{T}' = [P_1, x_2x_3 + 1], \quad \mathbb{T}'' = [x_1, \ldots, x_4].$$

Then, by SimSer, $\mathbb{P}$ can be decomposed into three reduced simple systems

$$[[P_1, x_3 - x_2, x_4 + x_2], [x_1(x_1 + 1)]], \quad [\mathbb{T}', [x_1]], \quad [\mathbb{T}'', \emptyset]. \tag{3.3.5}$$

The procedure proceeds roughly as follows. Let

$$[\mathbb{T}, \tilde{\mathbb{T}}] \leftarrow [\{P_1, P_2, P_3\}, \{x_2, I_3\}] = \mathfrak{T}.$$

P_3 is linear and thus squarefree with respect to x_4. To make P_2 squarefree with respect to x_3, compute the SRS of P_2 and $\partial P_2/\partial x_3$ with respect to x_3, which is $\partial P_2/\partial x_3$, $2x_2H_1$, and $4x_2H_2$, where H_1 is a polynomial of degree 1 in x_3 and H_2 a polynomial of class 2. Observe that $x_2 \in \tilde{\mathbb{T}}$, so there are two cases: (i) $H_2 \neq 0$ and P_2 is squarefree with respect to x_3, and (ii) $H_2 = 0$, $I = \text{ini}(H_1) \neq 0$ and P_2 is replaced by pquo(P_2, H_1, x_3) which is squarefree with respect to x_3. For the sake of simplicity, we point out that H_2 contains P_1 as a factor. Hence, by following the procedure the first case will be discarded and for the second case H_2 need not be added to $\mathbb{T}$. Therefore, set

$$[\mathbb{T}, \tilde{\mathbb{T}}] \leftarrow [\{P_1, H_3, P_3\}, [x_2, I_3, I]],$$

in which $H_3 = \mathrm{pquo}(P_2, H_1, x_3)$ has 42 terms and degree 2 in x_3 and I has 5 terms and degree 2 in x_2.

Next we want to eliminate I_3 from $\tilde{\mathbb{T}}$ by H_3. For this purpose, compute the SRS of H_3 and I_3 with respect to x_3: I_3 and H_4, where H_4 is a polynomial of 20 terms, also containing P_1 as a factor, so $\gcd(H_3, I_3, x_3) = I_3$ when $x_1 \neq 0$. Thus, set

$$[\mathbb{T}, \tilde{\mathbb{T}}] \leftarrow [\{P_1, H_5, P_3\}, \{x_1, x_2, I\}],$$

in which $H_5 = \mathrm{pp}(\mathrm{pquo}(H_3, I_3, x_3), x_3)$ consists of 11 terms.

Now P_1 is squarefree with respect to x_2 and both $\gcd(P_1, x_2, x_2)$ and $\gcd(P_1, I, x_2)$ are constants when $x_1(x_1 + 1) \neq 0$. Therefore, a simple system $[\{P_1, H_5, P_3\}, \{x_1(x_1 + 1)\}]$ is obtained. Finally, replacing H_5 and P_3 respectively by

$$\mathrm{pp}(\mathrm{prem}(H_5, P_1, x_2), x_3) = x_3 - x_2,$$
$$\mathrm{pp}(\mathrm{prem}(P_3, [P_1, x_3 - x_2]), x_3) = x_4 + x_2,$$

we arrive at the first reduced primitive simple system in (3.3.5).

Considering the polynomial sets obtained from $\mathbb{P}$ by replacing P_2 and P_3 respectively with their initials and reductums and following the same procedure, one will get the two other reduced simple systems.

Remark incidentally that by algorithm TriSerS $\mathbb{P}$ may be decomposed into three fine triangular systems $\mathfrak{T}$, $[\mathbb{T}', \{x_2\}]$, $[\mathbb{T}'', \emptyset]$.

Example 3.3.4. Let $\mathbb{P}$ be as in Example 2.4.1 and the polynomials H, G, P_2 there be renamed T_1, T_2, T_3:

$$T_1 = z^3 - z^2 + r^2 - 1,$$
$$T_2 = x^4 + z^2x^2 - r^2x^2 + z^4 - 2z^2 + 1,$$
$$T_3 = xy + z^2 - 1.$$

In addition, let

$$T = r^8 - 6r^6 + 71r^4 - 62r^2 - 67.$$

A simple series of $\mathbb{P}$ computed by SimSer consists of 9 simple systems $[\mathbb{T}_1, \tilde{\mathbb{T}}_1], \ldots, [\mathbb{T}_9, \tilde{\mathbb{T}}_9]$ with

$$\mathbb{T}_1 = [T_1, T_2, T_3],$$
$$\mathbb{T}_2 = [r^2 - 1, z - 1, x, y],$$
$$\mathbb{T}_3 = [r^2 - 1, z, x^4 - x^2 + 1, xy - 1],$$
$$\mathbb{T}_4 = [r^2 - 3, z + 1, x^2 - 2, y],$$

$$\begin{aligned}
\mathbb{T}_5 &= [r^2-3, z+1, x, y^2-2],\\
\mathbb{T}_6 &= [r^2-3, z^2-2z+2, T_2, T_3],\\
\mathbb{T}_7 &= [27r^2-31, 9z^2-3z-2, 27x^4+(9z-25)x^2-13z+17,\\
&\qquad 9xy+3z-7],\\
\mathbb{T}_8 &= [T, (r^4+14r^2+15)z+3r^4+13r^2-4,\\
&\qquad (z^2+z+1)x^2+z^5+z^4-z^3-3z^2+z+1, T_3],\\
\mathbb{T}_9 &= [T, (34r^6+155r^4+482r^2+292)z^2\\
&\qquad -(107r^6+165r^4+807r^2+433)z+205r^6-484r^4\\
&\qquad +779r^2+760, T_2, T_3];\\
\tilde{\mathbb{T}}_1 &= [(r^2-1)(r^2-3)(27r^2-31)T],\\
\tilde{\mathbb{T}}_2 &= \cdots = \tilde{\mathbb{T}}_9 = \emptyset.
\end{aligned}$$

In computing the series, we did not make use of polynomial factorization. The output is somewhat simpler when the occurring polynomials are factorized.

Example 3.3.5. A simple series of the polynomial set $\mathbb{P}$ given in Example 2.4.3 computed by SimSer with respect to the same variable ordering consists of 13 simple systems $[\mathbb{T}_1, \tilde{\mathbb{T}}_1], \ldots, [\mathbb{T}_{13}, \tilde{\mathbb{T}}_{13}]$, where $\mathbb{T}_1, \ldots, \mathbb{T}_7$ are as in Example 2.4.3 and

$$\begin{aligned}
\mathbb{T}_8 &= [H_1, 36z^3-8c^2z^2-42cz+81, H_4, P_3],\\
\mathbb{T}_9 &= [H_1, 2cz+3, 2c^2y^2-3cy-9, 3yx+2c],\\
\mathbb{T}_{10} &= [2c^3-27, 2c^2z^2+3cz-9, y-z, 2y^2x-xc+1],\\
\mathbb{T}_{11} &= [H_2, H_3, H_4, P_3],\\
\mathbb{T}_{12} &= [H_2, H_3, zy-z^2+c, x-z],\\
\mathbb{T}_{13} &= [H_2, 54(1938466c^3+138253)z^3-16c^2(440494c^3\\
&\qquad +31419)z^2-9c(4103430c^3+292663)z\\
&\qquad -3(7980362c^3+569169), (cz+1)y+cz^2-z, P_3];\\
\tilde{\mathbb{T}}_1 &= \tilde{\mathbb{T}}_2 = [cH_2],\ \ \tilde{\mathbb{T}}_3 = [H_1],\ \ \tilde{\mathbb{T}}_4 = [cH_1H_2],\\
\tilde{\mathbb{T}}_5 &= [2c^3-27],\ \ \tilde{\mathbb{T}}_6 = \cdots = \tilde{\mathbb{T}}_{13} = \emptyset;\\
H_1 &= 4c^3-27,\\
H_2 &= 8c^6-378c^3-27,\\
H_3 &= 36(18c^3+1)z^3+8c^2(10c^3+3)z^2-2c(250c^3+9)z\\
&\qquad -9(290c^3+21),\\
H_4 &= (z^3-cz+1)y+z^4-2cz^2+c^2.
\end{aligned}$$

For obtaining the simple series, factorization over $\mathbf{Q}$ has been done for some of the intermediate polynomials.

Computing simple series is expensive in general, mainly because of the high price that has to be carried to make polynomials squarefree and to eliminate inequation polynomials. In practice, it is even preferable to compute irreducible triangular series instead, making use of powerful routines available for polynomial factorization. This will be explained in Chap. 4.

3.4 Properties of simple systems

The significance of introducing simple systems may be seen partially from the properties that are stated and proved in this section. Let $\bar{\boldsymbol{K}}$ denote an *algebraic closure* of the ground field $\boldsymbol{K}$.

Theorem 3.4.1. Let $\mathfrak{S}$ be a simple system in $\boldsymbol{K}[\boldsymbol{x}]$. Then for any $1 < k \leq n$ and $\bar{\boldsymbol{x}}^{\{k-1\}} \in \mathrm{Zero}(\mathfrak{S}^{(k-1)})$ there exist $\bar{x}_k, \ldots, \bar{x}_l \in \bar{\boldsymbol{K}}$ such that $\bar{\boldsymbol{x}}^{\{l\}} \in \mathrm{Zero}(\mathfrak{S}^{(l)})$ for all $k \leq l \leq n$. In particular, $\mathfrak{S}$ is perfect over $\bar{\boldsymbol{K}}$.

Proof. Let $\mathfrak{S} = [\mathbb{T}, \breve{\mathbb{T}}]$ and $\breve{\mathfrak{S}}$ be reordered as a triangular set $[T_1, \ldots, T_r]$, with

$$p_i = \mathrm{cls}(T_i), \quad d_i = \mathrm{ldeg}(T_i), \quad I_i = \mathrm{ini}(T_i), \quad 1 \leq i \leq r.$$

Clearly, for every pair $k \leq l$ there exist i and $s \geq 0$ such that

$$p_{i-1} < k \leq p_i, \quad p_{i+s-1} < l \leq p_{i+s}.$$

Let

$$\bar{\boldsymbol{x}}^{\{k-1\}} \in \mathrm{Zero}(\mathfrak{S}^{(k-1)}).$$

If $s = 0$ and $l < p_i$, then take arbitrary $\bar{x}_k, \ldots, \bar{x}_l \in \boldsymbol{K}$. In this case, we have

$$\bar{\boldsymbol{x}}^{\{l\}} \in \mathrm{Zero}(\mathfrak{S}^{(l)})$$

and the theorem is already proved. Otherwise, take any $\bar{x}_k, \ldots, \bar{x}_{p_i-1} \in \boldsymbol{K}$. By definition,

$$I_i(\bar{\boldsymbol{x}}^{\{p_i-1\}}) \neq 0 \text{ and } \bar{T}_i = T_i(\bar{\boldsymbol{x}}^{\{p_i-1\}}, x_{p_i}) \text{ is squarefree.}$$

Thus, $\bar{T}_i$ has d_i distinct zeros in $\bar{\boldsymbol{K}}$ for x_{p_i}. If $T_i \in \mathbb{T}$, then take any of the d_i zeros for x_{p_i}. If $T_i \in \breve{\mathbb{T}}$, then take an element of $\boldsymbol{K}$ other than the d_i zeros of $\bar{T}_i$ for x_{p_i}.

If $s = 1$ and $l < p_{i+1}$, then take arbitrary $\bar{x}_{p_i+1}, \ldots, \bar{x}_l \in \boldsymbol{K}$; we have

$$\bar{\boldsymbol{x}}^{\{l\}} \in \mathrm{Zero}(\mathfrak{S}^{(l)}).$$

Otherwise, take arbitrary $\bar{x}_{p_i+1}, \ldots, \bar{x}_{p_{i+1}-1} \in \boldsymbol{K}$ for $x_{p_i+1}, \ldots, x_{p_{i+1}-1}$, respectively. Similarly,

$$I_{i+1}(\bar{\boldsymbol{x}}^{\{p_{i+1}-1\}}) \neq 0 \text{ and}$$
$$\bar{T}_{i+1} = T_{i+1}(\bar{\boldsymbol{x}}^{\{p_{i+1}-1\}}, x_{p_{i+1}}) \text{ is squarefree.}$$

Accordingly, $\bar{T}_{i+1}$ is a polynomial of degree d_{i+1} in $x_{p_{i+1}}$ and has d_{i+1} distinct zeros in $\bar{\boldsymbol{K}}$ for $x_{p_{i+1}}$.

Proceeding in this way, we shall construct a zero $\bar{\boldsymbol{x}}^{\{l\}}$ of $\mathfrak{S}^{(l)}$, and the theorem is proved. □

Corollary 3.4.2. Every simple system possesses the strong projection property.

Therefore, SimSer provides another method for solving parametric algebraic systems.

Theorem 3.4.3. Let $\mathfrak{P}$ be any polynomial system in $\boldsymbol{K}[\boldsymbol{x}]$ and Ψ a simple series of $\mathfrak{P}$. Then

a. $\text{Zero}(\mathfrak{P}) = \emptyset$ if and only if $\Psi = \emptyset$;

b. $\text{Zero}(\mathfrak{P})$ is finite if and only if $|\mathbb{T}| = n$ and $\tilde{\mathbb{T}} = \emptyset$ for every $[\mathbb{T}, \tilde{\mathbb{T}}] \in \Psi$.

Proof. Part a follows from (3.3.4) and Theorem 3.4.1.

b. For any $[\mathbb{T}, \tilde{\mathbb{T}}] \in \Psi$, if $|\mathbb{T}| = n$, then $\tilde{\mathbb{T}} = \emptyset$ and $\mathbb{T}$ can be written as $[T_1, \ldots, T_n]$ with $\text{cls}(T_i) = i$. Let $d_i = \text{ldeg}(T_i)$. Then, T_1 has d_1 distinct zeros in $\tilde{\boldsymbol{K}}$ for x_1, and for any of these d_1 zeros T_2 has d_2 distinct zeros in $\tilde{\boldsymbol{K}}$ for x_2, and so on. Therefore, $\mathbb{T}$ has a finite set of $d_1 \cdots d_n$ distinct zeros. If $|\mathbb{T}| < n$, then there exists a k such that $\mathbb{T}^{\langle k \rangle} = \emptyset$. Thus, the scope of x_k in $\text{Zero}(\mathbb{T}/\tilde{\mathbb{T}})$ is $\tilde{\boldsymbol{K}}$ when $\tilde{\mathbb{T}}^{\langle k \rangle} = \emptyset$ and is $\tilde{\boldsymbol{K}}$ minus a finite number of elements otherwise. In any case, $\text{Zero}(\mathbb{T}/\tilde{\mathbb{T}})$ is infinite. By (3.3.4), part b is proved. □

According to Theorem 3.4.3, one can apply algorithm SimSer to determine the solvability of any system of polynomial equations and inequations (with no need of polynomial factorization). In other words, the algorithm gives a solution to the decision problem in elementary algebra and geometry over algebraically closed fields. It is clear from the above proof that, when $\text{Zero}(\mathfrak{P})$ is finite, the exact number of zeros can be counted according to the leading degrees of the polynomials in $\mathbb{T}$; all the zeros can be successively computed from $\mathbb{T}$.

Theorem 3.4.4. For any simple system $[\mathbb{T}, \tilde{\mathbb{T}}]$ and polynomial P in $\boldsymbol{K}[\boldsymbol{x}]$,

$$\text{Zero}(\mathbb{T}/\tilde{\mathbb{T}}) \subset \text{Zero}(P) \iff \text{prem}(P, \mathbb{T}) = 0.$$

Proof. Let $\text{prem}(P, \mathbb{T}) = 0$ and $\bar{\boldsymbol{x}} \in \text{Zero}(\mathbb{T}/\tilde{\mathbb{T}})$. By definition, $\text{ini}(T)(\bar{\boldsymbol{x}}) \neq 0$ for any $T \in \mathbb{T}$. Hence, according to the pseudo-remainder formula (2.1.2) we have $P(\bar{\boldsymbol{x}}) = 0$. The "$\Longleftarrow$" part of the theorem is proved.

Now suppose that $\text{Zero}(\mathbb{T}/\tilde{\mathbb{T}}) \subset \text{Zero}(P)$. We want to show that

$$R = \text{prem}(P, \mathbb{T}) = 0.$$

For this purpose, let $\mathfrak{S} = [\mathbb{T}, \tilde{\mathbb{T}}]$ and $\breve{\mathfrak{S}}$ be reordered as a triangular set $[T_1, \ldots, T_r]$ with

$$\text{cls}(T_i) = p_i, \quad d_i = \text{ldeg}(T_i), \qquad 1 \le i \le r.$$

For any $\bar{\boldsymbol{x}}^{\{p_r-1\}} \in \text{Zero}(\mathfrak{S}^{(p_r-1)})$ and arbitrary $\bar{x}_{p_r+1}, \ldots, \bar{x}_n \in \tilde{\boldsymbol{K}}$, let

$$\hat{\boldsymbol{x}}_{p_r} = (\bar{\boldsymbol{x}}^{\{p_r-1\}}, x_{p_r}, \bar{x}_{p_r+1}, \ldots, \bar{x}_n).$$

Then $T_r(\hat{\boldsymbol{x}}_{p_r})$ has d_r distinct zeros for x_{p_r}. By the pseudo-remainder formula (2.1.2), $\text{Zero}(\mathfrak{S}) \subset \text{Zero}(R)$. Thus, $R(\hat{\boldsymbol{x}}_{p_r})$ also has d_r distinct zeros for x_{p_r} when $T_r \in \mathbb{T}$;

and any $\bar{x}_{p_r} \in \tilde{\boldsymbol{K}}$ other than the d_r zeros of $T_r(\hat{\boldsymbol{x}}_{p_r})$ is a zero of $R(\hat{\boldsymbol{x}}_{p_r})$ when $T_r \in \tilde{\mathbb{T}}$. As $\deg(R, x_{p_r}) < d_r$ when $T_r \in \mathbb{T}$, the coefficients R_i of R, considered as a polynomial in x_{p_r}, must be all zero for $\bar{\boldsymbol{x}}^{\{p_r-1\}} \in \text{Zero}(\mathfrak{S}^{(p_r-1)})$ and arbitrary $\bar{x}_{p_r+1}, \ldots, \bar{x}_n \in \tilde{\boldsymbol{K}}$. Namely, $\text{Zero}(\mathfrak{S}^{(p_r-1)}) \subset \text{Zero}(R_i)$ for each i. As $T_{r-1}(\bar{\boldsymbol{x}}^{\{p_{r-1}-1\}}, x_{p_{r-1}})$ has d_{r-1} distinct zeros for $x_{p_{r-1}}$ and $\deg(R_i, x_{p_{r-1}}) < d_{r-1}$ when $T_{r-1} \in \mathbb{T}$, the coefficients of every R_i, considered as a polynomial in $x_{p_{r-1}}$, are all zero for any

$$\bar{\boldsymbol{x}}^{\{p_{r-1}-1\}} \in \text{Zero}(\mathfrak{S}^{(p_{r-1}-1)})$$

and arbitrary $\bar{x}_{p_{r-1}+1}, \ldots, \bar{x}_{p_r-1}, \bar{x}_{p_r+1}, \ldots, \bar{x}_n \in \tilde{\boldsymbol{K}}$.

Continuing the argument for $T_{r-2}, \ldots, T_1$, we shall see that the coefficients of R, considered as a polynomial in $x_{p_1}, \ldots, x_{p_r}$, are all zero when any set of values is substituted for the other (parametric) variables. This implies that $R \equiv 0$, and the proof is complete. □

As a corollary of the above theorem, we have the following result.

Corollary 3.4.5. For any simple set $\mathbb{T}$ and polynomial P in $\boldsymbol{K}[\boldsymbol{x}]$,

$$\text{Zero}(\mathbb{T}/\text{ini}(\mathbb{T})) \subset \text{Zero}(P) \iff \text{prem}(P, \mathbb{T}) = 0.$$

Proof. From the remainder formula, it is easy to see that $\text{prem}(P, \mathbb{T}) = 0$ implies that $\text{Zero}(\mathbb{T}/\mathbb{I}) \subset \text{Zero}(P)$, where $\mathbb{I} = \text{ini}(\mathbb{T})$. As $\mathbb{T}$ is a simple set, there exists a $\tilde{\mathbb{T}}$ such that $[\mathbb{T}, \tilde{\mathbb{T}}]$ is a simple system. From the definition of simple systems, one knows that

$$\text{Zero}(\mathbb{T}/\tilde{\mathbb{T}}) \subset \text{Zero}(\mathbb{T}/\mathbb{I}).$$

Hence, by Theorem 3.4.4, if $\text{Zero}(\mathbb{T}/\mathbb{I}) \subset \text{Zero}(P)$ then $\text{prem}(P, \mathbb{T}) = 0$. □

Theorem 3.4.4 together with algorithm SimSer provides a solution to the radical ideal membership problem. It can also be used to prove the following properties about simple series.

Theorem 3.4.6. Let $[\mathbb{P}, \mathbb{Q}]$ be a polynomial system in $\boldsymbol{K}[\boldsymbol{x}]$ and Ψ a simple series of $[\mathbb{P}, \mathbb{Q}]$. Then

a. $\text{prem}(\mathbb{P}, \mathbb{T}) = \{0\}$ and $0 \notin \text{prem}(\mathbb{Q}, \mathbb{T})$ for every $[\mathbb{T}, \tilde{\mathbb{T}}] \in \Psi$;

b.
$$\text{Zero}(\mathbb{P}/\mathbb{Q}) = \bigcup_{[\mathbb{T},\tilde{\mathbb{T}}]\in\Psi} \text{Zero}(\mathbb{T}/\text{ini}(\mathbb{T}) \cup \mathbb{Q}). \tag{3.4.1}$$

Proof. a. Let $[\mathbb{T}, \tilde{\mathbb{T}}] \in \Psi$; then $\text{Zero}(\mathbb{T}/\tilde{\mathbb{T}}) \subset \text{Zero}(\mathbb{P}/\mathbb{Q})$. It follows that $\text{Zero}(\mathbb{T}/\tilde{\mathbb{T}}) \subset \text{Zero}(\mathbb{P})$ and $\text{Zero}(\mathbb{T}/\tilde{\mathbb{T}}) \not\subset \text{Zero}(Q)$ for any $Q \in \mathbb{Q}$. Hence, by Theorem 3.4.4 we have $\text{prem}(\mathbb{P}, \mathbb{T}) = \{0\}$ and $\text{prem}(Q, \mathbb{T}) \neq 0$ for any $Q \in \mathbb{Q}$.

b. By part a just proved and the pseudo-remainder formula, the right-hand side is contained in the left-hand side of (3.4.1). On the contrary, let $\bar{\boldsymbol{x}} \in \text{Zero}(\mathbb{P}/\mathbb{Q})$. Then there is a $[\mathbb{T}, \tilde{\mathbb{T}}] \in \Psi$ such that $\bar{\boldsymbol{x}} \in \text{Zero}(\mathbb{T}/\tilde{\mathbb{T}})$. Clearly, $\bar{\boldsymbol{x}}$ is not a zero of any polynomial in $\text{ini}(\mathbb{T})$. Hence $\bar{\boldsymbol{x}} \in \text{Zero}(\mathbb{T}/\text{ini}(\mathbb{T}) \cup \mathbb{Q})$, i.e., $\bar{\boldsymbol{x}}$ belongs to the right-hand side of (3.4.1). □

Corollary 3.4.7. Any simple series of a polynomial system $\mathfrak{P}$ is a W-characteristic series of $\mathfrak{P}$.

Theorem 3.4.8. Let $\mathfrak{S}_1 = [\mathbb{T}_1, \tilde{\mathbb{T}}_1]$ and $\mathfrak{S}_2 = [\mathbb{T}_2, \tilde{\mathbb{T}}_2]$ be two simple systems in $\boldsymbol{K}[\boldsymbol{x}]$ with $\text{Zero}(\mathfrak{S}_1) \subset \text{Zero}(\mathfrak{S}_2)$.

a. Then $\text{prem}(T_2, \mathbb{T}_1) = 0$ for all $T_2 \in \mathbb{T}_2$.
   For any $1 \le k \le n$:
b. If $\breve{\mathfrak{S}}_1^{\langle k\rangle} = \emptyset$, then $\breve{\mathfrak{S}}_2^{\langle k\rangle} = \emptyset$.
c. Assume that $\tilde{\mathbb{T}}_i^{\langle k\rangle} \neq \emptyset$ and let $T_i \in \tilde{\mathbb{T}}_i^{\langle k\rangle}$ for $i = 1, 2$. Then

$$\text{prem}(T_1, \mathbb{T}_1^{(k-1)} \cup [T_2]) = 0.$$

Proof. a. The result follows from Theorem 3.4.4.

b. Note that for any $1 \le k \le n$, $\text{Zero}(\mathfrak{S}_1^{(k)}) \subset \text{Zero}(\mathfrak{S}_2^{(k)})$, and the scope of x_k in $\text{Zero}(\mathfrak{S}_i^{(k)})$ is $\tilde{\boldsymbol{K}}$ for any fixed $\bar{\boldsymbol{x}}^{\{k-1\}} \in \text{Zero}(\mathfrak{S}_i^{(k-1)})$ if and only if $\breve{\mathfrak{S}}_i^{\langle k\rangle} = \emptyset$ for $i = 1, 2$. Hence, $\breve{\mathfrak{S}}_1^{\langle k\rangle} = \emptyset$ implies that $\breve{\mathfrak{S}}_2^{\langle k\rangle} = \emptyset$.

c. Let $\mathbb{T}_1^{*(k)} = \mathbb{T}_1^{(k-1)} \cup [T_2]$; then $[\mathbb{T}_1^{*(k)}, \tilde{\mathbb{T}}_1^{(k-1)}]$ is a simple system. And any zero of $[\mathbb{T}_1^{*(k)}, \tilde{\mathbb{T}}_1^{(k-1)}]$ for which $T_1 \neq 0$, if exists, is also a zero of $\mathfrak{S}_1^{(k)}$ and thus of $\mathfrak{S}_2^{(k)}$. The existence of such a zero would lead to a contradiction. Therefore, $\text{Zero}(\mathbb{T}_1^{*(k)}/\tilde{\mathbb{T}}_1^{(k-1)}) \subset \text{Zero}(T_1)$ and the conclusion follows from part a. □

Theorem 3.4.9. Let $[\mathbb{T}_1, \tilde{\mathbb{T}}_1]$ and $[\mathbb{T}_2, \tilde{\mathbb{T}}_2]$ be two simple systems in $\boldsymbol{K}[\boldsymbol{x}]$. Then $\text{Zero}(\mathbb{T}_1/\tilde{\mathbb{T}}_1) = \text{Zero}(\mathbb{T}_2/\tilde{\mathbb{T}}_2)$ if and only if the polynomials in $\mathbb{T}_1 \cup \tilde{\mathbb{T}}_1$ and in $\mathbb{T}_2 \cup \tilde{\mathbb{T}}_2$ can be put in a one-to-one correspondence such that for any corresponding polynomials T_1 and T_2 either $T_1 \in \mathbb{T}_1$ and $T_2 \in \mathbb{T}_2$, or $T_1 \in \tilde{\mathbb{T}}_1$ and $T_2 \in \tilde{\mathbb{T}}_2$, and

$$\text{prem}(I_2T_1 - I_1T_2, \mathbb{T}_1) = \text{prem}(I_2T_1 - I_1T_2, \mathbb{T}_2) = 0,$$

where $I_i = \text{ini}(T_i)$ for $i = 1, 2$.

Proof. We only need to prove the necessity. First of all, the leading variables must be exactly the same for the two systems $[\mathbb{T}_1, \tilde{\mathbb{T}}_1]$ and $[\mathbb{T}_2, \tilde{\mathbb{T}}_2]$. For the scope of a leading variable x_k in $\text{Zero}(\mathbb{T}_1^{(k)}/\tilde{\mathbb{T}}_1^{(k)})$ is a proper subset of $\tilde{\boldsymbol{K}}$ for any fixed $\bar{\boldsymbol{x}}^{\{k-1\}} \in \text{Zero}(\mathbb{T}_1^{(k-1)}/\tilde{\mathbb{T}}_1^{(k-1)})$, whereas in $\text{Zero}(\mathbb{T}_2^{(k)}/\tilde{\mathbb{T}}_2^{(k)})$ a free variable x_k may take any element of $\tilde{\boldsymbol{K}}$. Therefore, any $T_1 \in \mathbb{T}_1^{\langle k\rangle} \cup \tilde{\mathbb{T}}_1^{\langle k\rangle}$ corresponds to a $T_2 \in \mathbb{T}_2^{\langle k\rangle} \cup \tilde{\mathbb{T}}_2^{\langle k\rangle}$ $(1 \le k \le n)$, and vice versa. Thus, for any k and

$$\bar{\boldsymbol{x}}^{\{k-1\}} \in \text{Zero}(\mathbb{T}_1^{(k-1)}/\tilde{\mathbb{T}}_1^{(k-1)}) = \text{Zero}(\mathbb{T}_2^{(k-1)}/\tilde{\mathbb{T}}_2^{(k-1)}),$$

$T_1(\bar{\boldsymbol{x}}^{\{k-1\}}, x_k)$ and $T_2(\bar{\boldsymbol{x}}^{\{k-1\}}, x_k)$ are squarefree and have the same set of zeros for x_k. This implies that

$$T_1 \in \mathbb{T}_1^{\langle k\rangle} \iff T_2 \in \mathbb{T}_2^{\langle k\rangle},$$
$$I_2(\bar{\boldsymbol{x}}^{\{k-1\}}) \cdot T_1(\bar{\boldsymbol{x}}^{\{k-1\}}, x_k) - I_1(\bar{\boldsymbol{x}}^{\{k-1\}}) \cdot T_2(\bar{\boldsymbol{x}}^{\{k-1\}}, x_k) = 0.$$

The result is established by Theorem 3.4.4. □

Lemma 3.4.10. From any simple system $\mathfrak{S}$ in $\boldsymbol{K}[\boldsymbol{x}]$, one can compute a reduced simple system $\mathfrak{S}^*$ such that $\mathrm{Zero}(\mathfrak{S}) = \mathrm{Zero}(\mathfrak{S}^*)$.

Proof. According to Remark 2.1.1, one can compute a reduced triangular system $\mathfrak{S}^*$ such that $\mathrm{Zero}(\mathfrak{S}) = \mathrm{Zero}(\mathfrak{S}^*)$. We need to show that $\mathfrak{S}^*$ is a simple system. Referring to the proof of Lemma 2.1.4 and Remark 2.1.1 and notations therein with $\tilde{\mathbb{T}} = \mathbb{U}$ and $\tilde{\mathbb{T}}^* = \mathbb{U}^*$, one knows that

$$\mathrm{cls}(T_i^*) = \mathrm{cls}(T_i) = p_i, \quad \mathrm{ldeg}(T_i^*) = \mathrm{ldeg}(T_i) = d_i, \quad 2 \le i \le r.$$

Hence, $\breve{\mathfrak{S}}^*$ can be ordered as a triangular set and $T_i^*(\bar{\boldsymbol{x}}^{\{p_i-1\}}, x_{p_i})$ has the same set of d_i distinct zeros as $T_i(\bar{\boldsymbol{x}}^{\{p_i-1\}}, x_{p_i})$ for x_{p_i} and thus is squarefree for any

$$\bar{\boldsymbol{x}}^{\{p_i-1\}} \in \mathrm{Zero}([T_1, T_2^*, \ldots, T_{i-1}^*]/\tilde{\mathbb{T}}^{(p_i-1)})$$

and $2 \le i \le r$. Similarly, for any $T \in \tilde{\mathbb{T}}$ of class p, let $T^* = \mathrm{prem}(T, \mathbb{T}^*)$; then $\mathrm{cls}(T^*) = p$ and $T^*(\bar{\boldsymbol{x}}^{\{p-1\}}, x_p)$ has the same set of distinct zeros as $T(\bar{\boldsymbol{x}}^{\{p-1\}}, x_p)$ for x_p, and is squarefree for any

$$\bar{\boldsymbol{x}}^{\{p-1\}} \in \mathrm{Zero}(\mathbb{T}^{*(p-1)}/\tilde{\mathbb{T}}^{(p-1)}).$$

Therefore, $[\mathbb{T}^*, \tilde{\mathbb{T}}^*]$ is a reduced simple system. □

4 Irreducible zero decomposition

Polynomial factorization is not required theoretically for the algorithms described in the previous two chapters. Nevertheless, available factoring programs have been efficient enough to be used to enhance the performance of elimination algorithms. It is a good strategy to incorporate polynomial factorization (even over algebraic-extension fields) in the implementation of such algorithms. In this chapter, we elaborate how triangular systems can be further decomposed by making use of factorization in order to compute zero decompositions possessing better properties. For our exposition some of the material from Wu (1984; 1994, chap. 4) will be used without explicit mention.

4.1 Irreducibility of triangular sets

Definition 4.1.1. A triangular set $\mathbb{T} \subset \boldsymbol{K}[\boldsymbol{x}]$ is said to be *quasi-irreducible* if every polynomial in $\mathbb{T}$ is irreducible over the ground field $\boldsymbol{K}$.

A triangular system $[\mathbb{T}, \mathbb{U}]$ in $\boldsymbol{K}[\boldsymbol{x}]$ is said to be *quasi-irreducible* if $\mathbb{T}$ is quasi-irreducible.

Using polynomial factorization over $\boldsymbol{K}$, one has no difficulty to compute zero decompositions of the forms (2.2.7) and (2.1.8) with all triangular sets quasi-irreducible. This is done by splitting the corresponding polynomial systems when polynomials are factorized. More concretely, for any polynomial system $[\mathbb{P}, \mathbb{Q}]$, if $P_1, \ldots, P_t$ are all the irreducible factors of some polynomial $P \in \mathbb{P}$, we have

$$\mathrm{Zero}(\mathbb{P}/\mathbb{Q}) = \bigcup_{j=1}^{t} \mathrm{Zero}(\mathbb{P}_j/\mathbb{Q}), \tag{4.1.1}$$

where $\mathbb{P}_j = \mathbb{P} \setminus \{P\} \cup \{P_j\}$ for $1 \leq j \leq t$.

As a subalgorithm of IrrTriSer to be presented in Sect. 4.2, let us modify algorithm TriSer to QuaIrrTriSer with the following specification.

Algorithm QuaIrrTriSer: $\Psi \leftarrow \mathrm{QuaIrrTriSer}(\mathbb{P}, \mathbb{Q}, \mathbb{T})$. Given a triplet $[\mathbb{P}, \mathbb{Q}, \mathbb{T}]$ with $[\mathbb{T}, \mathbb{Q}]$ constituting a quasi-irreducible triangular system and all the polynomials in $\mathbb{Q}$ reduced with respect to $\mathbb{T}$, this algorithm computes a finite set Ψ of fine quasi-irreducible triangular systems $[\mathbb{T}_1, \mathbb{U}_1], \ldots, [\mathbb{T}_e, \mathbb{U}_e]$ such that

$$\mathrm{Zero}(\mathbb{P} \cup \mathbb{T}/\mathbb{Q}) = \bigcup_{i=1}^{e} \mathrm{Zero}(\mathbb{T}_i/\mathbb{U}_i). \tag{4.1.2}$$

As before, $\Psi = \emptyset$ when $\mathrm{Zero}(\mathbb{P} \cup \mathbb{T}/\mathbb{Q}) = \emptyset$ is detected. In the case $\mathbb{T} = \emptyset$, QuaIrrTriSer decomposes any polynomial system $[\mathbb{P}, \mathbb{Q}]$ into fine quasi-irreducible triangular systems. Algorithm QuaIrrTriSer is obtained from TriSer by replacing T1 with

T1′. Set $\Psi \leftarrow \emptyset$, $\Phi \leftarrow \{[\mathbb{P}, \mathbb{Q}, \mathbb{T}]\}$.

and P2.3 in PriTriSys with

P2.3′. Compute all the irreducible factors $F_1, \ldots, F_t$ of T over $\boldsymbol{K}$ and set $\bar{\mathbb{G}} \leftarrow \mathbb{G}$.

P2.3″. For $j = 1, \ldots, t$ do:

P2.3.1. Compute $\bar{\mathbb{G}}' \leftarrow \text{prem}(\bar{\mathbb{G}}, F_j)$.

P2.3.2. If $j = 1$, then set $\mathbb{G} \leftarrow \bar{\mathbb{G}}'$, $T \leftarrow F_j$. Otherwise, if $0 \notin \bar{\mathbb{G}}'$, then set $\Omega \leftarrow \Omega \cup \{[\mathbb{F}, \bar{\mathbb{G}}', [F_j] \cup \mathbb{T}]\}$.

Proof. For the modification of step T1 to T1′, we note that $\mathbb{P} \cup \mathbb{T}$ here corresponds to the set $\mathbb{P}$ in the input of TriSer, while the cases in which the initials of the polynomials in $\mathbb{T}$ happen to be zero need not be considered because $[\mathbb{T}, \mathbb{Q}]$ is a triangular system. Actually, any triplet from Φ in TriSer is of the same form as the input triplet to QuaIrrTriSer. For the modification of step P2.3 to P2.3′ and P2.3″, the polynomial T produced by Elim is factorized over the ground field $\boldsymbol{K}$ and the polynomial system is then split into subsystems by replacing T with its factors. One sees that for any triplet – say, $[\mathbb{F}^*, \mathbb{G}^*, \mathbb{T}^*]$ – produced in step P2.3.2, $\text{level}(\mathbb{F}^*) < \text{level}(\mathbb{F})$, where $\mathbb{F}$ is the first component of the corresponding triplet taken from Φ in step T2.1 (see the termination proof of TriSer). Hence, algorithm QuaIrrTriSer terminates as well.

To see the correctness of this algorithm, one only needs to be aware of the zero relation (4.1.1) for splitting of polynomial systems via factorization. The relation (4.1.2) is proved by the same argument as for the proof of (2.1.8) in algorithm TriSer. Since the corresponding T is replaced by its irreducible factors, by definition $\mathbb{T}_i$ is quasi-irreducible and thus so is $[\mathbb{T}_i, \mathbb{U}_i]$ for each i. $[\mathbb{T}_i, \mathbb{U}_i]$ is fine because all the polynomials in $\mathbb{U}_i$ are actually the pseudo-remainders of some polynomials (and thus are reduced) with respect to $\mathbb{T}_i$. □

In passing, those F_j whose classes are $< i$ are factors of the initial of T and thus need not be considered. Consequently, the corresponding triplets can be deleted from the set Ω.

Example 4.1.1. Recall Examples 2.3.1 and 2.3.2, and apply algorithm QuaIrrTriSer to the triplet $[\mathbb{P}, \emptyset, \emptyset]$ of level 4. It is easy to verify that all the polynomials in the triangular sets $\mathbb{T}_1$ and $\mathbb{T}_2$ produced by algorithm TriSer are irreducible. However, the first polynomial $t^3 + 1$ in $\mathbb{T}_3$ is reducible and factors as the product of two polynomials

$$t - 1 \text{ and } T_1 = t^2 + t + 1.$$

Hence, in QuaIrrTriSer $[\mathbb{T}_3, \mathbb{U}_3]$ is split into two triangular systems $[\mathbb{T}_3', \mathbb{U}_3']$ and $[\mathbb{T}_3'', \mathbb{U}_3'']$ with

$$\begin{aligned}
\mathbb{T}_3' &= [T_1, -z^5 + t^4, -z^3 y - t^3, zx^2 - t], \\
\mathbb{T}_3'' &= [t - 1, -z^5 + t^4, -z^3 y - t^3, zx^2 - t], \\
\mathbb{U}_3' &= \mathbb{U}_3'' = \{z\}.
\end{aligned}$$

Let a triangular set $\mathbb{T}$ be written in the form (2.1.1) and the leading variables $x_{p_1}, \ldots, x_{p_r}$ be renamed $y_1, \ldots, y_r$. Denote all the x_i in $\{x_1, \ldots, x_n\} \setminus \{x_{p_1}, \ldots, x_{p_r}\}$ by $u_1, \ldots, u_d$, abbreviated to $\boldsymbol{u}$. Clearly, $d + r = n$; we call $u_1, \ldots, u_d$ the *parameters* and $y_1, \ldots, y_r$ the *dependents* of $\mathbb{T}$. Then $\mathbb{T}$ can be written as

$$\mathbb{T} = \begin{bmatrix} T_1(\boldsymbol{u}, y_1), \\ T_2(\boldsymbol{u}, y_1, y_2), \\ \quad \cdots\cdots \\ T_r(\boldsymbol{u}, y_1, y_2, \ldots, y_r) \end{bmatrix}. \tag{4.1.3}$$

Let $\boldsymbol{K}_0$ be the transcendental-extension field $\boldsymbol{K}(\boldsymbol{u}) = \boldsymbol{K}(u_1, \ldots, u_d)$ of $\boldsymbol{K}$ acquired by adjoining $u_1, \ldots, u_d$. We define inductively the *irreducibility* and *generic zeros* of $\mathbb{T}$ as follows.

Definition 4.1.2. A fine triangular set $\mathbb{T}$ containing only one polynomial $T_1(\boldsymbol{u}, y_1)$ is said to be *irreducible* if T_1 is irreducible as a polynomial in $\boldsymbol{K}_0[y_1]$. In this case, let η_1 be a zero of T_1 in some algebraic-extension field of $\boldsymbol{K}_0$; then $(\boldsymbol{u}, \eta_1)$ is called a *generic zero* of $\mathbb{T}$.

Suppose that the irreducibility and generic zeros of any fine triangular set with length $<r$ have already been defined.

A fine triangular set $\mathbb{T}$ of length $r > 1$ as in (4.1.3) is said to be *irreducible* if the fine triangular set

$$\mathbb{T}^{\{r-1\}} = [T_1, \ldots, T_{r-1}]$$

is irreducible with a generic zero $(\boldsymbol{u}, \eta_1, \ldots, \eta_{r-1})$, and the polynomial

$$\bar{T}_r = T_r(\boldsymbol{u}, \eta_1, \ldots, \eta_{r-1}, y_r) \in \boldsymbol{K}_{r-1}[y_r]$$

is irreducible over $\boldsymbol{K}_{r-1}$, where $\boldsymbol{K}_{r-1} = \boldsymbol{K}_0(\eta_1, \ldots, \eta_{r-1})$ is the algebraic-extension field acquired from $\boldsymbol{K}_0$ by adjoining $\eta_1, \ldots, \eta_{r-1}$. In this case, let η_r be a zero of $\bar{T}_r$ in some algebraic-extension field of $\boldsymbol{K}_{r-1}$; then $(\boldsymbol{u}, \eta_1, \ldots, \eta_r)$ is called a *generic zero* of $\mathbb{T}$.

A fine triangular system $[\mathbb{T}, \mathbb{U}]$ is said to be *irreducible* if $\mathbb{T}$ is irreducible.

Let $\mathbb{T}$ as in (4.1.3) be an irreducible triangular set with $(\boldsymbol{u}, \eta_1, \ldots, \eta_r)$ as a generic zero. For the sake of brevity, we sometimes write $\boldsymbol{\xi}^{\{i\}}$ for $(\boldsymbol{u}, \eta_1, \ldots, \eta_i)$ with $\boldsymbol{\xi} = \boldsymbol{\xi}^{\{r\}}$. It is convenient to call $T_1, \ldots, T_r$ *adjoining polynomials* and $\mathbb{T}$ an *adjoining triangular set* of the extension field $\boldsymbol{K}_r = \boldsymbol{K}(\boldsymbol{\xi})$. Evidently, any generic zero $\boldsymbol{\xi}$ of $\mathbb{T}$ can be considered as a point of the linear space $\tilde{\boldsymbol{K}}^n$. The above $d = |\boldsymbol{u}|$, the number of parameters, is called the *dimension* of $\mathbb{T}$, denoted by $\dim(\mathbb{T})$.

If a fine triangular set $\mathbb{T}$ as above is reducible, then there is a k such that $\mathbb{T}^{\{k-1\}}$ is irreducible with a generic zero

$$\boldsymbol{\xi}^{\{k-1\}} = (\boldsymbol{u}, \eta_1, \ldots, \eta_{k-1})$$

and the polynomial

$$\bar{T}_k = T_k(\boldsymbol{\xi}^{\{k-1\}}, y_k) \in \boldsymbol{K}_{k-1}[y_k]$$

is reducible over $\boldsymbol{K}_{k-1} = \boldsymbol{K}(\boldsymbol{\xi}^{\{k-1\}})$. Let an irreducible factorization of $\bar{T}_k$ in $\boldsymbol{K}_{k-1}[y_k]$ be given by

$$\bar{T}_k = H_1 \cdots H_t,$$

in which each $H_i \in \boldsymbol{K}_{k-1}[y_k]$ is irreducible over $\boldsymbol{K}_{k-1}$ and $t \geq 2$. As the coefficients $\operatorname{coef}(H_i, y_k^j)$ are all elements of $\boldsymbol{K}_{k-1}$ and thus can be expressed as the quotients of polynomials in $\boldsymbol{\xi}^{\{k-1\}}$. By reducing fractions to a common denominator, one gets an expression of the form

$$\bar{D}\bar{T}_k = \bar{F}_1 \cdots \bar{F}_t,$$

where

$$\begin{aligned} &D \in \boldsymbol{K}[\boldsymbol{u}, y_1, \ldots, y_{k-1}], \quad F_i \in \boldsymbol{K}[\boldsymbol{u}, y_1, \ldots, y_k], \\ &\bar{D} = D(\boldsymbol{\xi}^{\{k-1\}}) \in \boldsymbol{K}_{k-1}, \quad \bar{F}_i = F_i(\boldsymbol{\xi}^{\{k-1\}}, y_k) \in \boldsymbol{K}_{k-1}[y_k]. \end{aligned}$$

The polynomial D may be assumed to be reduced with respect to $\mathbb{T}^{\{k-1\}}$, and so may each F_i with respect to $\mathbb{T}^{\{k\}}$.

Consider y_k as a free variable, renamed v. Then

$$\boldsymbol{\xi}^{*\{k-1\}} = (v, \boldsymbol{u}, \eta_1, \ldots, \eta_{k-1})$$

is a generic zero of $\mathbb{T}^{\{k-1\}} \subset \boldsymbol{K}[v, \boldsymbol{u}, y_1, \ldots, y_{k-1}]$. Let

$$G = F_1 \cdots F_t - DT_k \in \boldsymbol{K}[v, \boldsymbol{u}, y_1, \ldots, y_{k-1}].$$

Since $\bar{D}\bar{T}_k = \bar{F}_1 \cdots \bar{F}_t$, we have $G(\boldsymbol{\xi}^{*\{k-1\}}) = 0$. It follows from Lemma 4.3.1 that $\operatorname{prem}(G, \mathbb{T}^{\{k-1\}}) = 0$, so there are nonnegative integers $s_1, \ldots, s_{k-1}$ and polynomials $Q_1, \ldots, Q_{k-1} \in \boldsymbol{K}[v, \boldsymbol{u}, y_1, \ldots, y_{k-1}]$ such that

$$I_1^{s_1} \cdots I_{k-1}^{s_{k-1}} G = I_1^{s_1} \cdots I_{k-1}^{s_{k-1}} (F_1 \cdots F_t - DT_k) = \sum_{i=1}^{k-1} Q_i T_i,$$

or

$$I_1^{s_1} \cdots I_{k-1}^{s_{k-1}} F_1 \cdots F_t = \sum_{i=1}^{k} Q_i T_i. \tag{4.1.4}$$

In the above, y_k is renamed to help understand the application of Lemma 4.3.1. The renaming does not have any actual effect. The polynomials Q_i are all in the variables $\boldsymbol{u}, y_1, \ldots, y_k$.

We summarize the discussions as the following lemma.

Lemma 4.1.1. There is an algorithm which determines

a. whether a fine triangular set $\mathbb{T} \subset \boldsymbol{K}[\boldsymbol{u}, \boldsymbol{y}]$ is irreducible or not; and, if $\mathbb{T}$ is reducible,
b. an integer k such that the triangular set $\mathbb{T}^{\{k-1\}}$ formed by the first $k-1$ terms of $\mathbb{T}$ is irreducible with $\boldsymbol{\xi}^{\{k-1\}}$ as a generic zero, while the polynomial $T_k(\boldsymbol{\xi}^{\{k-1\}}, y_k)$ is reducible over $\boldsymbol{K}_{k-1} = \boldsymbol{K}(\boldsymbol{\xi}^{\{k-1\}})$;

c. an irreducible factorization of T_k of the form

$$DT_k \doteq F_1 \cdots F_t \tag{4.1.5}$$

over $\boldsymbol{K}_{k-1}$, where the polynomials

$$D \in \boldsymbol{K}[\boldsymbol{u}, y_1, \ldots, y_{k-1}],$$
$$F_i \in \boldsymbol{K}[\boldsymbol{u}, y_1, \ldots, y_k], \quad 1 \leq i \leq t,$$

are all reduced with respect to $\mathbb{T}^{\{k-1\}}$ and the dot equality means that $\operatorname{prem}(DT_k - F_1 \cdots F_t, \mathbb{T}^{\{k-1\}}) = 0$.

Let the algorithm indicated in Lemma 4.1.1 be specified as follows.

Algorithm Factor: $[k, D, \mathbb{F}] \leftarrow \operatorname{Factor}(\mathbb{T})$. Given a fine triangular set $\mathbb{T} \subset \boldsymbol{K}[\boldsymbol{x}]$, this algorithm computes an integer k, a polynomial D, and a finite set $\mathbb{F}$ of polynomials in $\boldsymbol{K}[\boldsymbol{x}]$ such that $0 \leq k \leq |\mathbb{T}|$ and

a. if $k = 0$, then $\mathbb{T}$ is irreducible;
b. if $k = 1$, then $\mathbb{T}$ is reducible, $|\mathbb{F}| > 1$, the first polynomial T_1 of class p_1 in $\mathbb{T}$ has a factorization $T_1 = \prod_{F \in \mathbb{F}} F$ over $\boldsymbol{K}_0 = \boldsymbol{K}(x_1, \ldots, x_{p_1-1})$, and each $F \in \mathbb{F} \subset \boldsymbol{K}_0[x_{p_1}]$ is irreducible over $\boldsymbol{K}_0$;
c. if $k > 1$, then $\mathbb{T}$ is reducible, $\mathbb{T}^{\{k-1\}}$ is irreducible, $|\mathbb{F}| > 1$, the kth polynomial T_k in $\mathbb{T}$ has a factorization $DT_k \doteq \prod_{F \in \mathbb{F}} F$ over the extension field $\boldsymbol{K}_{k-1}$ of $\boldsymbol{K}$ with adjoining triangular set $\mathbb{T}^{\{k-1\}}$, and each $F \in \mathbb{F} \subset \boldsymbol{K}_{k-1}[x_{p_k}]$ is irreducible over $\boldsymbol{K}_{k-1}$.

In specification c, the extension field $\boldsymbol{K}_{k-1}$ is obtained from $\boldsymbol{K}$ in a slightly different way: $\boldsymbol{K}_{k-1} = \boldsymbol{K}(x_1, \ldots, x_{p_k-1})$, where $x_{p_j} = \operatorname{lv}(T_j)$ is considered as an algebraic element with adjoining polynomial T_j for $1 \leq j \leq k-1$, and the other x_i are adjoined as transcendental elements. We shall refer to polynomial factorization over algebraic-extension fields as *algebraic factorization* for short. See Sect. 7.5 for a brief introduction to two algorithms of algebraic factorization.

4.2 Decomposition into irreducible triangular systems

From the formula (4.1.4) the following decomposition lemma may be easily established.

Lemma 4.2.1. Let a polynomial set $\mathbb{P}$ have a medial set

$$\mathbb{T} = [T_1, \ldots, T_r]$$

with

$$\operatorname{cls}(T_1) > 0, \quad I_i = \operatorname{ini}(T_i), \quad 1 \leq i \leq r.$$

Assume that $\mathbb{T}$ is reducible, so there is a k such that T_k has an irreducible factorization into polynomials $F_1, \ldots, F_t$ as of the form (4.1.5). Then the following zero decomposition holds

$$\operatorname{Zero}(\mathbb{P}) = \bigcup_{i=1}^{k-1} \operatorname{Zero}(\mathbb{P}_i) \cup \bigcup_{j=1}^{t} \operatorname{Zero}(\mathbb{Q}_j), \tag{4.2.1}$$

where $\mathbb{P}_i = \mathbb{P} \cup \{I_i\}$ and $\mathbb{Q}_j = \mathbb{P} \cup \{F_j\}$ for each i and j.

Proof. Any zero of either $\mathbb{P}_i$ or $\mathbb{Q}_j$ is obviously a zero of $\mathbb{P}$. Conversely, any zero of $\mathbb{P}$ is a zero of the T_i. By (4.1.4), it is also a zero of some I_i or F_j, and thus a zero of some $\mathbb{P}_i$ or $\mathbb{Q}_j$. □

As in Lemma 4.2.1 each I_i is already reduced with respect to $\mathbb{T}$ and each F_j is assumed to be reduced with respect to $\mathbb{T}^{\{k\}}$ and hence also reduced with respect to $\mathbb{T}$, any medial set of the polynomial set $\mathbb{P}_i \cup \mathbb{C}$ or $\mathbb{Q}_j \cup \mathbb{C}$ has rank lower than that of $\mathbb{T}$ by Lemma 2.2.4. Therefore, in proceeding with each $\mathbb{P}_i \cup \mathbb{C}$ or $\mathbb{Q}_j \cup \mathbb{C}$ as $\mathbb{P}$ to get further zero decomposition of the form (4.2.1), we shall arrive at a decomposition of the same form as (2.2.7) with all $\mathbb{C}_i$ irreducible.

A characteristic series or triangular series Ψ is said to be *irreducible* if every ascending set or triangular system in Ψ is irreducible. The following algorithm points out how to construct an irreducible characteristic series from any given polynomial set $\mathbb{P}$.

Algorithm IrrCharSer: $\Psi \leftarrow \text{IrrCharSer}(\mathbb{P})$. Given a nonempty polynomial set $\mathbb{P} \subset \boldsymbol{K}[\boldsymbol{x}]$, this algorithm computes an irreducible characteristic series Ψ of $\mathbb{P}$.

I1. Set $\Phi \leftarrow \{\mathbb{P}\}$, $\Psi \leftarrow \emptyset$.

I2. While $\Phi \neq \emptyset$, do:

- I2.1. Let $\mathbb{F}$ be an element of Φ and set $\Phi \leftarrow \Phi \setminus \{\mathbb{F}\}$.
- I2.2. Compute $\mathbb{C} \leftarrow \text{CharSet}(\mathbb{F})$.
- I2.3. If $\mathbb{C}$ is noncontradictory, then:
  - I2.3.1. Compute $[k, D, \mathbb{G}] \leftarrow \text{Factor}(\mathbb{C})$.
  - I2.3.2. If $k = 0$, then set

$$\Psi \leftarrow \Psi \cup \{\mathbb{C}\},$$
$$\Phi \leftarrow \Phi \cup \{\mathbb{F} \cup \mathbb{C} \cup \{I\}\colon\ I \in \text{ini}(\mathbb{C}) \setminus \boldsymbol{K}\};$$

else set

$$\Phi \leftarrow \Phi \cup \{\mathbb{F} \cup \mathbb{C} \cup \{I\}\colon\ I \in \text{ini}(\mathbb{C}^{\{k-1\}}) \setminus \boldsymbol{K}\}$$
$$\cup\ \{\mathbb{F} \cup \mathbb{C} \cup \{G\}\colon\ G \in \mathbb{G}\}.$$

Example 4.2.1. Refer to Example 2.2.3. It is easy to check that the first polynomial C_1 in the characteristic set $\mathbb{C}$ therein is irreducible over $\mathbf{Q}(x_1)$. To decide whether $\mathbb{C}$ is irreducible, one needs to verify whether the second polynomial C_2 in $\mathbb{C}$ is irreducible over the extension field $\mathbf{Q}(x_1, \eta)$ with η an extended zero of C_1. Application of any method of algebraic factorization should confirm that

$$C_2 \doteq (x_1 + 1)(x_3 - 2x_1x_2 + x_1)(x_3 + x_1x_2 - x_1)$$

over $\mathbf{Q}(x_1, \eta)$. Let

$$\mathbb{P}_1 = \mathbb{P} \cup \{x_1\}, \qquad \mathbb{P}_3 = \mathbb{P} \cup \{x_3 - 2x_1x_2 + x_1\},$$
$$\mathbb{P}_2 = \mathbb{P} \cup \{x_1 + 1\}, \quad \mathbb{P}_4 = \mathbb{P} \cup \{x_3 + x_1x_2 - x_1\}.$$

By Lemma 4.2.1, we have the following decomposition

$$\mathrm{Zero}(\mathbb{P}) = \bigcup_{i=1}^{4} \mathrm{Zero}(\mathbb{P}_i).$$

The characteristic sets $\mathbb{C}_1$ and $\mathbb{C}_2$ of $\mathbb{P}_1 \cup \mathbb{C}$ and $\mathbb{P}_2 \cup \mathbb{C}$ have already been given in Example 2.2.4. $\mathbb{P}_3 \cup \mathbb{C}$ and $\mathbb{P}_4 \cup \mathbb{C}$ have their characteristic sets

$$\begin{aligned}
\mathbb{C}_3 &= [C_1, x_3 - 2x_1x_2 + x_1, x_1(x_4 + x_2 - 1)],\\
\mathbb{C}_4 &= [C_1, x_3 + x_1x_2 - x_1, -x_1(x_4 - 2x_2 + 1)],
\end{aligned}$$

respectively. The factor x_1 of C_1 and the third polynomials in $\mathbb{C}_3$ and $\mathbb{C}_4$ can be simply removed; let the obtained ascending sets be denoted by $\mathbb{C}_3$ and $\mathbb{C}_4$ still.

Let us check whether the four ascending sets $\mathbb{C}_1, \ldots, \mathbb{C}_4$ are irreducible; both $\mathbb{C}_3$ and $\mathbb{C}_4$ are indeed so because all of their polynomials other than C_1 are linear in their leading variables. One can find that the third polynomial in $\mathbb{C}_1$ factors as

$$x_3^2 - 1 = (x_3 - 1)(x_3 + 1),$$

and so does the fourth polynomial in $\mathbb{C}_2$ as

$$x_4^2 - x_2x_4 + 3x_2 \doteq (x_4 + x_2 - 1)(x_4 - 2x_2 + 1)$$

over the algebraic-extension field $\mathbf{Q}(x_2)$ with adjoining polynomial $2x_2^2 + 1$ for x_2. By Lemma 4.2.1 again, we have further decompositions with the corresponding irreducible ascending sets as follows

$$\begin{aligned}
\mathbb{C}_1' &= [x_1 + 1, x_2, x_3 + 1, x_4 + 1],\\
\mathbb{C}_1'' &= [x_1 + 1, x_2, x_3 - 1, x_4 - 1],\\
\mathbb{C}_2' &= [x_1, 2x_2^2 + 1, x_3, x_4 + x_2 - 1],\\
\mathbb{C}_2'' &= [x_1, 2x_2^2 + 1, x_3, x_4 - 2x_2 + 1].
\end{aligned}$$

Thus, an irreducible characteristic series $\{\mathbb{C}_1', \mathbb{C}_1'', \mathbb{C}_2', \mathbb{C}_2'', \mathbb{C}_3, \mathbb{C}_4\}$ of $\mathbb{P}$ is finally obtained, with associated zero decomposition

$$\begin{aligned}
\mathrm{Zero}(\mathbb{P}) = {} & \mathrm{Zero}(\mathbb{C}_1') \cup \mathrm{Zero}(\mathbb{C}_1'') \cup \mathrm{Zero}(\mathbb{C}_2')\\
& \cup \mathrm{Zero}(\mathbb{C}_2'') \cup \mathrm{Zero}(\mathbb{C}_3/x_1 + 1) \cup \mathrm{Zero}(\mathbb{C}_4/x_1 + 1).
\end{aligned}$$

Remark 4.2.1. Irreducible weak-ascending sets can be defined as well, but irreducible quasi-ascending sets cannot. Algorithm IrrCharSer can also be used to compute irreducible weak-characteristic series of polynomial sets by modifying the corresponding notions.

Remark 4.2.2. A triangular set in which all the polynomials other than the first are linear in their leading variables is said to be *quasilinear*. The characteristic set of a general polynomial set happens quite often to be quasilinear. This may be observed from the feature of the characteristic-set algorithm, in which pseudo-division is the principal operation. Let $R = \text{prem}(G, F, x)$; normally, $\deg(R, x) = \deg(F, x) - 1$, i.e., the divided polynomial G is reduced to a remainder polynomial R of degree one less than that of the dividing polynomial F. The frequent occurrence of quasilinearity allows us to argue that, for computing irreducible characteristic series, algebraic factorization is not needed for the first characteristic set in the normal case. This gives one explanation of why irreducible decomposition is practically feasible, noting that in general the first characteristic set is the most complex one in terms of size. During the computation of characteristic series the adjunction of initials often destroys the quasilinearity of characteristic sets of the enlarged polynomial sets, unfortunately. Therefore, algebraic factorization is often required for verifying the irreducibility of these characteristic sets.

Lemma 4.2.2. Let $[\mathbb{T}, \mathbb{U}]$ be a fine triangular system in $\boldsymbol{K}[\boldsymbol{x}]$. Assume that $\mathbb{T}$ is reducible, so there exists a k such that the kth term T_k of $\mathbb{T}$ has an irreducible factorization into polynomials $F_1, \ldots, F_t$ as of the form (4.1.5). Then the following zero decomposition holds:

$$\text{Zero}(\mathbb{T}/\mathbb{U}) = \bigcup_{i=1}^{t} \text{Zero}(\mathbb{T}_i/\mathbb{U} \cup \{D\}) \cup \text{Zero}(\{D\} \cup \mathbb{T}/\mathbb{U}), \qquad (4.2.2)$$

where $\mathbb{T}_i = \mathbb{T} \setminus \{T_k\} \cup \{F_i\}$ for each i.

Proof. For any $\bar{\boldsymbol{x}} \in \text{Zero}(\mathbb{T}/\mathbb{U})$, we have $T_k(\bar{\boldsymbol{x}}) = 0$, so there must be an i such that $F_i(\bar{\boldsymbol{x}}) = 0$. If $D(\bar{\boldsymbol{x}}) \neq 0$, then $\bar{\boldsymbol{x}} \in \text{Zero}(\mathbb{T}_i/\mathbb{U} \cup \{D\})$. Otherwise, $\bar{\boldsymbol{x}} \in \text{Zero}(\{D\} \cup \mathbb{T}/\mathbb{U})$. Hence, in any case $\bar{\boldsymbol{x}}$ belongs to the right-hand side of (4.2.2).

On the other hand, let $\bar{\boldsymbol{x}}$ be contained in the right-hand side of (4.2.2). If $\bar{\boldsymbol{x}} \in \text{Zero}(\{D\} \cup \mathbb{T}/\mathbb{U})$, then $\bar{\boldsymbol{x}} \in \text{Zero}(\mathbb{T}/\mathbb{U})$ obviously. Otherwise, there is an i such that $\bar{\boldsymbol{x}} \in \text{Zero}(\mathbb{T}_i/\mathbb{U} \cup \{D\})$, so $F_i(\bar{\boldsymbol{x}}) = 0$ and $D(\bar{\boldsymbol{x}}) \neq 0$. It follows from (4.1.5) that $T_k(\bar{\boldsymbol{x}}) = 0$. Therefore $\bar{\boldsymbol{x}} \in \text{Zero}(\mathbb{T}/\mathbb{U})$. □

Remark 4.2.3. If, in particular, $D \in \boldsymbol{K}$ or $\dim(\mathbb{T}^{\{k-1\}}) = 0$, then (4.2.2) may be simplified to $\text{Zero}(\mathbb{T}/\mathbb{U}) = \bigcup_{i=1}^{t} \text{Zero}(\mathbb{T}_i/\mathbb{U})$. This is trivial for $D \in \boldsymbol{K}$. If $\dim(\mathbb{T}^{\{k-1\}}) = 0$, then by Proposition 4.3.10, we have

$$\text{Zero}(\{D\} \cup \mathbb{T}/\mathbb{U}) = \emptyset, \quad \text{Zero}(\mathbb{T}_i/\mathbb{U} \cup \{D\}) = \text{Zero}(\mathbb{T}_i/\mathbb{U}).$$

The following algorithm generalizes algorithm IrrCharSer. The strategy it employs is adapted from Wu (1986a) and is somewhat different from that used in IrrCharSer.

Algorithm IrrCharSerE: $\Psi \leftarrow \text{IrrCharSerE}(\mathbb{P}, \mathbb{Q})$. Given a polynomial system $[\mathbb{P}, \mathbb{Q}]$ in $\boldsymbol{K}[\boldsymbol{x}]$, this algorithm computes an irreducible characteristic series Ψ of $[\mathbb{P}, \mathbb{Q}]$.

I1. Set $\Phi \leftarrow \{[\mathbb{P}, \mathbb{Q}]\}$, $\Psi \leftarrow \emptyset$.
I2. While $\Phi \neq \emptyset$, do:
  I2.1. Let $[\mathbb{F}, \mathbb{G}]$ be an element of Φ and set $\Phi \leftarrow \Phi \setminus \{[\mathbb{F}, \mathbb{G}]\}$.
  I2.2. Compute $\mathbb{C} \leftarrow \text{CharSet}(\mathbb{F})$.

I2.3. If $\mathbb{C}$ is noncontradictory, then:

I2.3.1. Set

$$\mathbb{I} \leftarrow \mathrm{ini}(\mathbb{C}) \setminus \boldsymbol{K}, \quad \Phi \leftarrow \Phi \cup \{[\mathbb{F} \cup \mathbb{C} \cup \{I\}, \mathbb{G}]\colon\ I \in \mathbb{I}\}.$$

I2.3.2. Compute $[k, D, \mathbb{H}] \leftarrow \mathrm{Factor}(\mathbb{C})$. If $k = 0$, then go to I2.3.3. Set

$$\Phi \leftarrow \Phi \cup \{[\mathbb{C} \setminus \{\mathrm{op}(k, \mathbb{C})\} \cup \{H\}, \mathbb{G} \cup \mathbb{I} \cup \{D\}]\colon\ H \in \mathbb{H}\} \\ \cup \{[\mathbb{F} \cup \{D\}, \mathbb{G} \cup \mathbb{I}]\}$$

and go to I2.

I2.3.3. Compute $\mathbb{D} \leftarrow \mathrm{prem}(\mathbb{G} \cup \mathbb{I}, \mathbb{C})$. If $0 \notin \mathbb{D}$, then set

$$\Psi \leftarrow \Psi \cup \{[\mathbb{C}, \mathbb{D}]\}.$$

Since for each branch of the decomposition tree the basic sets of the successively adjoined polynomial sets are of steadily decreasing ranks, the above algorithm terminates obviously. Its correctness follows from the previous discussions.

Let the notations be as in Lemma 4.2.2 and $\mathbb{U}_i = \mathrm{prem}(\mathbb{U} \cup \{D\}, \mathbb{T}_i)$ (where the pseudo-division needs to be performed actually only with respect to $\mathbb{T}_i^{\{k\}} = [T_1, \dots, T_{k-1}, F_i]$). If $0 \in \mathbb{U}_i$ for some i, then the corresponding component in (4.2.2) can be simply removed. For those components in which $\mathbb{U}_i$ does not contain 0, it is easy to see that $[\mathbb{T}_i, \mathbb{U}_i]$ is still a fine triangular system and, in particular, $\mathbb{T}_i^{\{k\}}$ is irreducible for each i. Moreover, all $\mathbb{T}_i$ have the same set of parameters as $\mathbb{T}$.

The polynomial set $\{D\} \cup \mathbb{T}$ may no longer be in triangular form, yet it can be further triangularized by applying algorithm QuaIrrTriSer to

$$[\{T_1, \dots, T_q, D\}, \mathbb{U}, [T_{q+1}, \dots, T_r]],$$

where q is the biggest index such that $\mathrm{cls}(T_q) \leq \mathrm{cls}(D)$.

In step D2.2.3 of the following algorithm, the ordering is preserved naturally for ordered set collection. For instance, if $\mathbb{S} = [1, \dots, 10]$, then $[i \in \mathbb{S}\colon\ 4 \leq i < 8, 2 \mid i] = [4, 6]$.

Algorithm Decom: $[\Psi, \Phi] \leftarrow \mathrm{Decom}(\mathbb{T}, \mathbb{U})$. Given a fine quasi-irreducible triangular system $[\mathbb{T}, \mathbb{U}]$ in $\boldsymbol{K}[\boldsymbol{x}]$, this algorithm computes two sets

$$\Psi = \{[\mathbb{T}_1, \mathbb{U}_1], \dots, [\mathbb{T}_e, \mathbb{U}_e]\}, \\ \Phi = \{[\mathbb{P}_1, \mathbb{Q}_1, \mathbb{T}_1^*], \dots, [\mathbb{P}_h, \mathbb{Q}_h, \mathbb{T}_h^*]\}$$

such that

$$\mathrm{Zero}(\mathbb{T}/\mathbb{U}) = \bigcup_{i=1}^{e} \mathrm{Zero}(\mathbb{T}_i/\mathbb{U}_i) \cup \bigcup_{j=1}^{h} \mathrm{Zero}(\mathbb{P}_j \cup \mathbb{T}_j^*/\mathbb{Q}_j), \tag{4.2.3}$$

where each $[\mathbb{T}_i, \mathbb{U}_i]$ is an irreducible triangular system, $\mathbb{T}_i$ has the same set of parameters as $\mathbb{T}$, and $[\mathbb{P}_j, \mathbb{Q}_j, \mathbb{T}_j^*]$ is a triplet with $[\mathbb{T}_j^*, \mathbb{Q}_j]$ constituting a fine quasi-irreducible triangular system. $\mathrm{Zero}(\mathbb{T}/\mathbb{U}) = \emptyset$ is detected when $\Psi = \Phi = \emptyset$.

D1. Set $\Phi \leftarrow \emptyset$, $r \leftarrow |\mathbb{T}|$. If $r = 1$, then set $\Psi \leftarrow \{[\mathbb{T}, \mathbb{U}]\}$ and the algorithm terminates; else set $\Omega \leftarrow \{[[\mathrm{op}(1, \mathbb{T})], \mathbb{T} \setminus [\mathrm{op}(1, \mathbb{T})], \mathbb{U}]\}$.

D2. For $\iota = 2, \ldots, r$ do:

D2.1. Set $\Psi \leftarrow \emptyset$.

D2.2. For each $[\mathbb{T}', \mathbb{T}'', \mathbb{U}'] \in \Omega$ do:

D2.2.1. Set $T \leftarrow \mathrm{op}(1, \mathbb{T}'')$, $\mathbb{T}'' \leftarrow \mathbb{T}'' \setminus [T]$.

D2.2.2. Compute $[k, D, \mathbb{F}] \leftarrow \text{Factor}(\mathbb{T}' \cup [T])$. If $k = 0$, then set $D \leftarrow 1, \mathbb{F} \leftarrow \{T\}$.

D2.2.3. Set

$$\begin{aligned} \mathbb{T}^- &\leftarrow [T' \in \mathbb{T}' \colon \ \mathrm{cls}(T') \leq \mathrm{cls}(D)], \\ \mathbb{T}^+ &\leftarrow [T' \in \mathbb{T}' \colon \ \mathrm{cls}(T') > \mathrm{cls}(D)]. \end{aligned}$$

If $D \notin \boldsymbol{K}$ and $\mathbb{T}^- = \emptyset$ or $\dim(\mathbb{T}^-) > 0$, then set

$$\begin{aligned} \Phi &\leftarrow \Phi \cup \{[\mathbb{T}^- \cup \{D\}, \mathbb{U}', \mathbb{T}^+ \cup [T] \cup \mathbb{T}'']\}, \\ \mathbb{U}' &\leftarrow \mathbb{U}' \cup \{D\}. \end{aligned}$$

D2.2.4. For each $F \in \mathbb{F}$ do:

D2.2.4.1. Set $\mathbb{U}'' \leftarrow \mathrm{prem}(\mathbb{U}', \mathbb{T}' \cup [F])$.

D2.2.4.2. If $0 \notin \mathbb{U}''$, then set $\Psi \leftarrow \Psi \cup \{[\mathbb{T}' \cup [F], \mathbb{T}'', \mathbb{U}'']\}$.

D2.3. Set $\Omega \leftarrow \Psi$.

D3. Set $\Psi \leftarrow \{[\mathbb{T}', \mathbb{U}'] \colon \ [\mathbb{T}', \emptyset, \mathbb{U}'] \in \Psi\}$.

Proof. There is no recursive loop involved in this algorithm, so the termination is trivial. The correctness of the algorithm follows from Lemma 4.2.2 and Remark 4.2.3. □

By the way, the integer k in the factorization step D2.2.2 is known to be 0 or ι because $\mathbb{T}'$ is irreducible of length $\iota - 1$.

Example 4.2.2. Consider the triangular system $[\mathbb{T}_3', \mathbb{U}_3']$ produced in Example 4.1.1. One may verify that the second polynomial in $\mathbb{T}_3'$ factors as

$$-z^5 + t \doteq (z + t + 1)T_2 \tag{4.2.4}$$

over the algebraic-extension field obtained from $\mathbf{Q}$ with T_1 as adjoining polynomial, where

$$T_2 = -z^4 + tz^3 + z^3 - tz^2 - z + t + 1$$

and $T_1 = t^2 + t + 1$ as in Example 4.1.1. By replacing the polynomial $-z^5 + t$ with its two factors respectively, one obtains two triangular systems $[\mathbb{T}_3^*, \mathbb{U}_3^*]$ and $[\mathbb{T}_3^{**}, \mathbb{U}_3^{**}]$ with

$$\begin{aligned} &\mathbb{T}_3^* = [T_1, z + t + 1, T_3, T_4], \quad \mathbb{T}_3^{**} = [T_1, T_2, T_3, T_4], \\ &\mathbb{U}_3^* = \{t + 1\}, \quad \mathbb{U}_3^{**} = \{z\}, \end{aligned}$$

where

$$T_3 = -z^3y - t^3, \quad T_4 = zx^2 - t.$$

Since T_3 is linear in y (and thus irreducible), we need only to test whether T_4 is irreducible over the successive algebraic-extension fields $\mathbf{Q}(t, z)$ obtained from $\mathbf{Q}$ with $[T_1, z + t + 1]$ and with $[T_1, T_2]$ as adjoining triangular sets, respectively. Using algebraic factorization, one may determine that it is reducible and can be factorized as

$$T_4 \doteq -(t+1)(x+t)(x-t), \tag{4.2.5}$$

$$T_4 \doteq \frac{z}{D} T_4' T_4'', \tag{4.2.6}$$

respectively, where

$$\begin{aligned} D &= 4tz^3 + 2z^3 + tz^2 + 2z^2 + tz - 2z + 3t, \\ T_4' &= z^3x + z^2x + tx + x + tz^3 + z^3 + z^2 - z + 2t + 1, \\ T_4'' &= tz^3x + 2z^3x - tz^2x + tzx + tx + x - tz^3 - z^3 - tz - t \end{aligned}$$

and the factors $t + 1$, z and the denominator are viewed as elements of $\mathbf{Q}(t, z)$. Replacing T_4 in $\mathbb{T}_3^*$ and $\mathbb{T}_3^{**}$ respectively by the two factors whose leading variables are x, we obtain four irreducible triangular systems $[\mathbb{T}_{3i}, \mathbb{U}_{3i}]$ with

$$\begin{aligned} &\mathbb{T}_{31} = [T_1, z+t+1, T_3, x+t], && \mathbb{T}_{32} = [T_1, z+t+1, T_3, x-t], \\ &\mathbb{T}_{33} = [T_1, T_2, T_3, T_4'], && \mathbb{T}_{34} = [T_1, T_2, T_3, T_4''], \\ &\mathbb{U}_{31} = \mathbb{U}_{32} = \{t+1\}, && \mathbb{U}_{33} = \mathbb{U}_{34} = \{z\}. \end{aligned}$$

Thus, $[\mathbb{T}_3', \mathbb{U}_3']$ is decomposed into a set Ψ of four irreducible triangular systems $[\mathbb{T}_{31}, \mathbb{U}_{31}], \ldots, [\mathbb{T}_{34}, \mathbb{U}_{34}]$.

The polynomial corresponding to D in (4.1.5) is equal to 1 for (4.2.4) and (4.2.5). For the factorization (4.2.6), since the irreducible triangular set $[T_1, T_2]$ corresponding to $\mathbb{T}^-$ is of dimension 0, by Proposition 4.3.10 the adjunction of D into the triangular set need not be considered. Therefore, $\Phi = \emptyset$.

Algorithm IrrTriSer: $\Psi \leftarrow \text{IrrTriSer}(\mathbb{P}, \mathbb{Q})$. Given a polynomial system $[\mathbb{P}, \mathbb{Q}]$ in $\boldsymbol{K}[\boldsymbol{x}]$, this algorithm computes an irreducible triangular series Ψ of $[\mathbb{P}, \mathbb{Q}]$.

I1. Set $\Psi \leftarrow \emptyset$, $\Phi \leftarrow \{[\mathbb{P}, \mathbb{Q}, \emptyset, 0]\}$.

I2. While $\Phi \neq \emptyset$, do:

I2.1. Let $[\mathbb{F}, \mathbb{G}, \mathbb{T}, m]$ be an element of Φ and set $\Phi \leftarrow \Phi \setminus \{[\mathbb{F}, \mathbb{G}, \mathbb{T}, m]\}$.

I2.2. Compute $\Psi' \leftarrow \text{QuaIrrTriSer}(\mathbb{F}, \mathbb{G}, \mathbb{T})$.

I2.3. For each $[\mathbb{T}, \mathbb{U}] \in \Psi'$ do:

If $|\mathbb{T}| > m$, then compute $[\bar{\Psi}, \bar{\Phi}] \leftarrow \text{Decom}(\mathbb{T}, \mathbb{U})$ and set

$$\Psi \leftarrow \Psi \cup \bar{\Psi}, \quad \Phi \leftarrow \Phi \cup \{[\bar{\mathbb{P}}, \bar{\mathbb{Q}}, \bar{\mathbb{T}}, |\mathbb{T}|] : [\bar{\mathbb{P}}, \bar{\mathbb{Q}}, \bar{\mathbb{T}}] \in \bar{\Phi}\}.$$

Proof. To see the termination of the while-loop I2, consider any $[\mathbb{F}, \mathbb{G}, \mathbb{T}, m]$ taken from Φ in step I2.1 and $[\bar{\mathbb{P}}, \bar{\mathbb{Q}}, \bar{\mathbb{T}}, \bar{m}]$ added to Φ in step I2.3. Then we have $\bar{m} > m$. Since $\bar{m}$ is the number of polynomials in a triangular set and thus cannot be greater than n, the while-loop must terminate.

Now we show that, for each $[\mathbb{T}, \mathbb{U}] \in \Psi'$ as in step I2.3, if $|\mathbb{T}| \leq m$, then $\text{Zero}(\mathbb{T}/\mathbb{U}) = \emptyset$. When this is done, the correctness of IrrTriSer follows from the zero relations (4.1.2) and (4.2.3).

Let $[\mathbb{T}, \mathbb{U}] \in \Psi$ as in step I2.3. Then for any triplet $[\bar{\mathbb{P}}, \bar{\mathbb{Q}}, \bar{\mathbb{T}}]$ generated in Decom from $[\mathbb{T}, \mathbb{U}]$, $\bar{\mathbb{P}}$ is enlarged from an irreducible triangular set $\mathbb{T}^-$ by adjoining a single polynomial D. Moreover, $[\mathbb{T}^-, \bar{\mathbb{Q}}]$ is a triangular system. From the formation of the triplet in D2.2.3 of Decom one sees that

$$\text{cls}(D) \begin{cases} < \text{cls}(T), & \forall T \in \bar{\mathbb{T}}, \\ \geq \text{cls}(T), & \forall T \in \mathbb{T}^-, \end{cases}$$

$|\mathbb{T}^-| + |\bar{\mathbb{T}}| = |\mathbb{T}|$ and D is reduced with respect to $\mathbb{T}^-$. Let the quasi-irreducible triangular systems computed by QualrrTriSer from $[\bar{\mathbb{P}}, \bar{\mathbb{Q}}, \bar{\mathbb{T}}]$ be $[\mathbb{T}_1^*, \mathbb{U}_1^*], \ldots, [\mathbb{T}_h^*, \mathbb{U}_h^*]$. Then each $\mathbb{T}_i^*$ can be written as $\mathbb{T}_i' \cup \bar{\mathbb{T}}$ such that

$$\text{Zero}(\bar{\mathbb{P}}/\bar{\mathbb{Q}}) = \bigcup_{i=1}^{h} \text{Zero}(\mathbb{T}_i'/\mathbb{U}_i^*).$$

According to Theorem 6.1.11, if $|\mathbb{T}_i^*| \leq |\mathbb{T}|$, then $[\mathbb{T}_i^*, \mathbb{U}_i^*]$ is not perfect, i.e., $\text{Zero}(\mathbb{T}_i^*/\mathbb{U}_i^*) = \emptyset$, for each i. This proves what we wanted and thus the correctness of the algorithm. □

Excluding the case $|\mathbb{T}| \leq m$ in step I2.3 is crucial for the termination of IrrTriSer. We guess that this case never happens, but we cannot find a proof. If it is indeed so, then the algorithm may be slightly simplified by not considering the fourth element m and the correctness becomes obvious. When the condition in I2.3 is not imposed, the termination of the algorithm may be proved by requiring that in the algebraic factorization of T in D2.2.2 of Decom the polynomial D does not involve any dependent of $\mathbb{T}'$. The requirement can be satisfied if some additional computation is performed for algebraic factorization.

Example 4.2.3. Let us look at the triangular systems in Examples 2.3.2 and 4.1.1. Trivially, $[\mathbb{T}_2, \mathbb{U}_2]$ is irreducible. Algebraic factorization shows that $[\mathbb{T}_1, \mathbb{U}_1]$ is also irreducible. As we have seen in Example 4.2.2, $[\mathbb{T}_3', \mathbb{U}_3']$ can be decomposed into four irreducible triangular systems. It is easy to see that $[\mathbb{T}_3'', \mathbb{U}_3'']$ is reducible, because substitution of $t = 1$ into the second polynomial of $\mathbb{T}_3''$ yields $z^5 - 1$, which is reducible. In fact, this triangular system can also be decomposed by algorithm Decom into four irreducible triangular systems $[\mathbb{T}_{35}, \mathbb{U}_{35}], \ldots, [\mathbb{T}_{38}, \mathbb{U}_{38}]$ with

$$\begin{aligned}
\mathbb{T}_{35} &= [t-1, z-1, y+1, x-1], \\
\mathbb{T}_{36} &= [t-1, z-1, y+1, x+1], \\
\mathbb{T}_{37} &= [t-1, z^4+z^3+z^2+z+1, z^3y+1, x-z^2], \\
\mathbb{T}_{38} &= [t-1, z^4+z^3+z^2+z+1, z^3y+1, x+z^2], \\
\mathbb{U}_{35} &= \mathbb{U}_{36} = \emptyset, \\
\mathbb{U}_{37} &= \mathbb{U}_{38} = \{z\}.
\end{aligned}$$

We omit the details for this decomposition.

In summary, the original polynomial set $\mathbb{P}$ is decomposed into a sequence of ten irreducible triangular systems $[\mathbb{T}_1, \mathbb{U}_1]$, $[\mathbb{T}_2, \mathbb{U}_2]$, $[\mathbb{T}_{31}, \mathbb{U}_{31}]$, ..., $[\mathbb{T}_{38}, \mathbb{U}_{38}]$ such that

$$\mathrm{Zero}(\mathbb{P}) = \mathrm{Zero}(\mathbb{T}_1/\mathbb{U}_1) \cup \mathrm{Zero}(\mathbb{T}_2/\mathbb{U}_2) \cup \bigcup_{j=1}^{8} \mathrm{Zero}(\mathbb{T}_{3j}/\mathbb{U}_{3j}).$$

By Theorem 4.3.11 b, each $\mathbb{U}_i$ in the above decomposition may be substituted by $\mathrm{ini}(\mathbb{T}_i)$. As $|\mathbb{T}_2| = |\mathbb{T}_{3j}| = 4$ (the number of variables) for $1 \le j \le 8$, we have

$$\mathrm{Zero}(\mathbb{T}_i/\mathrm{ini}(\mathbb{T}_i)) = \mathrm{Zero}(\mathbb{T}_i), \quad i = 2, 31, \ldots, 38,$$

according to Proposition 4.3.10. Therefore,

$$\mathrm{Zero}(\mathbb{P}) = \mathrm{Zero}(\mathbb{T}_1/\mathrm{ini}(\mathbb{T}_1)) \cup \mathrm{Zero}(\mathbb{T}_2) \cup \bigcup_{j=1}^{8} \mathrm{Zero}(\mathbb{T}_{3j}). \tag{4.2.7}$$

Example 4.2.4. As further illustration, let us take a more complicated polynomial system $\mathfrak{P} = [\{P_1, P_2, P_3\}, \{x_3\}]$, where

$$\begin{aligned}
P_1 &= x_3(x_5^2 - x_4^2 + 2x_1x_4 - x_1^2) + 2x_1(x_1 - x_4)x_5,\\
P_2 &= x_3(x_5^2 - x_4^2 + 2x_2x_4 - x_2^2) + 2x_2(x_2 - x_4)x_5,\\
P_3 &= x_3[(x_1 - x_6)(x_2x_6 + x_3^2) + (x_2 - x_6)(x_1x_6 + x_3^2)].
\end{aligned}$$

With respect to the variable ordering $x_1 \prec \cdots \prec x_6$, $\mathfrak{P}$ may be decomposed into seven (reduced) irreducible triangular sets $\mathbb{T}_i$ such that

$$\mathrm{Zero}(\mathfrak{P}) = \bigcup_{i=1}^{7} \mathrm{Zero}(\mathbb{T}_i/\mathrm{ini}(\mathbb{T}_i) \cup \{x_3\}), \tag{4.2.8}$$

where

$$\begin{aligned}
\mathbb{T}_1 &= [T_1, T_2, T_3],\\
\mathbb{T}_2 &= [T_1, T_2, T_3'],\\
\mathbb{T}_3 &= [x_2 + x_1, x_3^2 + x_1^2, x_4, x_5 - x_3],\\
\mathbb{T}_4 &= [x_2 + x_1, x_4^2 - x_3^2 - x_1^2, x_5 - x_3, x_6],\\
\mathbb{T}_5 &= [x_2 + x_1, x_4, x_3x_5^2 + 2x_1^2x_5 - x_1^2x_3, x_6],\\
\mathbb{T}_6 &= [x_2 - x_1, T_2', x_6 - x_1],\\
\mathbb{T}_7 &= [x_2 - x_1, T_2', x_1x_6 + x_3^2];\\
T_1 &= 4x_4^4 - 8(x_2 + x_1)x_4^3 - 4(x_3^2 - x_2^2 - 3x_1x_2 - x_1^2)x_4^2\\
&\quad + 4(x_2x_3^2 + x_1x_3^2 - x_1x_2^2 - x_1^2x_2)x_4 - (x_2^2 + 2x_1x_2 + x_1^2)x_3^2,\\
T_2 &= 2(x_4 - x_2 - x_1)x_5 - 2x_3x_4 + (x_2 + x_1)x_3,\\
T_2' &= x_3x_5^2 - 2x_1(x_4 - x_1)x_5 - x_3x_4^2 + 2x_1x_3x_4 - x_1^2x_3,\\
T_3 &= (x_2 + x_1)x_6 + 2x_4^2 - 2(x_2 + x_1)x_4,\\
T_3' &= (x_2 + x_1)x_6 - 2x_4^2 + 2(x_2 + x_1)x_4 + 2x_3^2 - 2x_1x_2.
\end{aligned}$$

4.3 Properties of irreducible triangular systems

In what follows, we write $\boldsymbol{z}^{\{i\}}$ for $(\boldsymbol{u}, y_1, \dots, y_i)$ and $\boldsymbol{\xi}^{\{i\}}$ for $(\boldsymbol{u}, \eta_1, \dots, \eta_i)$ with $\boldsymbol{z} = \boldsymbol{z}^{\{r\}}$ and $\boldsymbol{\xi} = \boldsymbol{\xi}^{\{r\}}$. Obviously, $\boldsymbol{z}$ is a permutation of $\boldsymbol{x}$. The following lemma is taken from Wu (1994, pp. 174–175).

Lemma 4.3.1. Let $\mathbb{T}$ be an irreducible triangular set in $\boldsymbol{K}[\boldsymbol{z}]$ with a generic zero $\boldsymbol{\xi}$. Then, for any polynomial $P \in \boldsymbol{K}[\boldsymbol{z}]$,

$$\operatorname{prem}(P, \mathbb{T}) = 0 \iff P(\boldsymbol{\xi}) = 0.$$

Proof. Let $\mathbb{T} = [T_1, \dots, T_r]$ as in (4.1.3) with

$$I_i = \operatorname{ini}(T_i),\ d_i = \operatorname{ldeg}(T_i), \quad 1 \le i \le r,$$

and $\boldsymbol{\xi}$ be of the form

$$\boldsymbol{\xi} = (\boldsymbol{u}, \eta_1, \dots, \eta_r).$$

As before, $\boldsymbol{K}_k = \boldsymbol{K}(\boldsymbol{\xi}^{\{k\}})$. We first prove the following assertion.

If $R \in \boldsymbol{K}[\boldsymbol{z}]$ is reduced with respect to $\mathbb{T}$ and $R(\boldsymbol{\xi}) = 0$, then $R \equiv 0$.

Note that η_r is an extended zero of the polynomials

$$\bar{R} = R(\boldsymbol{\xi}^{\{r-1\}}, y_r), \quad \bar{T}_r = T_r(\boldsymbol{\xi}^{\{r-1\}}, y_r) \in \boldsymbol{K}_{r-1}[y_r].$$

As $\bar{T}_r$ is irreducible over $\boldsymbol{K}_{r-1}$ and $\deg(R, y_r) < d_r$, $\bar{R} \equiv 0$. Hence, all the coefficients of $\bar{R}$ as a polynomial in y_r are identically equal to 0, viz.,

$$R_i(\boldsymbol{\xi}^{\{r-1\}}) = \operatorname{coef}(\bar{R}, y_r^i) \equiv 0,\ 0 \le i < d_r.$$

Similarly, η_{r-1} is an extended zero of the polynomials

$$\bar{R}_i = R_i(\boldsymbol{\xi}^{\{r-2\}}, y_{r-1}), \quad \bar{T}_{r-1} = T_{r-1}(\boldsymbol{\xi}^{\{r-2\}}, y_{r-1}) \in \boldsymbol{K}_{r-2}[y_{r-1}].$$

Since R is reduced with respect to $\mathbb{T}$, so is each R_i. Therefore, $\deg(R_i, y_{r-1}) < d_{r-1}$. This and the irreducibility of $\bar{T}_{r-1}$ over $\boldsymbol{K}_{r-2}$ imply that $\bar{R}_i \equiv 0$ for every i. It follows that the coefficients of $\bar{R}_i$ in y_{r-1} are all identically 0, and so are the coefficients of R_i in y_{r-1} when $\boldsymbol{z}^{\{r-2\}}$ is substituted by $\boldsymbol{\xi}^{\{r-2\}}$.

The above argument may be continued for $T_{r-2}, \dots, T_1$. In this way, we shall see that all the coefficients of R as a polynomial in $\boldsymbol{K}_0[y_1, \dots, y_r]$ must be identically 0. Therefore, $R \equiv 0$ and the assertion is proved.

To complete the proof of Lemma 4.3.1, let $R = \operatorname{prem}(P, \mathbb{T})$. Then there are integers $s_i \ge 0$ and polynomials Q_i such that

$$I_1^{s_1} \cdots I_r^{s_r} P = \sum_{i=1}^{r} Q_i T_i + R. \tag{4.3.1}$$

As $T_i(\boldsymbol{\xi}) = 0$, plunging $\boldsymbol{\xi}$ into formula (4.3.1) yields

$$I_1(\boldsymbol{\xi})^{s_1} \cdots I_r(\boldsymbol{\xi})^{s_r} P(\boldsymbol{\xi}) = R(\boldsymbol{\xi}).$$

Since each I_i is a nonzero polynomial reduced with respect to $\mathbb{T}$, $I_i(\boldsymbol{\xi}) \neq 0$ by the above assertion. Hence,

$$P(\boldsymbol{\xi}) = 0 \iff R(\boldsymbol{\xi}) = 0 \iff R = 0.$$

The second " $\iff$ " is ensured by the above assertion because R is reduced with respect to $\mathbb{T}$. The proof is complete. □

Definition 4.3.1. Let P be any polynomial and $\mathbb{T} = [T_1, \ldots, T_r]$ a triangular set in $\boldsymbol{K}[\boldsymbol{x}]$. The polynomial

$$\operatorname{res}(P, \mathbb{T}) \triangleq \operatorname{res}(\ldots \operatorname{res}(P, T_r, \operatorname{lv}(T_r)), \ldots, T_1, \operatorname{lv}(T_1))$$

is called the *resultant* of P with respect to $\mathbb{T}$.

Clearly, $R = \operatorname{res}(P, \mathbb{T})$ does not involve $\operatorname{lv}(T_i)$ for any i. When the variables $\boldsymbol{x}$ are renamed $\boldsymbol{u}$ and $\boldsymbol{y}$ with $y_i = \operatorname{lv}(T_i)$ as before, we have $R \in \boldsymbol{K}[\boldsymbol{u}]$.

Lemma 4.3.2. Let $\mathbb{T} = [T_1, \ldots, T_r]$ be a triangular set and P a polynomial in $\boldsymbol{K}[\boldsymbol{z}]$, and $R = \operatorname{res}(P, \mathbb{T})$. Then in $\boldsymbol{K}[\boldsymbol{z}]$ one can determine polynomials Q and $Q_1, \ldots, Q_r$ such that

$$QP = Q_1 T_1 + \cdots + Q_r T_r + R. \tag{4.3.2}$$

If $\mathbb{T}$ is irreducible with a generic zero $\boldsymbol{\xi} = (\boldsymbol{u}, \eta_1, \ldots, \eta_r)$ and $\operatorname{prem}(P, \mathbb{T}) \neq 0$, then $R(\boldsymbol{u}) \neq 0$ and $Q(\boldsymbol{\xi}) \neq 0$.

Proof. The first half of the lemma is a direct consequence of Lemma 1.3.1.

To prove the second half, let

$$R_r = \operatorname{res}(P, T_r, y_r), \quad R_i = \operatorname{res}(R_{i+1}, T_i, y_i), \; i = r-1, \ldots, 1,$$

where $y_i = \operatorname{lv}(T_i)$ for each i and $R_1 = R$. Since $\mathbb{T}$ is irreducible and $\operatorname{prem}(P, \mathbb{T}) \neq 0$, $P(\boldsymbol{\xi}) \neq 0$ by Lemma 4.3.1. On the other hand,

$$\bar{T}_r = T_r(\boldsymbol{\xi}^{\{r-1\}}, y_r)$$

is irreducible over $\boldsymbol{K}(\boldsymbol{\xi}^{\{r-1\}})$ and $T_r(\boldsymbol{\xi}) = \bar{T}_r(\eta_r) = 0$. Thus, the two polynomials $P(\boldsymbol{\xi}^{\{r-1\}}, y_r)$ and $\bar{T}_r$ cannot have a common zero for y_r in any extension field of $\boldsymbol{K}(\boldsymbol{\xi}^{\{r-1\}})$. Therefore,

$$R_r(\boldsymbol{\xi}^{\{r-1\}}) \neq 0.$$

As $T_{r-1}(\boldsymbol{\xi}^{\{r-2\}}, y_{r-1})$ is irreducible over $\boldsymbol{K}(\boldsymbol{\xi}^{\{r-2\}})$ and $T_{r-1}(\boldsymbol{\xi}^{\{r-1\}}) = 0$, we have

$$R_{r-1}(\boldsymbol{\xi}^{\{r-2\}}) \neq 0$$

for the same reason. Continuing this argument, finally we shall have

$$R(\boldsymbol{u}) = R_1(\boldsymbol{u}) \neq 0.$$

Plunging $\boldsymbol{\xi}$ into the polynomials in (4.3.2), one immediately gets $Q(\boldsymbol{\xi}) \neq 0$. The lemma is proved. □

See Wu (1994, pp. 175–177) for another proof of Lemma 4.3.2. The following theorem and its proof are also adapted from Wu (1994, pp. 189 f).

Theorem 4.3.3. Every irreducible triangular system in $\boldsymbol{K}[\boldsymbol{x}]$ is perfect over the algebraic closure $\bar{\boldsymbol{K}}$ of $\boldsymbol{K}$.

Proof. Let $[\mathbb{T}, \mathbb{U}]$ be an irreducible triangular system with $\mathbb{T} = [T_1, \ldots, T_r]$ written in the form (4.1.3), and let

$$I_i = \mathrm{ini}(T_i),\ 1 \le i \le r, \quad \text{and} \quad V = \prod_{U \in \mathbb{U}} U.$$

As $\mathrm{prem}(I_i, \mathbb{T}^{\{i-1\}}) \neq 0$, by Lemma 4.3.2 there exist polynomials $Q_i, Q_{ij} \in \boldsymbol{K}[\boldsymbol{z}^{\{i-1\}}]$ such that

$$R_i = Q_i I_i - \sum_{j=1}^{i-1} Q_{ij} T_j \in \boldsymbol{K}[\boldsymbol{u}]$$

and $R_i \neq 0$ for each i. Since $\mathrm{prem}(U, \mathbb{T}) \neq 0$ for any $U \in \mathbb{U}$, $\mathrm{prem}(V, \mathbb{T}) \neq 0$ according to Lemma 4.3.1. Again, by Lemma 4.3.2 there are polynomials $H, H_i \in \boldsymbol{K}[\boldsymbol{z}]$ such that

$$R = HV - \sum_{i=1}^{r} H_i T_i \in \boldsymbol{K}[\boldsymbol{u}], \tag{4.3.3}$$

and $R \neq 0$. Hence, there exists a point

$$\bar{\boldsymbol{u}} = (\bar{u}_1, \ldots, \bar{u}_d) \in \boldsymbol{K}^d$$

such that

$$R_1(\bar{\boldsymbol{u}}) \cdots R_r(\bar{\boldsymbol{u}}) R(\bar{\boldsymbol{u}}) \neq 0.$$

Such $\bar{\boldsymbol{u}}$ may be chosen as a rational point.

Now we proceed to determine numbers $\bar{y}_i \in \bar{\boldsymbol{K}}$ by induction such that the point

$$\bar{\boldsymbol{z}} = (\bar{\boldsymbol{u}}, \bar{y}_1, \ldots, \bar{y}_r) \in \bar{\boldsymbol{K}}^{d+r}$$

satisfies the relations

$$T_i(\bar{\boldsymbol{z}}^{\{i\}}) = 0, \quad I_{i+1}(\bar{\boldsymbol{z}}^{\{i\}}) \neq 0. \tag{4.3.4}$$

First of all, let

$$\bar{T}_1 = T_1(\bar{\boldsymbol{u}}, y_1) \in \boldsymbol{K}[y_1], \quad \bar{I}_1 = I_1(\bar{\boldsymbol{u}}) \in \boldsymbol{K}.$$

Since

$$Q_1(\bar{\boldsymbol{u}}) I_1(\bar{\boldsymbol{u}}) = R_1(\bar{\boldsymbol{u}}) \neq 0,$$

$\bar{I}_1 \neq 0$ and $\bar{T}_1$ is a polynomial in y_1 of degree ≥ 1. Thus, one can take a number $\bar{y}_1$ from some algebraic-extension field of $\boldsymbol{K}$ such that

$$\bar{T}_1(\bar{y}_1) = 0, \quad \text{or} \quad T_1(\bar{\boldsymbol{z}}^{\{1\}}) = 0.$$

As

$$R_2 = Q_2 I_2 - Q_{21} T_1, \quad R_2(\bar{z}^{\{1\}}) = R_2(\bar{u}) \neq 0,$$

we have $I_2(\bar{z}^{\{1\}}) \neq 0$. So (4.3.4) holds for $i = 1$.

Suppose that we have already found $\bar{y}_1, \dots, \bar{y}_i$ satisfying (4.3.4) and want to find $\bar{y}_{i+1}$.

Let

$$\bar{T}_{i+1} = T_{i+1}(\bar{z}^{\{i\}}, y_{i+1}) \in \boldsymbol{K}'[y_{i+1}],$$

where $\boldsymbol{K}'$ is some algebraic extension of $\boldsymbol{K}$ containing $\bar{y}_1, \dots, \bar{y}_i$. The leading coefficient of $\bar{T}_{i+1}$ as a polynomial in y_{i+1} is

$$I_{i+1}(\bar{z}^{\{i\}}) \neq 0.$$

Hence, one can choose a number $\bar{y}_{i+1}$ in some algebraic extension of $\boldsymbol{K}'$ and thus of $\boldsymbol{K}$ such that $\bar{T}_{i+1}(\bar{y}_{i+1}) = 0$ or $T_{i+1}(\bar{z}^{\{i+1\}}) = 0$. Therefore,

$$R_{i+2} = Q_{i+2} I_{i+2} - \sum_{j=1}^{i+1} Q_{i+2\ j} T_j,$$

$$R_{i+2}(\bar{z}^{\{i+1\}}) = R_{i+2}(\bar{u}) \neq 0,$$

and

$$T_1(\bar{z}^{\{i+1\}}) = T_1(\bar{z}^{\{1\}}) = 0, \dots, T_{i+1}(\bar{z}^{\{i+1\}}) = 0$$

imply immediately that

$$I_{i+2}(\bar{z}^{\{i+1\}}) \neq 0.$$

Finally, plunging the above-constructed $\bar{z}$ into (4.3.3) one sees that $V(\bar{z}) \neq 0$, and thus $\bar{z}$ is a zero of $[\mathbb{T}, \mathbb{U}]$. This completes the proof of the theorem. □

Corollary 4.3.4. Every irreducible triangular set in $\boldsymbol{K}[\boldsymbol{x}]$ is perfect over the algebraic closure $\bar{\boldsymbol{K}}$ of $\boldsymbol{K}$.

Corollary 4.3.5. Any irreducible triangular set and system in $\boldsymbol{K}[\boldsymbol{x}]$ are perfect.

As a matter of fact, Corollary 4.3.5 can be established without using Theorem 4.3.3. For any generic zero of an irreducible triangular set $\mathbb{T}$ is a zero of $[\mathbb{T}, \mathrm{ini}(\mathbb{T})]$ and any fine triangular system $[\mathbb{T}, \mathbb{U}]$ in some extension field of $\boldsymbol{K}$.

Corollary 4.3.6. Let Ψ be an irreducible triangular series of any polynomial system $\mathfrak{P}$ in $\boldsymbol{K}[\boldsymbol{x}]$. Then $\mathrm{Zero}(\mathfrak{P}) = \emptyset \iff \Psi = \emptyset$.

Proposition 4.3.7. Any irreducible triangular set is a simple set in $\boldsymbol{K}[\boldsymbol{x}]$.

Proof. Let $\mathbb{T} = [T_1, \dots, T_r]$ be an irreducible triangular set written in the form (4.1.3) with

$$I_i = \mathrm{ini}(T_i), \quad T_i' = \partial T_i / \partial y_i, \quad 1 \leq i \leq r,$$

and let

$$D = I_1 \cdots I_r T_1' \cdots T_r'.$$

As $\mathrm{prem}(I_i, \mathbb{T}) \neq 0$ and $\mathrm{prem}(T_i', \mathbb{T}) \neq 0$ for each i, $\mathrm{prem}(D, \mathbb{T}) \neq 0$. By Lemma 4.3.2, there are polynomials $Q, Q_i \in \boldsymbol{K}[\boldsymbol{z}]$ such that

$$R = \mathrm{res}(D, \mathbb{T}) = QD - \sum_{i=1}^{r} Q_i T_i \neq 0 \tag{4.3.5}$$

and $R \in \boldsymbol{K}[\boldsymbol{u}]$. Let

$$\tilde{T}_t = \mathrm{sqfr}(R),$$

where $\mathrm{sqfr}(R)$ denotes the product of all the distinct irreducible factors of R over $\boldsymbol{K}$ (i.e., the greatest squarefree divisor of R) and the index t is to be determined as follows. Construct $t - 1$ polynomials

$$\tilde{T}_{i-1} = \mathrm{sqfr}(\mathrm{ini}(\tilde{T}_i)\, \mathrm{res}(\tilde{T}_i, \partial \tilde{T}_i / \partial u_{p_i}, u_{p_i})), \quad i = t, \ldots, 2,$$

such that

$$\tilde{T}_0 = \mathrm{ini}(\tilde{T}_1)\, \mathrm{res}(\tilde{T}_1, \partial \tilde{T}_1 / \partial u_{p_1}, u_{p_1}) \in \boldsymbol{K},$$

where $u_{p_i} = \mathrm{lv}(\tilde{T}_i)$ and $\tilde{T}_i \neq 0$ for each i. Let $\tilde{\mathbb{T}} = [\tilde{T}_1, \ldots, \tilde{T}_t]$. We want to show that $[\mathbb{T}, \tilde{\mathbb{T}}]$ is a simple system. From the construction of $\tilde{T}_i$, it is easy to see that $\mathrm{ini}(\tilde{T}_i)(\bar{\boldsymbol{u}}^{\{p_i-1\}}) \neq 0$ and $\tilde{T}_i(\bar{\boldsymbol{u}}^{\{p_i-1\}}, u_{p_i})$ is squarefree for any $\bar{\boldsymbol{u}}^{\{p_i-1\}} \in \mathrm{Zero}(\emptyset / \tilde{\mathbb{T}}^{\{i-1\}})$.

Now let

$$\bar{\boldsymbol{z}}^{\{i-1\}} = (\bar{\boldsymbol{u}}, \bar{\boldsymbol{y}}^{\{i-1\}}) \in \mathrm{Zero}(\mathbb{T}^{\{i-1\}} / \tilde{\mathbb{T}}).$$

Clearly, $R(\bar{\boldsymbol{z}}^{\{i-1\}}) = R(\bar{\boldsymbol{u}}) \neq 0$. To see the squarefreeness of $T_i(\bar{\boldsymbol{z}}^{\{i-1\}}, y_i)$, let us proceed to derive a contradiction by supposing the opposite: $T_i(\bar{\boldsymbol{z}}^{\{i-1\}}, y_i)$ and $T_i'(\bar{\boldsymbol{z}}^{\{i-1\}}, y_i)$ have a common divisor of degree ≥ 1 in y_i. Then there exists a $\bar{y}_i \in \tilde{\boldsymbol{K}}$ such that

$$T_i(\bar{\boldsymbol{z}}^{\{i\}}) = T_i'(\bar{\boldsymbol{z}}^{\{i\}}) = 0.$$

It follows that

$$D(\bar{\boldsymbol{z}}^{\{i\}}, \bar{y}_{i+1}, \ldots, \bar{y}_r) = 0$$

for any $\bar{y}_{i+1}, \ldots, \bar{y}_r \in \tilde{\boldsymbol{K}}$. Clearly, this is also true if $I_i(\bar{\boldsymbol{z}}^{\{i-1\}}) = 0$.

On the other hand, since $\mathbb{T}$ is irreducible, by Corollary 4.3.5 there exist $\bar{y}_{i+1}, \ldots, \bar{y}_r \in \tilde{\boldsymbol{K}}$ such that

$$I_j(\bar{\boldsymbol{z}}) \neq 0, \;\; T_j(\bar{\boldsymbol{z}}) = 0, \quad j > i.$$

Plunging $\bar{\boldsymbol{z}}$ into (4.3.5), one sees that $D(\bar{\boldsymbol{z}}) \neq 0$. This leads to a contradiction. Hence,

$$I_i(\bar{\boldsymbol{z}}^{\{i-1\}}) \neq 0 \text{ and } T_i(\bar{\boldsymbol{z}}^{\{i-1\}}, y_i) \text{ is squarefree.}$$

Thus $[\mathbb{T}, \tilde{\mathbb{T}}]$ is a simple system, and the proposition is proved. □

Another simpler proof of this proposition is provided by Lemma 4.4.1.

Roughly speaking, a simple set is a triangular set $\mathbb{T}$ in which each polynomial of class p is squarefree with respect to x_p over every extension field obtained from $\boldsymbol{K}$ with an irreducible component of $\mathbb{T}^{\{p-1\}}$ as adjoining triangular set. Note that an irreducible triangular system is not necessarily a simple system. This can be seen from the triangular system $[\mathbb{T}_1, \{T\}]$ in Example 3.3.2: it is not a simple system, though $\mathbb{T}_1$ is irreducible.

As a consequence of Corollary 3.4.5 and Proposition 4.3.7, we have the following corollary.

Corollary 4.3.8. For any irreducible triangular set $\mathbb{T}$ and polynomial P in $\boldsymbol{K}[\boldsymbol{x}]$, $\operatorname{Zero}(\mathbb{T}/\operatorname{ini}(\mathbb{T})) \subset \operatorname{Zero}(P) \iff \operatorname{prem}(P, \mathbb{T}) = 0$.

The following corollary corresponds to Theorem 3.4.4.

Corollary 4.3.9. For any irreducible triangular system $[\mathbb{T}, \mathbb{U}]$ and polynomial P in $\boldsymbol{K}[\boldsymbol{x}]$, $\operatorname{Zero}(\mathbb{T}/\mathbb{U}) \subset \operatorname{Zero}(P) \iff \operatorname{prem}(P, \mathbb{T}) = 0$.

Proof. As $\operatorname{Zero}(\mathbb{T}/\mathbb{U}) \subset \operatorname{Zero}(\mathbb{T}/\operatorname{ini}(\mathbb{T}))$, the direction "$\Longleftarrow$" follows from Corollary 4.3.8.

For the other direction, let $\boldsymbol{\xi}$ be a generic zero of $\mathbb{T}$. For any $U \in \mathbb{U}$, as $\operatorname{prem}(U, \mathbb{T}) \neq 0$, by Lemma 4.3.1 $U(\boldsymbol{\xi}) \neq 0$. This implies that

$$\boldsymbol{\xi} \in \operatorname{Zero}(\mathbb{T}/\mathbb{U}) \subset \operatorname{Zero}(P)$$

and thus $P(\boldsymbol{\xi}) = 0$. Applying Lemma 4.3.1 again, we have $\operatorname{prem}(P, \mathbb{T}) = 0$. . □

Proposition 4.3.10. Let $\mathbb{T}$ be an irreducible triangular set and P a polynomial in $\boldsymbol{K}[\boldsymbol{x}]$ with $\operatorname{prem}(P, \mathbb{T}) \neq 0$. If $\dim(\mathbb{T}) = 0$, then

$$\operatorname{Zero}(\{P\} \cup \mathbb{T}) = \emptyset, \quad \operatorname{Zero}(\mathbb{T}/\mathbb{I}) = \operatorname{Zero}(\mathbb{T}),$$

where $\mathbb{I} = \operatorname{ini}(\mathbb{T})$.

Proof. The first equality follows from Lemma 4.3.2, and the second is obvious by noting that $\operatorname{Zero}(\mathbb{T}) = \operatorname{Zero}(\mathbb{T}/\mathbb{I}) \cup \bigcup_{I \in \mathbb{I}} \operatorname{Zero}(\{I\} \cup \mathbb{T})$. □

In zero decompositions of the form (2.2.8) computed with characteristic sets, $\operatorname{Zero}(\mathbb{C}_i/\operatorname{ini}(\mathbb{C}_i) \cup \mathbb{Q})$ is placed instead of $\operatorname{Zero}(\mathbb{T}_i/\mathbb{U}_i)$ in the zero decomposition associated to a triangular series, where each $\mathbb{C}_i$ is an ascending set having the properties that $\operatorname{prem}(\mathbb{P}, \mathbb{C}_i) = \{0\}$ and $0 \notin \operatorname{prem}(\mathbb{Q}, \mathbb{C}_i)$. In general there is no guarantee that $\operatorname{prem}(\mathbb{P}, \mathbb{T}_i) = \{0\}$, however. And each $\mathbb{U}_i$ may contain many more polynomials than $\operatorname{ini}(\mathbb{C}_i) \cup \mathbb{Q}$ does. It is remarkable that the property $\operatorname{prem}(\mathbb{P}, \mathbb{T}_i) = \{0\}$ is recovered when the triangular series is irreducible or simple.

Parallel to Theorem 3.4.6 for simple series, let us state the properties for irreducible triangular series as the following theorem. Here property a is easily proved by applying Corollary 4.3.9, while the proof of b is an analogy to that of Theorem 3.4.6 b.

Theorem 4.3.11. Let Ψ be an irreducible triangular series of any polynomial system $[\mathbb{P}, \mathbb{Q}]$ in $\boldsymbol{K}[\boldsymbol{x}]$. Then

a. $\operatorname{prem}(\mathbb{P}, \mathbb{T}) = \{0\}$ and $0 \notin \operatorname{prem}(\mathbb{Q}, \mathbb{T})$ for any $[\mathbb{T}, \mathbb{U}] \in \Psi$;

b. $$\mathrm{Zero}(\mathbb{P}/\mathbb{Q}) = \bigcup_{[\mathbb{T},\mathbb{U}]\in\Psi} \mathrm{Zero}(\mathbb{T}/\mathrm{ini}(\mathbb{T}) \cup \mathbb{Q}). \tag{4.3.6}$$

If $\dim(\mathbb{T}) = 0$, then $\mathrm{Zero}(\mathbb{T}/\mathrm{ini}(\mathbb{T})\cup\mathbb{Q})$ in (4.3.6) can be simplified to $\mathrm{Zero}(\mathbb{T}/\mathbb{Q})$.

Proof. a. Let $[\mathbb{T}, \mathbb{U}] \in \Psi$; then $\mathrm{Zero}(\mathbb{T}/\mathbb{U}) \subset \mathrm{Zero}(\mathbb{P}/\mathbb{Q})$. Hence, for all $P \in \mathbb{P}$ and $Q \in \mathbb{Q}$: $\mathrm{Zero}(\mathbb{T}/\mathbb{U}) \subset \mathrm{Zero}(P)$ and $\mathrm{Zero}(\mathbb{T}/\mathbb{U}) \not\subset \mathrm{Zero}(Q)$; and it thus follows from Corollary 4.3.9 that $\mathrm{prem}(P, \mathbb{T}) = 0$ and $\mathrm{prem}(Q, \mathbb{T}) \neq 0$.

b. By part a and the pseudo-remainder formula, any $\boldsymbol{x}$ belonging to the right-hand side of (4.3.6) is contained in the left-hand side. On the contrary, let $\bar{\boldsymbol{x}} \in \mathrm{Zero}(\mathbb{P}/\mathbb{Q})$. By definition there is a $[\mathbb{T}, \mathbb{U}] \in \Psi$ such that $\bar{\boldsymbol{x}} \in \mathrm{Zero}(\mathbb{T}/\mathbb{U})$. Since $[\mathbb{T}, \mathbb{U}]$ is a triangular system, $I(\bar{\boldsymbol{x}}) \neq 0$ for any $I \in \mathrm{ini}(\mathbb{T})$. Hence $\bar{\boldsymbol{x}} \in \mathrm{Zero}(\mathbb{T}/\mathrm{ini}(\mathbb{T}) \cup \mathbb{Q})$, i.e., $\bar{\boldsymbol{x}}$ belongs to the right-hand side of (4.3.6). If $\dim(\mathbb{T}) = 0$, by Proposition 4.3.10 $\mathrm{Zero}(\mathbb{T}/\mathrm{ini}(\mathbb{T}) \cup \mathbb{Q})$ may be simplified to $\mathrm{Zero}(\mathbb{T}/\mathbb{Q})$. □

Property a in Theorem 4.3.11 is satisfied by each irreducible triangular system $[\mathbb{T}, \mathbb{U}] \in \Psi$, no matter whether or not the other triangular systems in Ψ are irreducible. It can be used to avoid some verifications of the 0 pseudo-remainder in decomposition algorithms based on characteristic sets.

Corollary 4.3.12. Any irreducible triangular series of a polynomial system $\mathfrak{P}$ in $\boldsymbol{K}[\boldsymbol{x}]$ is an irreducible W-characteristic series of $\mathfrak{P}$.

Some of the results stated in this section are consequences of the properties about simple systems shown in Sect. 3.4. Most of the other results newly proved for irreducible triangular sets or systems also hold or can be generalized for simple sets or systems when the corresponding notions are appropriately substituted. These include the properties in Lemmas 4.3.1 and 4.3.2, Theorem 4.3.3, and Proposition 4.3.10. A generalization of Theorem 4.3.3 will be given as Theorem 5.1.12. The generalization of other results will be discussed somewhere else.

4.4 Irreducible simple systems

A simple system is said to be *irreducible* or *prime* if it is irreducible as a triangular system. We want to decompose any polynomial system $\mathfrak{P}$ into irreducible simple systems. This may be achieved by first decomposing $\mathfrak{P}$ into irreducible triangular systems $\mathfrak{T}_i$ and then computing simple systems from each $\mathfrak{T}_i$.

To explain the process in detail, consider an irreducible triangular system $[\mathbb{T}, \mathbb{U}]$ and let

$$\mathbb{U}' = \{\partial T/\partial \mathrm{lv}(T):\ T \in \mathbb{T}\}$$

and

$$\mathbb{R} = \{\mathrm{sqfr}(\mathrm{res}(U, \mathbb{T})):\ U \in \mathbb{U} \cup \mathbb{U}'\}.$$

Since $\mathbb{T}$ is irreducible and $\mathrm{prem}(U, \mathbb{T}) \neq 0$ for every $U \in \mathbb{U} \cup \mathbb{U}'$, any polynomial $R \in \mathbb{R}$ is nonzero and does not involve the dependents of $\mathbb{T}$ and

$$\mathrm{Zero}(\mathbb{T}/\mathbb{U}) = \mathrm{Zero}(\mathbb{T}/\mathbb{R}) \cup \bigcup_{R\in\mathbb{R}} \mathrm{Zero}(\mathbb{T} \cup \{R\}/\mathbb{U}).$$

Compute a simple series $[\mathbb{T}_1, \tilde{\mathbb{T}}_1], \ldots, [\mathbb{T}_q, \tilde{\mathbb{T}}_q]$ of $[\emptyset, \mathbb{R}]$. There must be some $\mathbb{T}_i$ which is empty. Suppose, otherwise, that all $\mathbb{T}_i$ are nonempty, and let y be a new variable. Then

$$\bigcup_{i=1}^{q} \mathrm{Zero}(\mathbb{T}_i \cup [y]/\tilde{\mathbb{T}}_i) = \mathrm{Zero}([y]/\mathbb{R}),$$

$\max_{1\le i\le q} \dim(\mathbb{T}_i \cup [y]) \le n-1$, and $\dim([y]) = n$ in $\boldsymbol{K}[\boldsymbol{x}, y]$. This leads to a contradiction with Corollary 6.1.6. So we may assume that $\mathbb{T}_1, \ldots, \mathbb{T}_l$ $(1 \le l \le q)$ are all those $\mathbb{T}_i$ which are empty. Then,

$$\mathrm{Zero}(\mathbb{T}/\mathbb{U}) = \bigcup_{i=1}^{l} \mathrm{Zero}(\mathbb{T}/\tilde{\mathbb{T}}_i) \cup \bigcup_{i=l+1}^{q} \mathrm{Zero}(\mathbb{T} \cup \mathbb{T}_i/\tilde{\mathbb{T}}_i) \cup \bigcup_{R\in\mathbb{R}} \mathrm{Zero}(\mathbb{T} \cup \{R\}/\mathbb{U}).$$

Note the fact that $\mathbb{T} \cup \mathbb{T}_i$ for $i > l$ and $\mathbb{T} \cup \{R\}$ for $R \in \mathbb{R}$ are all enlarged from $\mathbb{T}$ by adjoining at least one polynomial which does not involve any dependent of $\mathbb{T}$.

We want to show that $[\mathbb{T}, \tilde{\mathbb{T}}_i]$ is an irreducible simple system for $1 \le i \le l$. For this purpose, consider a fixed i (≥ 1 and $\le l$) and a polynomial $T \in \mathbb{T}$ of class p. Let

$$\bar{\boldsymbol{x}}^{\{p-1\}} \in \mathrm{Zero}(\mathbb{T}^{(p-1)}/\tilde{\mathbb{T}}_i^{(p-1)});$$

then $R(\bar{\boldsymbol{x}}^{\{p-1\}}, x_p, \ldots, x_n) \ne 0$ for all $R \in \mathbb{R}$. It follows from the construction of $\mathbb{R}$ that $\mathrm{ini}(T)(\bar{\boldsymbol{x}}^{\{p-1\}}) \ne 0$ and

$$T(\bar{\boldsymbol{x}}^{\{p-1\}}, x_p), \quad \frac{\partial T}{\partial x_p}(\bar{\boldsymbol{x}}^{\{p-1\}}, x_p)$$

do not have any common divisor of degree ≥ 1 in x_p. Therefore, $T(\bar{\boldsymbol{x}}^{\{p-1\}}, x_p)$ is squarefree. Note that $[\emptyset, \tilde{\mathbb{T}}_i]$ is simple and any polynomial in $\tilde{\mathbb{T}}_i$ does not involve the dependents of $\mathbb{T}$. Hence $[\mathbb{T}, \tilde{\mathbb{T}}_i]$ is simple.

What has been explained above may be summarized as the following lemma. One of its consequences is Proposition 4.3.7.

Lemma 4.4.1. From any irreducible triangular system $[\mathbb{T}, \mathbb{U}]$ in $\boldsymbol{K}[\boldsymbol{x}]$, one can compute a finite number of triangular or empty sets $\tilde{\mathbb{T}}_1, \ldots, \tilde{\mathbb{T}}_l$ and polynomial systems $[\mathbb{F}_1, \mathbb{U}_1], \ldots, [\mathbb{F}_m, \mathbb{U}_m]$ with $\mathbb{F}_j \ne \emptyset$ such that each $[\mathbb{T}, \tilde{\mathbb{T}}_i]$ is an irreducible simple system, every polynomial in $\mathbb{F}_i$ does not involve the dependents of $\mathbb{T}$ and

$$\mathrm{Zero}(\mathbb{T}/\mathbb{U}) = \bigcup_{i=1}^{l} \mathrm{Zero}(\mathbb{T}/\tilde{\mathbb{T}}_i) \cup \bigcup_{j=1}^{m} \mathrm{Zero}(\mathbb{T} \cup \mathbb{F}_j/\mathbb{U}_j).$$

Now consider an arbitrary polynomial system $\mathfrak{P}$ and let $[\mathbb{T}_1, \mathbb{U}_1], \ldots, [\mathbb{T}_r, \mathbb{U}_r]$ be an irreducible triangular series of $\mathfrak{P}$. For each $[\mathbb{T}_i, \mathbb{U}_i]$, one can determine triangular or empty sets $\tilde{\mathbb{T}}_{i1}, \ldots, \tilde{\mathbb{T}}_{il_i}$ and polynomial systems $[\mathbb{F}_{i1}, \mathbb{U}_{i1}], \ldots, [\mathbb{F}_{im_i}, \mathbb{U}_{im_i}]$ with $\mathbb{F}_{ik} \neq \emptyset$, according to Lemma 4.4.1, such that

$$\mathrm{Zero}(\mathbb{T}_i/\mathbb{U}_i) = \bigcup_{j=1}^{l_i} \mathrm{Zero}(\mathbb{T}_i/\tilde{\mathbb{T}}_{ij}) \cup \bigcup_{k=1}^{m_i} \mathrm{Zero}(\mathbb{T}_i \cup \mathbb{F}_{ik}/\mathbb{U}_{ik}),$$

where each $[\mathbb{T}_i, \tilde{\mathbb{T}}_{ij}]$ is simple and $\deg(F, \mathrm{lv}(T)) = 0$ for every $F \in \mathbb{F}_{ik}$ and $T \in \mathbb{T}_i$.

One may decompose each polynomial system $[\mathbb{T}_i \cup \mathbb{F}_{ik}, \mathbb{U}_{ik}]$ into irreducible triangular systems $[\mathbb{T}^*_{ij}, \mathbb{U}^*_{ij}]$ and apply Lemma 4.4.1 to each obtained $[\mathbb{T}^*_{ij}, \mathbb{U}^*_{ij}]$, and so on. As $\mathbb{T}$ is irreducible and $\deg(F, \mathrm{lv}(T)) = 0$ for any $F \in \mathbb{F}_{ik}$ and $T \in \mathbb{T}_i$, $|\mathbb{T}^*_{ij}| > |\mathbb{T}_i|$. Hence, the recursive process must terminate. Finally, $\mathfrak{P}$ will be decomposed into finitely many irreducible simple systems. In other words, we have the following theorem.

Theorem 4.4.2. There is an algorithm which computes, from any given polynomial system $\mathfrak{P}$ in $\boldsymbol{K}[\boldsymbol{x}]$, a finite number of irreducible simple systems $\mathfrak{S}_1, \ldots, \mathfrak{S}_e$ such that $\mathrm{Zero}(\mathfrak{P}) = \bigcup_{i=1}^{e} \mathrm{Zero}(\mathfrak{S}_i)$.

The above theoretical approach may have undesirable performance. It has been so explained mainly for simplicity and ease of termination proof. In practice, one may compute directly a simple series of each irreducible triangular system $[\mathbb{T}_i, \mathbb{U}_i]$ and then examine which of the obtained simple systems are already irreducible. For the reducible ones, one decomposes them further into irreducible triangular systems, and so forth. In this way, $\mathfrak{P}$ should also be decomposed into irreducible simple systems, but the termination is not evident.

Example 4.4.1. Consider the irreducible triangular systems in (4.2.7). As $\dim(\mathbb{T}_2) = \dim(\mathbb{T}_{3j}) = 0$ for $1 \leq j \leq 8$, it is easy to see that each $[\mathbb{T}_i, \emptyset]$ is a simple system for $i = 2, 31, \ldots, 38$. Now recall the triangular set

$$\mathbb{T}_1 = \begin{bmatrix} -z^5 + t^4, \\ z^6y^2 + 2t^3z^3y - t^7z^5 + 2t^4z^5 - tz^5 + t^6, \\ (t^3 - 1)z^3x - z^3y - t^3 \end{bmatrix},$$

where $t \prec z \prec y \prec x$. The factors of the initials and derivatives of the three polynomials which need to be considered are $t^3 - 1$, z and $z^3y + t^3$. As

$$\mathrm{sqfr}(\mathrm{res}(z, \mathbb{T}_1)) = t, \quad \mathrm{sqfr}(\mathrm{res}(z^3y + t^3, \mathbb{T}_1)) = t(t^3 - 1),$$

we can take $\mathbb{R} = \{t, t^3 - 1\}$. A simple series of $[\emptyset, \mathbb{R}]$ consists of a single simple system $[\emptyset, \tilde{\mathbb{T}}_1]$, where $\tilde{\mathbb{T}}_1 = [t(t^3 - 1)]$. Therefore, an irreducible simple system $[\mathbb{T}_1, \tilde{\mathbb{T}}_1]$ is obtained. Computing directly a simple series of $[\mathbb{T}_1, \mathrm{ini}(\mathbb{T}_1)]$ yields the same result. In any case, we have

$$\mathrm{Zero}(\mathbb{P}) = \mathrm{Zero}(\mathbb{T}_1/\tilde{\mathbb{T}}_1) \cup \mathrm{Zero}(\mathbb{T}_2) \cup \bigcup_{j=1}^{8} \mathrm{Zero}(\mathbb{T}_{3j}).$$

As an alternative to decompose $\mathfrak{P}$ into irreducible simple systems, one can compute a simple series of $\mathfrak{P}$ first. Each of the obtained simple systems may be further decomposed into irreducible triangular systems by algorithm Decom. However, these triangular systems are not necessarily simple, and from them simple systems have to be determined by a technique similar to the one demonstrated above. This approach has obvious disadvantages. The computation of simple series is very expensive, due to the high price of making polynomials squarefree. Apparently, the cost is spent in vain when the polynomials finally have to be factorized. Therefore, we do not pursue any further in this direction.

5 Various elimination algorithms

It is somewhat unusual to postpone the presentation of important elimination methods based on resultants and Gröbner bases to this later chapter. The main reason for this is that these methods are already well-known, fully described in standard textbooks, and widely accessible. In order to reduce overlap with existing materials in the literature, we shall not introduce the methods in detail and be satisfied by only giving them a brief review. Most formal proofs will be omitted.

As the reader may have been aware, our emphasis is placed mainly on a systematic treatment of elimination techniques based on pseudo-division. The objective is to establish various decompositions of zero sets (rather than ideals) of multivariate polynomials. This attempt is continued in part of this chapter.

5.1 Regular systems

Roughly speaking, a regular system is a simple system without the requirement on squarefreeness. We want to modify the subresultant-based algorithms described in Chaps. 2 and 3 to decompose any polynomial system into regular systems. It will also be shown that the decomposition can be computed by an alternative algorithm.

Definition 5.1.1. A triangular system $[\mathbb{T}, \mathbb{U}]$ in $\boldsymbol{K}[\boldsymbol{x}]$ is said to be *regular* or called a *regular system* if for any $1 \leq k \leq n$:

a. either $\mathbb{T}^{\langle k \rangle} = \emptyset$ or $\mathbb{U}^{\langle k \rangle} = \emptyset$;

b. $I(\bar{\boldsymbol{x}}^{\{k-1\}}) \neq 0$ for any $I \in \mathrm{ini}(\mathbb{U}^{\langle k \rangle})$ and $\bar{\boldsymbol{x}}^{\{k-1\}} \in \mathrm{Zero}(\mathbb{T}^{(k-1)}/\mathbb{U}^{(k-1)})$.

A triangular set $\mathbb{T}$ is said to be *regular* or called a *regular set* if there exists a polynomial set $\mathbb{U}$ such that $[\mathbb{T}, \mathbb{U}]$ is a regular system.

A triangular series Ψ is called a *regular series* if every $\mathfrak{T} \in \Psi$ is a regular system.

Ψ is called a *regular series* of a polynomial system $\mathfrak{P}$ if it is a regular series and $\mathrm{Zero}(\mathfrak{P}) = \bigcup_{\mathfrak{T} \in \Psi} \mathrm{Zero}(\mathfrak{T})$.

A regular series of $[\mathbb{P}, \emptyset]$ is also called a *regular series* of the polynomial set $\mathbb{P}$.

In the above definition, condition b is also satisfied for every $I \in \mathrm{ini}(\mathbb{T}^{\langle k \rangle})$ as $[\mathbb{T}, \mathbb{U}]$ is a triangular system. For example, with respect to the ordering $x \prec y$, $[xy - 1]$ is a regular set because $[[xy - 1], \{x\}]$ is a regular system; but neither is $\mathbb{T} = [x^2 - 1, (x + 1)y - 1]$. For $[\mathbb{T}, \emptyset]$ is not a triangular system by definition, while $\mathbb{U} = \emptyset$ is the only possible set such that condition a holds.

For convenience, sometimes $\emptyset$ is also regarded as a regular set. Refer to Sect. 3.1 for regular systems, projection is rather easy.

Subresultant-based algorithm

The following algorithm RegSer is an extension of TriSer. It may also be considered as simplified from SimSer. The algorithm decomposes any polynomial system into finitely many regular systems, where the elimination strategy for the equation-polynomials is almost the same as that employed in TriSer. The main new ingredient is step R2.2.3 in which the polynomial P_2 of class k obtained in step R2.2.2 is used to eliminate the inequation-polynomials from $\mathbb{U}^{\langle k\rangle} \neq \emptyset$. Roughly speaking, the elimination is realized by computing SRS and removing GCDs.

Algorithm RegSer: $\Psi \leftarrow \mathrm{RegSer}(\mathbb{P}, \mathbb{Q})$. Given a polynomial system $[\mathbb{P}, \mathbb{Q}]$ in $\boldsymbol{K}[\boldsymbol{x}]$, this algorithm computes a regular series Ψ of $[\mathbb{P}, \mathbb{Q}]$.

R1. Set $\Phi \leftarrow \{[\mathbb{P}, \mathbb{Q}, n]\}$, $\Psi \leftarrow \emptyset$.

R2. While $\Phi \neq \emptyset$, do:

R2.1. Let $[\mathbb{T}, \mathbb{U}, l]$ be an element of Φ and set $\Phi \leftarrow \Phi \setminus \{[\mathbb{T}, \mathbb{U}, l]\}$.

R2.2. For $k = l, \ldots, 1$, do:

R2.2.1. Set $\mathbb{T} \leftarrow \mathbb{T} \setminus \{0\}$, $\mathbb{U} \leftarrow \mathbb{U} \setminus (\boldsymbol{K} \setminus \{0\})$. If $\mathbb{T} \cap \boldsymbol{K} \neq \emptyset$ or $0 \in \mathbb{U}$, then go to R2. If $\mathbb{T}^{\langle k\rangle} = \emptyset$, then go to R2.2.4.

R2.2.2. Repeat:

R2.2.2.1. Let P_2 be an element of $\mathbb{T}^{\langle k\rangle}$ with minimal degree in x_k and set

$$\Phi \leftarrow \Phi \cup \{[\mathbb{T} \setminus \{P_2\} \cup \{\mathrm{ini}(P_2), \mathrm{red}(P_2)\}, \mathbb{U}, k]\},$$
$$\mathbb{U} \leftarrow \mathbb{U} \cup \{\mathrm{ini}(P_2)\}.$$

If $|\mathbb{T}^{\langle k\rangle}| = 1$, then go to R2.2.3; else take a polynomial P_1 from $\mathbb{T}^{\langle k\rangle} \setminus \{P_2\}$.

R2.2.2.2. Compute the SRS $H_2, \ldots, H_r$ of P_1 and P_2 with respect to x_k and set $I_i \leftarrow \mathrm{lc}(H_i, x_k)$ for $2 \le i \le r$. If $\mathrm{cls}(H_r) < k$, then set $\bar{r} \leftarrow r - 1$; else set $\bar{r} \leftarrow r$.

R2.2.2.3. Set

$$\Phi \leftarrow \Phi \cup \{[\mathbb{T} \setminus \{P_1, P_2\} \cup \{H_i, I_{i+1}, \ldots, I_r\}, \mathbb{U} \cup \{I_i\}, k]\colon 2 \le i \le \bar{r} - 1\},$$
$$\mathbb{T} \leftarrow \mathbb{T} \setminus \{P_1, P_2\} \cup \{H_r, H_{\bar{r}}\},$$
$$\mathbb{U} \leftarrow \mathbb{U} \cup \{I_{\bar{r}}\}.$$

R2.2.3. While $\mathbb{U}^{\langle k\rangle} \neq \emptyset$ and $\mathrm{cls}(P_2) = k$, do:

R2.2.3.1. Let P_1 be a polynomial in $\mathbb{U}^{\langle k\rangle}$; compute the SRS $H_2, \ldots, H_r$ of P_1 and P_2 if $\deg(P_1, x_k) \ge \deg(P_2, x_k)$, or of P_2 and P_1 otherwise, with respect to x_k, and set $I_i \leftarrow \mathrm{lc}(H_i, x_k)$ for $2 \le i \le r$.

R2.2.3.2. Set

$$\Phi \leftarrow \Phi \cup \{[\mathbb{T} \setminus \{P_2\} \cup \{\mathrm{pquo}(P_2, H_i, x_k), I_{i+1}, \ldots, I_r\}, \mathbb{U} \cup \{I_i\}, k]\colon 2 \le i \le r - 1\},$$
$$\mathbb{T} \leftarrow \mathbb{T} \setminus \{P_2\} \cup \{\mathrm{pquo}(P_2, H_r, x_k)\},$$
$$P_2 \leftarrow \mathrm{pquo}(P_2, H_r, x_k).$$

If $\text{cls}(H_r) < k$, then set $\mathbb{U} \leftarrow \mathbb{U} \setminus \{P_1\} \cup \{I_r\}$; else set $\mathbb{U} \leftarrow \mathbb{U} \cup \{I_r\}$.

R2.2.4. If $\mathbb{U}^{\langle k \rangle} \neq \emptyset$, then for each $P_1 \in \mathbb{U}^{\langle k \rangle}$ do:

$$\begin{aligned}&\Phi \leftarrow \Phi \cup \{[\mathbb{T} \cup \{\text{ini}(P_1)\}, \mathbb{U} \setminus \{P_1\} \cup \{\text{red}(P_1)\}, k]\},\\&\mathbb{U} \leftarrow \mathbb{U} \cup \{\text{ini}(P_1)\}.\end{aligned}$$

R2.3. Set $\Psi \leftarrow \Psi \cup \{[\mathbb{T}, \mathbb{U}]\}$, with $\mathbb{T}$ ordered as a triangular set.

The termination and correctness of RegSer may be proved by an argument similar to the proof of those of SimSer. We only need to note the following. Recall Lemma 3.3.2 and drop the assumption that $P_2(\bar{x}^{\{k-1\}}, x_k)$ is squarefree for $\bar{x}^{\{k-1\}} \in \text{Zero}(\mathbb{P}/\mathbb{Q})$. Corresponding to (3.3.1) therein is the zero relation

$$\text{Zero}(\mathbb{P} \cup \{P_2\}/\mathbb{Q} \cup \{P_1\}) = \bigcup_{i=2}^{r} \text{Zero}(\mathbb{P} \cup \mathbb{P}_i/\mathbb{Q} \cup \{P_1, I_i\}).$$

Clearly, $\text{cls}(H_i) = k$ holds for $2 \leq i \leq r-1$ but not necessarily for $i = r$. If $\text{cls}(H_r) < k$, then $I_r = H_r$ and

$$\begin{aligned}\text{Zero}(\mathbb{P} \cup \mathbb{P}_r/\mathbb{Q} \cup \{P_1, I_r\}) &= \text{Zero}(\mathbb{P} \cup \{\text{pquo}(P_2, I_r, x_k)\}/\mathbb{Q} \cup \{I_r\})\\&= \text{Zero}(\mathbb{P} \cup \{P_2\}/\mathbb{Q} \cup \{I_r\}),\end{aligned}$$

i.e., the polynomial P_1 may be eliminated. Otherwise, the process may continue, for example, by computing the SRS of $\text{pquo}(P_2, H_i, x_k)$ and P_1 with respect to x_k for each i. This procedure will terminate eventually because the degree of $\text{pquo}(P_2, H_i, x_k)$ is less than that of P_2 in x_k when $\text{cls}(H_i) = k$. Roughly speaking, the conditional GCD of P_2 and P_1 is removed from P_2 by pquo recursively until no such factors can be removed; then P_1 is eliminated.

Example 5.1.1. The polynomial set $\mathbb{P}$ in Example 2.4.1 may be decomposed by RegSer into four regular systems $[\mathbb{T}_i, \mathbb{U}_i]$ such that

$$\text{Zero}(\mathbb{P}) = \bigcup_{i=1}^{4} \text{Zero}(\mathbb{T}_i/\mathbb{U}_i),$$

where

$$\begin{aligned}&\mathbb{T}_3 = [r^4 - 4r^2 + 3, -z^2 + r^2 z - z - r^2 + 1, F, P_2],\\&\mathbb{U}_1 = \{r^4 - 4r^2 + 3\}, \quad \mathbb{U}_2 = \mathbb{U}_3 = \mathbb{U}_4 = \emptyset,\end{aligned}$$

$\mathbb{T}_1, \mathbb{T}_2$ and F, P_2 are as in Example 2.4.1, and $\mathbb{T}_4$ as in Example 3.3.4.

To give more details, let T_1, T_2, T_3 denote the three polynomials in $\mathbb{T}_1$ successively. Compute the SRS of $x = \text{ini}(T_3)$ and T_2 with respect to x; let R be the last polynomial in the subchain (which is identical to the resultant of x and T_2 with respect to x). The inequation-polynomial in $\mathbb{U}_1$ is acquired as the last in the SRS of squarefreed R and T_1 with respect to z. In splitting according to the SRS some new polynomial systems are generated from which the two regular sets $\mathbb{T}_3$ and $\mathbb{T}_4$ are obtained.

Example 5.1.2. Recall the polynomial set $\mathbb{P}$ and variable ordering given in Example 3.2.2. A regular series of $\mathbb{P}$ computed by RegSer consists of 6 regular systems $[\mathbb{T}_1, \mathbb{U}_1], [\mathbb{T}_2, \emptyset], \ldots, [\mathbb{T}_6, \emptyset]$, where the triangular sets $\mathbb{T}_i$ are either the same as or very similar to those listed in Example 3.2.2 and $\mathbb{U}_1$ contains x and two other univariate polynomials that are T_{31} and T_{41} in Example 3.2.2.

Algorithm based on generalized GCD

Definition 5.1.2. Let $\mathfrak{T} = [\mathbb{T}, \mathbb{U}]$ be an arbitrary triangular system in $\boldsymbol{K}[\boldsymbol{x}]$. A zero $(\xi_1, \ldots, \xi_n)$ of $\mathfrak{T}$ is said to be *regular* if either $\xi_i = x_i$ or x_i is a dependent of $\mathbb{T}$ for any $1 \leq i \leq n$.

When $\mathfrak{T}$ is regular, any regular zero of $\mathfrak{T}$ is also called a *regular zero* of $\mathbb{T}$.

As usual, we write $\boldsymbol{\xi}^{\{i\}}$ for $\xi_1, \ldots, \xi_i$ or $(\xi_1, \ldots, \xi_i)$ with $\boldsymbol{\xi} = \boldsymbol{\xi}^{\{n\}}$. The set of all regular zeros of $\mathfrak{T}$ or $\mathbb{T}$ is denoted RegZero($\mathfrak{T}$) or RegZero($\mathbb{T}$). Apparently, RegZero($\mathfrak{T}$) $\subset$ Zero($\mathfrak{T}$).

Proposition 5.1.1. The regular zeros of any regular set are well-defined. In other words, for any two regular systems $[\mathbb{T}, \mathbb{U}_1]$ and $[\mathbb{T}, \mathbb{U}_2]$,

$$\mathrm{RegZero}(\mathbb{T}/\mathbb{U}_1) = \mathrm{RegZero}(\mathbb{T}/\mathbb{U}_2).$$

Proof. Let $\boldsymbol{\xi} \in \mathrm{RegZero}(\mathbb{T}/\mathbb{U}_1)$. First, consider any $U \in \mathbb{U}_2$ of smallest class p. Clearly x_p is a parameter of $\mathbb{T}$ by definition, so $\xi_p = x_p$ is an indeterminate. Therefore, $U(\boldsymbol{\xi}^{\{p\}}) = 0$ implies that $\mathrm{ini}(U)(\boldsymbol{\xi}^{\{p-1\}}) = 0$. Since $[\mathbb{T}, \mathbb{U}_2]$ is a regular system, by definition $\mathrm{ini}(U)(\boldsymbol{\xi}^{\{p-1\}}) \neq 0$. It follows that $U(\boldsymbol{\xi}^{\{p\}}) \neq 0$.

Now suppose that $\mathbb{U}_2^{\langle i\rangle} \neq \emptyset$, and $U(\boldsymbol{\xi}^{\{i-1\}}) \neq 0$ for all $U \in \mathbb{U}_2^{(i-1)}$. Then

$$\boldsymbol{\xi}^{\{i-1\}} \in \mathrm{Zero}(\mathbb{T}^{(i-1)}/\mathbb{U}_2^{(i-1)}).$$

Consider any $U \in \mathbb{U}_2^{\langle i\rangle}$. By definition, x_i is a parameter of $\mathbb{T}$ and $\xi_i = x_i$. As $[\mathbb{T}, \mathbb{U}_2]$ is regular, $\mathrm{ini}(U)(\boldsymbol{\xi}^{\{i-1\}}) \neq 0$. For the same reason as above, we have $U(\boldsymbol{\xi}^{\{i\}}) \neq 0$. Hence, by induction $U(\boldsymbol{\xi}) \neq 0$ for all $U \in \mathbb{U}_2$. This shows that $\boldsymbol{\xi} \in \mathrm{RegZero}(\mathbb{T}/\mathbb{U}_2)$; thereby $\mathrm{RegZero}(\mathbb{T}/\mathbb{U}_1) \subset \mathrm{RegZero}(\mathbb{T}/\mathbb{U}_2)$. The other direction is proved by the same argument. □

Corollary 5.1.2. For any regular system $[\mathbb{T}, \mathbb{U}]$ and regular zero $\boldsymbol{\xi}$ of $\mathbb{T}$, $U(\boldsymbol{\xi}) \neq 0$ for all $U \in \mathbb{U}$.

If $\mathbb{T}$ is written as

$$\mathbb{T} = [T_1(\boldsymbol{u}, y_1), \ldots, T_r(\boldsymbol{u}, y_1, \ldots, y_r)], \tag{5.1.1}$$

then any regular zero of $\mathfrak{T}$ has the form

$$\boldsymbol{\xi} = (\boldsymbol{u}, \eta_1, \ldots, \eta_r) \in \mathrm{Zero}(\mathfrak{T}), \tag{5.1.2}$$

where $\eta_i \in \tilde{\boldsymbol{K}} \supset \boldsymbol{K}(\boldsymbol{u})$ for each i.

Lemma 5.1.3. Every perfect triangular system in $\boldsymbol{K}[\boldsymbol{x}]$ has a regular zero.

Proof. Let $\mathfrak{T} = [\mathbb{T}, \mathbb{U}]$ be a perfect triangular system and write $\mathbb{T}$ as

$$\mathbb{T} = [T_1(\boldsymbol{u}, y_1), \ldots, T_r(\boldsymbol{u}, y_1, \ldots, y_r)]$$

as before with

$$I_i(\boldsymbol{u}, y_1, \ldots, y_{i-1}) = \mathrm{ini}(T_i), \; 1 \le i \le r, \quad V = \prod_{U \in \mathbb{U}} U.$$

Since $I_1(\boldsymbol{u}) \neq 0$ in $\boldsymbol{K}(\boldsymbol{u})$, $T_1(\boldsymbol{u}, y_1)$ must have zeros for y_1 in some suitably chosen algebraic-extension field $\tilde{\boldsymbol{K}}$ of $\boldsymbol{K}(\boldsymbol{u})$. Because $\mathfrak{T}$ is perfect, V can vanish only at some but not all of these zeros. For, otherwise, any zero of T_1 for specialized values of $\boldsymbol{u}$ is also a zero of V and thus $\mathfrak{T}$ is not perfect. Therefore, the zero set

$$\mathcal{Z}_1 = \{(\boldsymbol{u}, \bar{y}_1) \colon \bar{y}_1 \in \tilde{\boldsymbol{K}}, T_1(\boldsymbol{u}, \bar{y}_1) = 0, V(\boldsymbol{u}, \bar{y}_1, y_2, \ldots, y_r) \neq 0\}$$

is not empty.

For any $(\boldsymbol{u}, \bar{y}_1) \in \mathcal{Z}_1$, by the definition of a triangular system $I_2(\boldsymbol{u}, \bar{y}_1) \neq 0$ and thus $T_2(\boldsymbol{u}, \bar{y}_1, y_2)$ has zeros for y_2 in some algebraic-extension field $\tilde{\boldsymbol{K}}$. For the same reason, V may vanish at $(\boldsymbol{u}, \bar{y}_1, \bar{y}_2)$ only for some but not all $(\boldsymbol{u}, \bar{y}_1) \in \mathcal{Z}_1$ and $\bar{y}_2 \in \mathrm{Zero}(T_2(\boldsymbol{u}, \bar{y}_1, y_2))$. In other words,

$$\mathcal{Z}_2 = \left\{ (\boldsymbol{u}, \bar{y}_1, \bar{y}_2) \colon \begin{array}{l} (\boldsymbol{u}, \bar{y}_1) \in \mathcal{Z}_1, \bar{y}_2 \in \tilde{\boldsymbol{K}}, T_2(\boldsymbol{u}, \bar{y}_1, \bar{y}_2) = 0, \\ V(\boldsymbol{u}, \bar{y}_1, \bar{y}_2, y_3, \ldots, y_r) \neq 0 \end{array} \right\} \neq \emptyset.$$

The above reasoning may continue for T_3, T_4 and so on. In this way, a regular zero of $\mathfrak{T}$ will finally be constructed and the lemma is proved. □

The algorithms presented below are adapted from Kalkbrener (1993). They are somewhat complicated by the cross-calling. The basic idea here is to compute GCDs modulo regular sets with splitting on demand.

Algorithm Split: $[\Delta, \Lambda] \leftarrow \mathrm{Split}(\mathbb{T}, P, k)$. Given an integer k ($1 \le k \le n$), a polynomial P and a regular set $\mathbb{T}$ in $\boldsymbol{K}[\boldsymbol{x}^{\{k\}}]$, this algorithm computes two sets Δ and Λ of regular sets in $\boldsymbol{K}[\boldsymbol{x}^{\{k\}}]$ such that

$$\mathrm{RegZero}(\mathbb{T}) \cap \mathrm{Zero}(P) = \bigcup_{\mathbb{T}^* \in \Delta} \mathrm{RegZero}(\mathbb{T}^*),$$

$$\mathrm{RegZero}(\mathbb{T}/P) = \bigcup_{\mathbb{T}^* \in \Lambda} \mathrm{RegZero}(\mathbb{T}^*).$$

S1. Compute $\Omega \leftarrow \mathrm{GenGCD}(\mathbb{T}^{(k-1)}, \mathbb{T}^{\langle k \rangle} \cup \{P\}, k)$.
S2. If $\mathbb{T}^{\langle k \rangle} = \emptyset$, then set

$$\Delta \leftarrow \{\mathbb{S} \colon [\mathbb{S}, G] \in \Omega, G = 0\}, \quad \Lambda \leftarrow \{\mathbb{S} \colon [\mathbb{S}, G] \in \Omega, G \neq 0\}$$

and the algorithm terminates.

S3. Let F be the only element of $\mathbb{T}^{\langle k \rangle}$ and set

$$\Gamma \leftarrow \left\{ \mathbb{S} \cup [\mathrm{pquo}(F, G, x_k)]\colon \begin{array}{l} [\mathbb{S}, G] \in \Omega, \mathrm{cls}(G) = k, \\ \deg(G, x_k) < \deg(F, x_k) \end{array} \right\},$$

$$\Delta \leftarrow \{\mathbb{S} \cup [G]\colon [\mathbb{S}, G] \in \Omega, \mathrm{cls}(G) = k\},$$

$$\Lambda \leftarrow \{\mathbb{S} \cup [F]\colon [\mathbb{S}, G] \in \Omega, \mathrm{cls}(G) < k\} \cup \{\mathrm{op}(2, \mathrm{Split}(\mathbb{S}, P, k))\colon \mathbb{S} \in \Gamma\}.$$

Refer to Definition 6.2.2 for the *saturation* sat($\mathbb{T}$) of any triangular set $\mathbb{T}$. Zero(sat($\mathbb{T}$)) represents the union of the irreducible algebraic varieties whose generic points are regular zeros of $[\mathbb{T}, \mathrm{ini}(\mathbb{T})]$.

Algorithm GenGCD: $\Omega \leftarrow \mathrm{GenGCD}(\mathbb{T}, \mathbb{P}, k)$. Given an integer k ($1 \leq k \leq n$), a polynomial set $\mathbb{P} \subset \boldsymbol{K}[\boldsymbol{x}^{\{k\}}]$ and a regular set $\mathbb{T} \subset \boldsymbol{K}[\boldsymbol{x}^{\{k-1\}}]$, this algorithm computes a finite set Ω of pairs $[\mathbb{T}_1, G_1], \ldots, [\mathbb{T}_l, G_l]$, with each $\mathbb{T}_i$ a regular set in $\boldsymbol{K}[\boldsymbol{x}^{\{k-1\}}]$ and G_i a polynomial in $\boldsymbol{K}[\boldsymbol{x}^{\{k\}}]$, such that

a. $\mathrm{RegZero}(\mathbb{T}) = \bigcup_{i=1}^{l} \mathrm{RegZero}(\mathbb{T}_i)$;

b. for any $1 \leq i \leq l$ and $\boldsymbol{\xi}^{\{k-1\}} \in \mathrm{RegZero}(\mathbb{T}_i)$,

$$G_i \neq 0 \Longrightarrow \mathrm{lc}(G_i, x_k)(\boldsymbol{\xi}^{\{k-1\}}) \neq 0$$

and $G_i(\boldsymbol{\xi}^{\{k-1\}}, x_k)$ is a GCD of the polynomials in $\mathbb{P}^{\langle \xi, k-1 \rangle}$ with respect to x_k;

c. $\mathrm{Zero}(\mathrm{sat}(\mathbb{T}_i)) \cap \mathrm{Zero}(\mathbb{P}) \subset \mathrm{Zero}(G_i)$ for any $1 \leq i \leq l$.

G1. If $k = 1$; or $\mathbb{P} = \emptyset$; or $k > 1$, $|\mathbb{P}| = 1$ and $\mathrm{op}(1, \mathrm{Split}(\mathbb{T}, \mathrm{lc}(\mathrm{op}(1, \mathbb{P}), x_k), k-1)) = \emptyset$, then set

$$\Omega \leftarrow \begin{cases} \{[\emptyset, 0]\} & \text{when } k = 1 \text{ and } \mathbb{P} = \emptyset, \\ \{[\emptyset, \mathrm{gcd}(\mathbb{P})]\} & \text{when } k = 1 \text{ and } \mathbb{P} \neq \emptyset, \\ \{[\mathbb{T}, 0]\} & \text{when } k > 1 \text{ and } \mathbb{P} = \emptyset, \\ \{[\mathbb{T}, \mathrm{op}(1, \mathbb{P})]\} & \text{when } k > 1 \text{ and } |\mathbb{P}| = 1, \end{cases}$$

and the algorithm terminates.

G2. Let P be an element of $\mathbb{P}$ with minimal degree in x_k, set

$$\mathbb{P}' \leftarrow \mathbb{P} \setminus \{P\} \cup \{\mathrm{red}(P, x_k)\} \setminus \{0\},$$

and compute

$$[\Delta, \Lambda] \leftarrow \mathrm{Split}(\mathbb{T}, \mathrm{lc}(P, x_k), k-1),$$

$$\mathbb{P}'' \leftarrow \{P\} \cup \mathrm{prem}(\mathbb{P}, P, x_k) \setminus \{0\},$$

$$\Omega \leftarrow \bigcup_{\mathbb{S} \in \Delta} \mathrm{GenGCD}(\mathbb{S}, \mathbb{P}', k) \cup \bigcup_{\mathbb{S} \in \Lambda} \mathrm{GenGCD}(\mathbb{S}, \mathbb{P}'', k).$$

Algorithm RegSer*: $\Psi \leftarrow \mathrm{RegSer}^*(\mathbb{T}, \mathbb{P}, k)$. Given an integer k ($1 \leq k \leq n$), a nonempty polynomial set $\mathbb{P} \subset \boldsymbol{K}[\boldsymbol{x}^{\{k\}}]$ and a regular set $\mathbb{T} \subset \boldsymbol{K}[\boldsymbol{x}^{\{k-1\}}]$, this algorithm computes a set Ψ of regular sets in $\boldsymbol{K}[\boldsymbol{x}^{\{k\}}]$ such that

a. $$\mathrm{Zero}(\mathrm{sat}(\mathbb{T})) \cap \mathrm{Zero}(\mathbb{P}) \subset \bigcup_{\mathbb{T}^* \in \Psi} \mathrm{Zero}(\mathrm{sat}(\mathbb{T}^*)) \subset \mathrm{Zero}(\mathbb{P}); \tag{5.1.3}$$

b. for any $\mathbb{T}^* \in \Psi$, either $\mathrm{RegZero}(\mathbb{T}^{*(k-1)}) \subset \mathrm{RegZero}(\mathbb{T})$, or $|\mathbb{T}^{*(k-1)}| < |\mathbb{T}|$.

R1. If $k = 1$ then set

$$\Psi \leftarrow \begin{cases} \emptyset & \text{when } \mathrm{gcd}(\mathbb{P}) \in \boldsymbol{K}, \\ \{[\mathrm{gcd}(\mathbb{P})]\} & \text{otherwise,} \end{cases}$$

and the procedure terminates.

R2. Compute

$$\begin{aligned} &\Omega \leftarrow \mathrm{GenGCD}(\mathbb{T}, \mathbb{P}, k),\\ &\Gamma \leftarrow \textstyle\bigcup_{\substack{[\mathbb{S},\,G] \in \Omega\\ G \neq 0}} \mathrm{RegSer}^*(\mathbb{S}^{(k-2)}, \mathbb{S}^{[k-2]} \cup \{\mathrm{lc}(G, x_k)\}, k-1),\\ &\Psi \leftarrow \{\mathbb{S}\colon\ [\mathbb{S}, G] \in \Omega, G = 0\} \cup \{\mathbb{S} \cup [G]\colon\ [\mathbb{S}, G] \in \Omega,\\ &\qquad \mathrm{cls}(G) = k\} \cup \textstyle\bigcup_{\mathbb{S} \in \Gamma} \mathrm{RegSer}^*(\mathbb{S}, \mathbb{P}, k). \end{aligned}$$

When $\mathbb{T} = \emptyset$, (5.1.3) leads to

$$\mathrm{Zero}(\mathbb{P}) = \bigcup_{\mathbb{T}^* \in \Psi} \mathrm{Zero}(\mathrm{sat}(\mathbb{T}^*)). \tag{5.1.4}$$

Hence, with $\mathbb{T} = \emptyset$ and $k = n$, algorithm RegSer* decomposes any polynomial set $\mathbb{P} \subset \boldsymbol{K}[\boldsymbol{x}]$ into a finite set Ψ of regular sets such that (5.1.4) holds. In general, (5.1.4) does not imply that

$$\mathrm{Zero}(\mathbb{P}) = \bigcup_{\mathbb{T}^* \in \Psi} \mathrm{Zero}(\mathbb{T}^*/\mathrm{ini}(\mathbb{T}^*)). \tag{5.1.5}$$

However, one may observe from the algorithms that (5.1.5) does hold for any Ψ computed by RegSer* from $\mathbb{T} = \emptyset$, $\mathbb{P}$ and $k = n$. Therefore, Ψ can be taken as a regular series of the polynomial set $\mathbb{P}$.

The correctness and termination proofs for the above algorithms involve some technical arguments for which new notations and terminologies may have to be introduced. We omit the details and refer to Kalkbrener (1993). The interested reader may also work out his own proofs. Kalkbrener (1994) extended the algorithm to decompose radicals of polynomial ideals into primes – the equivalent problem of decomposing algebraic varieties into irreducible components will be discussed in Sect. 6.2.

Properties

When a regular zero $\boldsymbol{\xi}$ is written in the form (5.1.2), $\boldsymbol{\xi}^{\{i\}}$ stands alternatively for $\boldsymbol{u}, \eta_1, \ldots, \eta_i$ or $(\boldsymbol{u}, \eta_1, \ldots, \eta_i)$ with $\boldsymbol{\xi} = \boldsymbol{\xi}^{\{r\}}$ as before.

Proposition 5.1.4. Let $\mathbb{T}$ as in (5.1.1) be a regular set. Then for any $1 \leq i \leq r-1$ and $\boldsymbol{\xi}^{\{i\}} \in \mathrm{RegZero}(\mathbb{T}^{\{i\}})$,

$$\mathrm{ini}(T_{i+1})(\boldsymbol{\xi}^{\{i\}}) \neq 0. \tag{5.1.6}$$

Proof. As $\mathbb{T}$ is regular, there exists a $\mathbb{U}$ such that $[\mathbb{T}, \mathbb{U}]$ is a regular system. In particular, $\mathbb{U} \subset \boldsymbol{K}[\boldsymbol{u}]$. For any $1 \leq i \leq r-1$, let $\boldsymbol{\xi}^{\{i\}} \in \text{RegZero}(\mathbb{T}^{\{i\}})$. Clearly, $U(\boldsymbol{\xi}^{\{i\}}) \neq 0$ for any $U \in \mathbb{U}$. As $[\mathbb{T}, \mathbb{U}]$ is a triangular system, (5.1.6) holds by definition. □

Proposition 5.1.5. For any regular set $\mathbb{T}$ and polynomial P in $\boldsymbol{K}[\boldsymbol{x}]$,

$$\text{res}(P, \mathbb{T}) \neq 0 \iff P(\boldsymbol{\xi}) \neq 0 \quad \text{for any } \boldsymbol{\xi} \in \text{RegZero}(\mathbb{T}).$$

Proof. ($\Longrightarrow$) Let the variables $\boldsymbol{x}$ be renamed so that $\mathbb{T}$ is written in the form (5.1.1). If there exists a $\boldsymbol{\xi} \in \text{RegZero}(\mathbb{T})$ such that $P(\boldsymbol{\xi}) = 0$, then plunging $\boldsymbol{\xi}$ into (4.3.2) in Lemma 4.3.2 yields $R = \text{res}(P, \mathbb{T}) = 0$. This contradicts the assumption that $R \neq 0$.

($\Longleftarrow$) Let $R_1 = R_1(\boldsymbol{z}^{\{r-1\}}) = \text{res}(P, T_r, y_r)$ and $\boldsymbol{\xi}^{\{r-1\}} \in \text{RegZero}(\mathbb{T}^{\{r-1\}})$. As $\mathbb{T}$ is regular, by Proposition 5.1.4 we have $\text{ini}(T_r)(\boldsymbol{\xi}^{\{r-1\}}) \neq 0$. If $R_1(\boldsymbol{\xi}^{\{r-1\}}) = 0$, then $P(\boldsymbol{\xi}^{\{r-1\}}, y_r)$ and $T_r(\boldsymbol{\xi}^{\{r-1\}}, y_r)$ have a common zero η_r for y_r. This is impossible because $\boldsymbol{\xi} \in \text{RegZero}(\mathbb{T})$ and $P(\boldsymbol{\xi}) = 0$ contradict the hypothesis that $P(\boldsymbol{\xi}) \neq 0$ for any $\boldsymbol{\xi} \in \text{RegZero}(\mathbb{T})$. Hence $R_1(\boldsymbol{\xi}^{\{r-1\}}) \neq 0$ for any $\boldsymbol{\xi}^{\{r-1\}} \in \text{RegZero}(\mathbb{T}^{\{r-1\}})$.

Next, consider $R_2 = \text{res}(R_1, T_{r-1}, y_{r-1})$ and use the same argument. We shall see that $R_2(\boldsymbol{\xi}^{\{r-2\}}) \neq 0$ for any $\boldsymbol{\xi}^{\{r-2\}} \in \text{RegZero}(\mathbb{T}^{\{r-2\}})$. In this way, one will finally arrive at $R(\boldsymbol{u}) = R_r(\boldsymbol{u}) \neq 0$. The proof is complete. □

Since any simple set is regular, Proposition 5.1.5 holds as well when $\mathbb{T}$ is a simple set. From Propositions 5.1.4 and 5.1.5, the following result is obtained.

Corollary 5.1.6. For any regular or simple set $\mathbb{T} \subset \boldsymbol{K}[\boldsymbol{x}]$ and any $I \in \text{ini}(\mathbb{T})$, $\text{res}(I, \mathbb{T}) \neq 0$.

The conclusion in the above corollary is also a sufficient condition for any triangular set to be regular. This is stated as follows.

Lemma 5.1.7. Let $\mathbb{T} = [T_1, \ldots, T_r]$ be a triangular set in $\boldsymbol{K}[\boldsymbol{x}]$ and assume that $\text{res}(\text{ini}(T_i), \mathbb{T}^{\{i-1\}}) \neq 0$, for $2 \leq i \leq r$. Then $\mathbb{T}$ is regular.

Proof. Let $R_1 = \text{ini}(T_1) \prod_{i=2}^{r} \text{res}(\text{ini}(T_i), \mathbb{T}^{\{i-1\}})$; then R_1 is not equal to 0 and does not involve any $\text{lv}(T_i)$. Let $R_i = \text{ini}(R_{i-1})$ for $i = 2, \ldots, t$ such that R_t is a constant. It is easy to verify by definition that $[\mathbb{T}, \{R_1, \ldots, R_t\}]$ is a regular system. The lemma follows immediately. □

Let $\mathbb{T}$ be any triangular set in $\boldsymbol{K}[\boldsymbol{x}]$. Summarizing the above results, we have the equivalence of the following conditions:

a. $\mathbb{T}$ is regular;
b. $\text{res}(I, \mathbb{T}) \neq 0$ for any $I \in \text{ini}(\mathbb{T})$;
c. either $|\mathbb{T}| = 1$, or $\mathbb{T}^{(n-1)}$ is regular and

$$I(\boldsymbol{\xi}^{\{n-1\}}) \neq 0 \quad \text{for } I \in \text{ini}(\mathbb{T}^{(n)}) \text{ and all } \boldsymbol{\xi}^{\{n-1\}} \in \text{RegZero}(\mathbb{T}^{(n-1)}).$$

Therefore, either of the conditions b and c may be taken for the definition of a regular set. In fact, they have been used respectively to define the equivalent concepts of *proper ascending chains* in Yang and Zhang (1994) and *regular chains* in Kalkbrener (1993). Condition b may be regarded as an effective criterion to check whether a given triangular set is regular. The results of Proposition 5.1.5, Corollary 5.1.6, and Lemma 5.1.7 are also given in Yang and Zhang (1994).

The following proposition follows from the specification of algorithm Split and the definition of saturation.

Proposition 5.1.8. Let $\mathbb{T}$ be a regular set and P a polynomial in $\boldsymbol{K}[\boldsymbol{x}]$. Then

a. $P(\boldsymbol{\xi}) \neq 0$ for any $\boldsymbol{\xi} \in \mathrm{RegZero}(\mathbb{T})$
$\iff \mathrm{RegZero}(\mathbb{T}) \cap \mathrm{Zero}(P) = \emptyset$
$\iff \mathrm{op}(1, \mathrm{Split}(\mathbb{T}, P, n)) = \emptyset$;

b. $\mathrm{Zero}(\mathrm{sat}(\mathbb{T})) \subset \mathrm{Zero}(P)$
$\iff \mathrm{RegZero}(\mathbb{T}) \subset \mathrm{Zero}(P)$
$\iff \mathrm{op}(2, \mathrm{Split}(\mathbb{T}, P, n)) = \emptyset$.

In contrast with Theorem 3.4.4 and Corollary 4.3.9, we have the following theorem. The proof of this theorem as well as Theorem 5.1.11 requires a result given late in Sect. 6.2 (see Definition 6.2.3 and Theorem 6.2.4).

Theorem 5.1.9. For any regular system $[\mathbb{T}, \mathbb{U}]$ and polynomial P in $\boldsymbol{K}[\boldsymbol{x}]$, $\mathrm{Zero}(\mathbb{T}/\mathbb{U}) \subset \mathrm{Zero}(P)$ if and only if there exists an integer $d > 0$ such that $\mathrm{prem}(P^d, \mathbb{T}) = 0$.

Proof. The sufficiency follows obviously from the pseudo-remainder formula and the definition of regular systems.

To show the necessity, suppose that $\mathrm{Zero}(\mathbb{T}/\mathbb{U}) \subset \mathrm{Zero}(P)$, let

$$V = \prod_{U \in \mathbb{U}} \mathrm{res}(U, \mathbb{T}),$$

and write $\mathbb{T}$ in the form (5.1.1) with $\mathrm{ini}(T_i) = I_i$ and $\mathrm{ldeg}(T_i) = d_i$ for $1 \leq i \leq r$. Then, $V \in \boldsymbol{K}[\boldsymbol{u}]$, $V \neq 0$ (according to Corollary 5.1.2 and Proposition 5.1.5), and

$$\mathrm{Zero}(\mathbb{T}/V) \subset \mathrm{Zero}(\mathbb{T}/\mathbb{U}) \subset \mathrm{Zero}(P)$$

(by Lemma 4.3.2). It follows that $\mathrm{Zero}(\mathbb{T}/VP) = \emptyset$. We complete the proof of the theorem by proving the following assertion with induction on r.

For any regular set $\mathbb{T}$ and nonzero polynomials $V \in \boldsymbol{K}[\boldsymbol{u}]$ and $P \in \boldsymbol{K}[\boldsymbol{u}, y_1, \ldots, y_r]$ as above, if $\mathrm{Zero}(\mathbb{T}/VP) = \emptyset$, then there exists an integer $d > 0$ such that $\mathrm{prem}(P^d, \mathbb{T}) = 0$.

Consider first the case $r = 1$ and let $R = \mathrm{prem}(P^{d_1}, T_1)$. Denote all the nonzero coefficients of R in y_1 by $R_1, \ldots, R_l$. According to Lemmas 3.1.1 and 3.1.2 b, $\mathrm{Zero}(\emptyset/VR_j) = \emptyset$ for all j. This implies that $R_j \equiv 0$ for $1 \leq j \leq l$; therefore, $R \equiv 0$ and the assertion is proved.

Now suppose that the assertion holds for any regular set $\mathbb{T}$ with $|\mathbb{T}| < r$; we proceed to prove the assertion for $|\mathbb{T}| = r > 1$. Let

$$\mathbb{T}^{\{r-1\}} = [T_1, \ldots, T_{r-1}],\ J_{r-1} = I_1 \cdots I_{r-1},\ R = \mathrm{prem}(P^{d_r}, T_r),$$

and denote all the nonzero coefficients of R in y_r by $R_1, \ldots, R_l$. Again by Lemmas 3.1.2 and 3.1.2b, $\mathrm{Zero}(\mathbb{T}^{\{r-1\}}/VR_j) = \emptyset$ for all j. By the induction hypothesis, there exists an integer $k_j > 0$ such that $\mathrm{prem}(R_j^{k_j}, \mathbb{T}^{\{r-1\}}) = 0$ for each j. Thus, there exists an integer $s_j \geq 0$ such that

$$J_{r-1}^{s_j} R_j^{k_j} \in \mathrm{Ideal}(\mathbb{T}^{\{r-1\}}), \quad 1 \leq j \leq l.$$

Set $k = \max_{1 \leq j \leq l} k_j$, and $s = \max_{1 \leq j \leq l} s_j$; then $J_{r-1}^s R^k \in \mathrm{Ideal}(\mathbb{T})$. On the other hand, $R = \mathrm{prem}(P^{d_r}, T_r)$ implies that there exists an integer $q_r \geq 0$ such that $I_r^{q_r} P^{d_r} - R \in \mathrm{Ideal}(\{T_r\})$. Hence

$$J_{r-1}^s I_r^{q_r k} P^{d_r k} = J_{r-1}^s R^k + J_{r-1}^s (I_r^{q_r} P^{d_r} - R)$$
$$[(I_r^{q_r} P^{d_r})^{k-1} + \cdots + R^{k-1}] \in \mathrm{Ideal}(\mathbb{T}).$$

Let $d = d_r k$ and $q = \max(s, q_r k)$. Then $(I_1 \cdots I_r)^q P^d \in \mathrm{Ideal}(\mathbb{T})$, so $P^d \in \mathrm{sat}(\mathbb{T})$. By Theorem 6.2.4, $P^d \in \text{p-sat}(\mathbb{T})$, wherefore $\mathrm{prem}(p^d, \mathbb{T}) = 0$. The proof is complete. □

Corollary 5.1.10. For any regular set $\mathbb{T}$ and polynomial P in $\boldsymbol{K}[\boldsymbol{x}]$, $\mathrm{Zero}(\mathbb{T}/\mathrm{ini}(\mathbb{T})) \subset \mathrm{Zero}(P)$ if and only if there exists an integer $d > 0$ such that $\mathrm{prem}(P^d, \mathbb{T}) = 0$.

Proof. The sufficient condition is obvious, so we only need to prove the necessity. As $\mathbb{T}$ is regular, there exists a polynomial set $\mathbb{U} \subset \boldsymbol{K}[\boldsymbol{x}]$ such that $[\mathbb{T}, \mathbb{U}]$ is a regular system and $\mathrm{Zero}(\mathbb{T}/\mathbb{U}) \subset \mathrm{Zero}(\mathbb{T}/\mathrm{ini}(\mathbb{T}))$. If $\mathrm{Zero}(\mathbb{T}/\mathrm{ini}(\mathbb{T})) \subset \mathrm{Zero}(P)$, then $\mathrm{Zero}(\mathbb{T}/\mathbb{U}) \subset \mathrm{Zero}(P)$. In view of Theorem 5.1.9, there exists an integer $d > 0$ such that $\mathrm{prem}(P^d, \mathbb{T}) = 0$. □

The reader should compare the following with Theorems 3.4.6 and 4.3.11.

Theorem 5.1.11. Let $[\mathbb{P}, \mathbb{Q}]$ be a polynomial system in $\boldsymbol{K}[\boldsymbol{x}]$ and $[\mathbb{T}_1, \mathbb{U}_1], \ldots, [\mathbb{T}_e, \mathbb{U}_e]$ a regular series of $[\mathbb{P}, \mathbb{Q}]$. Then:

a. there exists an integer $d > 0$ such that $\mathrm{prem}(P^d, \mathbb{T}_i) = 0$ for all $P \in \mathbb{P}$ and $1 \leq i \leq e$;

b. for any integers $m > 0$, $1 \leq i \leq e$ and polynomial $Q \in \mathbb{Q}$, $\mathrm{prem}(Q^m, \mathbb{T}_i) \neq 0$;

c.
$$\mathrm{Zero}(\mathbb{P}/\mathbb{Q}) = \bigcup_{i=1}^{e} \mathrm{Zero}(\mathbb{T}_i/\mathrm{ini}(\mathbb{T}_i) \cup \mathbb{Q}). \tag{5.1.7}$$

Proof. a. From Definition 5.1.1, we know that $\mathrm{Zero}(\mathbb{P}/\mathbb{Q}) = \bigcup_{i=1}^{e} \mathrm{Zero}(\mathbb{T}_i/\mathbb{U}_i)$, so $\mathrm{Zero}(\mathbb{T}_i/\mathbb{U}_i) \subset \mathrm{Zero}(\mathbb{P}/\mathbb{Q}) \subset \mathrm{Zero}(\mathbb{P})$ for each i. By Theorem 5.1.9, there exists an integer $d_{Pi} > 0$ such that $\mathrm{prem}(P^{d_{Pi}}, \mathbb{T}_i) = 0$ for any $P \in \mathbb{P}$ and $1 \leq i \leq e$. It follows that $P^{d_{Pi}} \in \mathrm{sat}(\mathbb{T}_i)$. Let

$$d = \max_{\substack{P \in \mathbb{P} \\ 1 \leq i \leq e}} d_{Pi}.$$

We have $P^d \in \mathrm{sat}(\mathbb{T}_i)$, and thus $\mathrm{prem}(P^d, \mathbb{T}_i) = 0$ for all $P \in \mathbb{P}$ and $1 \leq i \leq e$ according to Theorem 6.2.4.

b. Suppose otherwise that there exist $m > 0$, $1 \le i \le e$ and $Q \in \mathbb{Q}$ such that $\mathrm{prem}(Q^m, \mathbb{T}_i) = 0$. Then

$$\mathrm{Zero}(\mathbb{T}_i/\mathbb{U}_i) \subset \mathrm{Zero}(\mathbb{T}_i/\mathrm{ini}(\mathbb{T}_i)) \subset \mathrm{Zero}(Q).$$

This contradicts the fact that $\mathrm{Zero}(\mathbb{T}_i/\mathbb{U}_i) \subset \mathrm{Zero}(\mathbb{P}/\mathbb{Q})$.

c. By part a and the pseudo-remainder formula, the right-hand side is clearly contained in the left-hand side of (5.1.7).

Now, consider any $\bar{\boldsymbol{x}} \in \mathrm{Zero}(\mathbb{P}/\mathbb{Q})$. Then there exists an i such that

$$\bar{\boldsymbol{x}} \in \mathrm{Zero}(\mathbb{T}_i/\mathbb{U}_i) \subset \mathrm{Zero}(\mathbb{T}_i/\mathrm{ini}(\mathbb{T}_i) \cup \mathbb{Q}).$$

Hence, $\bar{\boldsymbol{x}}$ belongs to the right-hand side of (5.1.7). The theorem is proved. □

In view of Theorem 5.1.11 c, it is proper to call $\mathbb{T}_1, \ldots, \mathbb{T}_e$ a *regular series* of $\mathbb{P}$ when $[\mathbb{T}_1, \mathbb{U}_1], \ldots, [\mathbb{T}_e, \mathbb{U}_e]$ is a regular series of $\mathbb{P}$.

Let $\mathfrak{T} = [\mathbb{T}, \mathbb{U}]$ be a regular system and write $\mathbb{T}$ in the form (5.1.1) with $\mathrm{ini}(T_i) = I_i$ for each i. Let

$$R = \prod_{U \in \mathbb{U}} \mathrm{res}(U, \mathbb{T}) \in \boldsymbol{K}[\boldsymbol{u}].$$

Then, $R \neq 0$ by Corollary 5.1.2 and Proposition 5.1.5, and

$$\mathrm{Zero}(\mathbb{T}/R) \subset \mathrm{Zero}(\mathfrak{T}).$$

Clearly, $I_1(\boldsymbol{u}) \neq 0$ and thus T_1 has a zero η_1 for y_1 in $\boldsymbol{K}(\boldsymbol{u})$. By Proposition 5.1.4, $I_2(\boldsymbol{u}, \eta_1) \neq 0$. Therefore $T_2(\boldsymbol{u}, \eta_1, y_2)$ has a zero η_2 for y_2 in $\boldsymbol{K}(\boldsymbol{u})(\eta_1)$. It follows from Proposition 5.1.4 that $I_3(\boldsymbol{u}, \eta_1, \eta_2) \neq 0$. Continuing in this way, one can obtain a regular zero $(\boldsymbol{u}, \eta_1, \ldots, \eta_r)$ of $[\mathbb{T}, \{R\}]$ and thus of $\mathfrak{T}$. Hence $\mathfrak{T}$ is perfect.

Furthermore, one can construct a zero of $\mathfrak{T}$ with specialized values $\bar{\boldsymbol{u}}$ of $\boldsymbol{u}$. In other words, we have the following.

Theorem 5.1.12. Any regular system in $\boldsymbol{K}[\boldsymbol{x}]$ is perfect over the algebraic closure $\bar{\boldsymbol{K}}$ of $\boldsymbol{K}$.

Proof. Let $[\mathbb{T}, \mathbb{U}]$ be a regular system with $\mathbb{T} = [T_1, \ldots, T_r]$ and

$$\mathrm{cls}(T_i) = p_i, \quad \mathrm{ini}(T_i) = I_i, \quad 1 \le i \le r.$$

Obviously, there exists an

$$\bar{\boldsymbol{x}}^{\{p_1-1\}} \in \mathrm{Zero}(\emptyset/\mathbb{U}^{(p_1-1)}).$$

As $[\mathbb{T}, \mathbb{U}]$ is a triangular system, $I_1(\bar{\boldsymbol{x}}^{\{p_1-1\}}) \neq 0$. Hence, $T_1(\bar{\boldsymbol{x}}^{\{p_1-1\}}, x_{p_1})$ has a zero $\bar{x}_{p_1}$ in some algebraic extension of $\boldsymbol{K}$ for x_{p_1}. Since $\mathbb{U}^{\langle p_1\rangle} = \emptyset$ and $\mathrm{ini}(U)(\bar{\boldsymbol{x}}^{\{j-1\}}) \neq 0$ for any $U \in \mathbb{U}^{\langle j\rangle}$, $\bar{\boldsymbol{x}}^{\{j-1\}} \in \mathrm{Zero}(T_1/\mathbb{U}^{(j-1)})$ and $j = p_1 + 1, \ldots, p_2 - 1$, one can choose $\bar{x}_{p_1+1}, \ldots, \bar{x}_{p_2-1}$ in $\bar{\boldsymbol{K}}$ such that

$$\bar{\boldsymbol{x}}^{\{p_2-1\}} \in \mathrm{Zero}(T_1/\mathbb{U}^{(p_2-1)}).$$

Thus, $I_2(\bar{x}^{\{p_2-1\}}) \neq 0$ because $[\mathbb{T}, \mathbb{U}]$ is a triangular system. Therefore, $T_2(\bar{x}^{\{p_2-1\}}, x_{p_2})$ has a zero $\bar{x}_{p_2}$ in some algebraic extension of $\boldsymbol{K}$ for x_{p_2}. Continuing in this way, we shall finally construct a zero $\bar{x}$ of $[\mathbb{T}, \mathbb{U}]$, so $\text{Zero}(\mathbb{T}/\mathbb{U}) \neq \emptyset$ in $\bar{\boldsymbol{K}}$. □

We may list some corollaries of this theorem as follows.

Corollary 5.1.13. Any regular set $\mathbb{T} \subset \boldsymbol{K}[\boldsymbol{x}]$ is perfect.

Proof. As $\mathbb{T}$ is regular, there exists a polynomial set $\mathbb{U}$ such that $[\mathbb{T}, \mathbb{U}]$ is regular and thus $\text{Zero}(\mathbb{T}/\mathbb{U}) \neq \emptyset$. The corollary is proved by observing that $\text{Zero}(\mathbb{T}/\mathbb{U}) \subset \text{Zero}(\mathbb{T}/\text{ini}(\mathbb{T}))$. □

Corollary 5.1.14. For any polynomial system $\mathfrak{P}$ in $\boldsymbol{K}[\boldsymbol{x}]$, $\text{Zero}(\mathfrak{P}) = \emptyset$ if and only if any regular series of $\mathfrak{P}$ is empty.

Corollary 5.1.15. Let $\mathfrak{P} = [\mathbb{P}, \mathbb{Q}]$ be a polynomial system and P a polynomial in $\boldsymbol{K}[\boldsymbol{x}]$, and let Ψ and Ψ^* be any regular series of $\mathfrak{P}$ and $[\mathbb{P}, \mathbb{Q} \cup \{P\}]$, respectively. The following are equivalent:

a. $\text{Zero}(\mathfrak{P}) \subset \text{Zero}(P)$;
b. $\Psi^* = \emptyset$;
c. $\text{op}(2, \text{Split}(\mathbb{T}, P, n)) = \emptyset$ for all $\mathbb{T} \in \Psi$.

Several results will be proved in the following chapter for arbitrary triangular sets. From those results, special properties such as unmixed dimensionality for regular systems may be obtained.

Let $\mathbb{T}$ as in (5.1.1) be a regular set with $d_i = \text{ldeg}(T_i)$ and $d = d_1 \cdots d_r$; $\mathbb{T}$ is perfect. If $\mathbb{T}$ is irreducible, then it has d distinct regular zeros which are also called *generic zeros* of $\mathbb{T}$ and generate the same extension field of $\boldsymbol{K}$. If $\mathbb{T}$ is simple and reducible, then it has d distinct regular zeros which generate more than one extension field of $\boldsymbol{K}$ of the same transcendence degree. If $\mathbb{T}$ is reducible but not simple, then it has less than d distinct regular zeros which generate one or more extension fields of $\boldsymbol{K}$ of the same transcendence degree.

The above remakrs may help understand the difference among regular set, simple set, and irreducible triangular set. The term "regular zero" which was introduced by Kalkbrener (1993) for a regular set is used here for an arbitrary triangular system. It can be understood as "generic zero," but this notion has been used in algebraic geometry exclusively for irreducible varieties and the corresponding irreducible triangular sets.

5.2 Canonical triangular sets

One gain of introducing regular sets is Corollary 5.1.13, which ensures the nonemptiness of $\text{Zero}(\mathbb{T}/\text{ini}(\mathbb{T}))$ for any triangular set $\mathbb{T}$ that is regular and may be reducible. Now, we want to impose more restrictions, but not irreducibility, on triangular sets in order to make them canonical.

Definition 5.2.1. A triangular system $[\mathbb{T}, \mathbb{U}]$ in $\boldsymbol{K}[\boldsymbol{x}]$ is said to be *normal* if

$$\deg(I, \text{lv}(T)) = 0 \quad \text{for any } T \in \mathbb{T} \text{ and } I \in \text{ini}(\mathbb{T} \cup \mathbb{U}).$$

A triangular set $\mathbb{T}$ is said to be *normal* if $[\mathbb{T}, \text{ini}(\mathbb{I})]$ is normal.

In other words, the initial of any polynomial in a triangular system $[\mathbb{T}, \mathbb{U}]$ does not involve the dependents of $\mathbb{T}$. A normal triangular set is called a *p-chain* in Gao and Chou (1992). When $\mathbb{T}$ is normal, it is quite trivial to perform projection for $[\mathbb{T}, \text{ini}(\mathbb{I})]$ (see Sect. 3.1). The following algorithm exhibits how to compute a normal simple set from any simple set.

Algorithm Norm: $[\mathbb{T}^*, \mathbb{F}] \leftarrow \text{Norm}(\mathbb{T})$. Given a simple set $\mathbb{T} \subset \boldsymbol{K}[\boldsymbol{x}]$, this algorithm computes a normal simple set $\mathbb{T}^*$ and a polynomial set $\mathbb{F}$ such that

$$\text{Zero}(\mathbb{T}/\tilde{\mathbb{T}}) = \text{Zero}(\mathbb{T}^*/\tilde{\mathbb{T}} \cup \mathbb{F}) \cup \bigcup_{F \in \mathbb{F}} \text{Zero}(\mathbb{T} \cup \{F\}/\tilde{\mathbb{T}})$$

and $\deg(F, \text{lv}(T)) = 0$ for any $F \in \mathbb{F}$ and $T \in \mathbb{T}$, where $\tilde{\mathbb{T}}$ is $\emptyset$ or any triangular set that makes $[\mathbb{T}, \tilde{\mathbb{T}}]$ a simple system.

N1. Let the polynomials in $\mathbb{T}$ be $T_1, \ldots, T_r$ and set $\mathbb{F} \leftarrow \emptyset$.
N2. For $i = r, \ldots, 2$ do:
  N2.1. Compute $R \leftarrow \text{res}(\text{ini}(T_i), [T_1, \ldots, T_{i-1}])$ and a polynomial Q such that

$$Q_1 T_1 + \cdots + Q_{i-1} T_{i-1} + Q \cdot \text{ini}(T_i) = R$$

  for some $Q_1, \ldots, Q_{i-1} \in \boldsymbol{K}[\boldsymbol{x}]$.
  N2.2. Compute $T_i^* \leftarrow R \cdot \text{lv}(T_i)^{\text{ldeg}(T_i)} + Q \cdot \text{red}(T_i)$. If $R \notin \boldsymbol{K}$ and $\text{sqfr}(R) \nmid \prod_{F \in \text{ini}(\mathbb{T}) \cup \mathbb{F}} F$, then set $\mathbb{F} \leftarrow \mathbb{F} \cup \{R\}$.
N3. Set $\mathbb{T}^* \leftarrow [T_1, T_2^*, \ldots, T_r^*]$.

Proof. Let $\mathbb{T} = [T_1, \ldots, T_r]$ with

$$p_i = \text{cls}(T_i), \quad I_i = \text{ini}(T_i), \quad d_i = \text{ldeg}(T_i), \quad 1 \le i \le r,$$

and

$$R_i = \text{res}(I_i, [T_1, \ldots, T_{i-1}]), \quad 2 \le i \le r.$$

Since $\mathbb{T}$ is simple, by Corollary 5.1.6 R_i is a nonzero polynomial not involving the variables $x_{p_1}, \ldots, x_{p_{i-1}}$ for each i. In other words, $\deg(R_i, x_{p_j}) = 0$ for any pair of i and j. By Lemma 4.3.2, there are polynomials Q_{ij} and Q_i such that

$$\sum_{j=1}^{i-1} Q_{ij} T_j + Q_i I_i = R_i, \quad 2 \le i \le r. \tag{5.2.1}$$

Let

$$\begin{aligned} T_i^* &= R_i x_{p_i}^{d_i} + Q_i \cdot \text{red}(T_i), \quad 2 \le i \le r, \\ \mathbb{T}^* &= [T_1, T_2^*, \ldots, T_r^*], \\ \mathbb{F} &= \{R_2, \ldots, R_r\}. \end{aligned}$$

If $R_i \in \boldsymbol{K}$ or every irreducible factor of R_i is a divisor of some polynomial in $\text{ini}(\mathbb{T})$ or another R_j for $j \neq i$, then R_i is not needed and can be deleted from $\mathbb{F}$.

Let $\tilde{\mathbb{T}}$ be $\emptyset$ or any triangular set such that $[\mathbb{T}, \tilde{\mathbb{T}}]$ makes up a simple system. We now show that

$$\operatorname{Zero}(\mathbb{T}/\tilde{\mathbb{T}}) = \operatorname{Zero}(\mathbb{T}^*/\tilde{\mathbb{T}} \cup \mathbb{F}) \cup \bigcup_{i=2}^{r} \operatorname{Zero}(\mathbb{T} \cup \{R_i\}/\tilde{\mathbb{T}}). \tag{5.2.2}$$

For this purpose, consider any i and let

$$\bar{\boldsymbol{x}}^{\{p_i-1\}} \in \operatorname{Zero}([T_1, \ldots, T_{i-1}]/\tilde{\mathbb{T}}^{(p_i-1)} \cup \mathbb{F}).$$

One knows from (5.2.1) that

$$Q_i(\bar{\boldsymbol{x}}^{\{p_i-1\}}) I_i(\bar{\boldsymbol{x}}^{\{p_i-1\}}) = R_i(\bar{\boldsymbol{x}}^{\{p_i-1\}}) \neq 0,$$

so after $\boldsymbol{x}^{\{p_i-1\}}$ is substituted by $\bar{\boldsymbol{x}}^{\{p_i-1\}}$

$$T_i^* = Q_i T_i = R_i x_{p_i}^{d_i} + Q_i \cdot \operatorname{red}(T_i)$$

has the same set of d_i distinct zeros as T_i for x_{p_i} (and thus is squarefree). It follows that

$$\operatorname{Zero}(\mathbb{T}/\tilde{\mathbb{T}} \cup \mathbb{F}) = \operatorname{Zero}(\mathbb{T}^*/\tilde{\mathbb{T}} \cup \mathbb{F})$$

and thus the zero relation (5.2.2) holds.

Apparently, $\mathbb{T}^*$ is normal (but $[\mathbb{T}^*, \tilde{\mathbb{T}} \cup \mathbb{F}]$ is not necessarily a simple system). It remains to show that $\mathbb{T}^*$ is a simple set. In fact, one can construct a triangular or empty set $\tilde{\mathbb{T}}^*$ from $\tilde{\mathbb{T}} \cup \mathbb{F}$ such that $[\mathbb{T}^*, \tilde{\mathbb{T}}^*]$ is a simple system. The construction proceeds as follows. Let $R = R_2 \cdots R_r$. We repeat the following until $R \in \boldsymbol{K}$:

1. If there exists a $T \in \tilde{\mathbb{T}}$ such that $\operatorname{cls}(T) = \operatorname{cls}(R)$, then set

$$R \leftarrow R\,T, \quad \tilde{\mathbb{T}} \leftarrow \tilde{\mathbb{T}} \setminus \{T\}.$$

2. Compute $\tilde{R} \leftarrow \operatorname{sqfr}(R)$ and set

$$\tilde{\mathbb{T}} \leftarrow \tilde{\mathbb{T}} \cup \{\tilde{R}\}, \quad R \leftarrow \operatorname{ini}(\tilde{R}) \cdot \operatorname{res}(\tilde{R}, \partial \tilde{R}/\partial \operatorname{lv}(\tilde{R}), \operatorname{lv}(\tilde{R})).$$

Let $\tilde{\mathbb{T}}^*$ be the final $\tilde{\mathbb{T}}$, ordered as a triangular set if it is nonempty. Then it is not difficult to verify that $[\mathbb{T}^*, \tilde{\mathbb{T}}^*]$ is a simple system by definition (see the proof of Proposition 4.3.7 for a similar verification). Therefore, $\mathbb{T}^*$ is a normal simple set. □

Lemma 5.2.1. From any normal simple set $\mathbb{T} \subset \boldsymbol{K}[\boldsymbol{x}]$, one can compute a normal, reduced and primitive simple set $\mathbb{T}^*$ such that

$$\operatorname{Zero}(\mathbb{T}/\operatorname{ini}(\mathbb{T})) = \operatorname{Zero}(\mathbb{T}^*/\operatorname{ini}(\mathbb{T})).$$

Proof. Let $\mathbb{T} = [T_1, \ldots, T_r]$ and

$$T_i^* = \operatorname{pp}(\operatorname{prem}(T_i, \mathbb{T}^{\{i-1\}}), \operatorname{lv}(T_i)), \quad 2 \le i \le r.$$

As $\mathbb{T}$ is normal, T_i^* is clearly well-defined and primitive with $\text{cls}(T_i^*) = \text{cls}(T_i)$. Set

$$\mathbb{T}^* = [T_1, T_2^*, \ldots, T_r^*].$$

Then $\mathbb{T}^*$ is reduced and primitive, and the zero relation is easily verified. □

Remark 5.2.1. The normal simple set $\mathbb{T}^*$ and the polynomial set $\mathbb{F}$ computed from a simple set $\mathbb{T}$ by algorithm Norm possess the following property: For any polynomial G and triangular or empty set $\tilde{\mathbb{T}}$ with $[\mathbb{T}, \tilde{\mathbb{T}}]$ a simple system,

$$\text{Zero}(\mathbb{T}^*/\tilde{\mathbb{T}} \cup \mathbb{F}) \subset \text{Zero}(G) \iff \text{prem}(G, \mathbb{T}) = 0.$$

The property holds still when $\mathbb{T}^*$ is made reduced and primitive according to Lemma 5.2.1. The proof is an analogy to the proof of Theorem 3.4.4. One needs to note that all the polynomials in $\mathbb{F}$ do not involve the dependents of $\mathbb{T}^*$.

In fact, algorithm Norm works as well for any regular set $\mathbb{T}$, with respect to which the resultant R of any $I \in \text{ini}(\mathbb{T})$ never vanishes identically. One can also try to normalize an arbitrary triangular or empty set $\mathbb{T}$, but there is no guarantee to succeed. The following alternative algorithm does the job and returns a normalized triangular set when successful. It always succeeds when $\mathbb{T}$ is regular, simple, or irreducible.

Algorithm NormG: $[\mathbb{T}^*, \mathbb{F}] \leftarrow \text{NormG}(\mathbb{T})$. Given a triangular set $\mathbb{T} \subset \boldsymbol{K}[\boldsymbol{x}]$, this algorithm computes a pair $[\mathbb{T}^*, \mathbb{F}]$ such that either $\mathbb{T}^* = \texttt{Fail}$ (in this case the algorithm fails), or $\mathbb{T}^*$ is a normal triangular set and $\mathbb{F}$ a polynomial set satisfying

$$\text{Zero}(\mathbb{T}/\mathbb{F}) \subset \text{Zero}(\mathbb{T}^*), \quad \text{Zero}(\mathbb{T}^*/\text{ini}(\mathbb{T}^*)) \subset \text{Zero}(\mathbb{T}/\text{ini}(\mathbb{T})). \quad (5.2.3)$$

N1. Let the polynomials in $\mathbb{T}$ be $T_1, \ldots, T_r$ and set $\mathbb{F} \leftarrow \emptyset$, $T_r^* \leftarrow T_r$. If $r = 1$, then set $\mathbb{T}^* \leftarrow [T_1^*]$ and the procedure terminates.

N2. For $i = r-1, \ldots, 1$ do:

N2.1. Set $I \leftarrow \text{ini}(T_r^*)$. If $\text{cls}(I) < \text{cls}(T_i)$, then go to N3; else set $y \leftarrow \text{lv}(T_i)$.

N2.2. Compute $R \leftarrow \text{gcd}(T_i, I, y)$ and a polynomial Q such that $R = PT_i + QI$ for some $P \in \boldsymbol{K}[\boldsymbol{x}]$.

N2.3. If $\text{cls}(R) < \text{cls}(T_i)$, then go to N2.4. Otherwise, compute

$$D \leftarrow \text{Remo}(T_i/R, R, y)$$

and set $\mathbb{F} \leftarrow \mathbb{F} \cup \{R\}$. If $\text{cls}(D) = \text{cls}(T_i)$, then set $T_i \leftarrow D$; else set $\mathbb{T}^* \leftarrow \texttt{Fail}$ and the procedure terminates.

N2.4. Set $T_r^* \leftarrow R \cdot \text{lv}(T_r^*)^{\text{ldeg}(T_r^*)} + Q \cdot \text{red}(T_r^*)$.

N3. Compute $[\mathbb{T}^\star, \mathbb{F}^\star] \leftarrow \text{NormG}([T_1, \ldots, T_{r-1}])$. If $\mathbb{T}^\star = \texttt{Fail}$, then set $\mathbb{T}^* \leftarrow \texttt{Fail}$; else set $\mathbb{F} \leftarrow \mathbb{F} \cup \mathbb{F}^\star$, $\mathbb{T}^* \leftarrow \mathbb{T}^\star \cup [T_r^*]$.

The simple subalgorithm Remo is given below.

Algorithm Remo: $H \leftarrow \text{Remo}(F, G, x_k)$. Given two polynomials F and G in $\boldsymbol{K}[\boldsymbol{x}]$ and a variable x_k, this algorithm computes a polynomial H such that $\text{gcd}(H, G, x_k)$ does not involve x_k.

Set $R \leftarrow \gcd(F, G, x_k)$.
If $\deg(R, x_k) = 0$, then set $H \leftarrow F$; else compute $H \leftarrow \mathrm{Remo}(F/R, G, x_k)$.

Proof. For NormG the termination is obvious, so we only need to show its correctness. As in the algorithm, let $|\mathbb{T}| = r$; then $r = 1$ is a trivial case.

For $r > 1$, assume that step N2 has iterated for $i = r - 1, \ldots, k + 1$ and let the current values of $\mathbb{F}$ and $\mathbb{T}$ be denoted $\tilde{\mathbb{F}}$ and

$$\tilde{\mathbb{T}} = [T_1(\boldsymbol{z}^{\{1\}}), \ldots, T_{r-1}(\boldsymbol{z}^{\{r-1\}}), T_r^*(\boldsymbol{z}^{\{r\}})]$$

respectively, where $\boldsymbol{z}^{\{i\}}$ stands for $(\boldsymbol{u}, y_1, \ldots, y_i)$ with $\boldsymbol{z} = \boldsymbol{z}^{\{r\}}$ as usual. Then (5.2.3) holds when $\mathbb{F}$ and $\mathbb{T}^*$ are replaced by $\tilde{\mathbb{F}}$ and $\tilde{\mathbb{T}}$ respectively.

Now consider N2 for iteration $i = k$. Let $I_j = \mathrm{ini}(T_j)$ for $1 \le j \le r - 1$ and $I = \mathrm{ini}(T_r^*)$; then $I \in \boldsymbol{K}[\boldsymbol{z}^{\{k\}}]$. If $\mathrm{cls}(I) < \mathrm{cls}(T_k)$, then proceed with the iteration for $i = k - 1$. Suppose, otherwise, that $\mathrm{cls}(I) = \mathrm{cls}(T_k)$. There are two cases.

Case 1. T_k and I are relatively prime with respect to $y_k = \mathrm{lv}(T_k)$, i.e., $R = \gcd(T_k, I, y_k) \in \boldsymbol{K}[\boldsymbol{z}^{\{k-1\}}]$. This is similar to the case handled by Norm. One can determine polynomials $P, Q \in \boldsymbol{K}[\boldsymbol{z}^{\{k\}}]$ such that

$$PT_k + QI = R \in \boldsymbol{K}[\boldsymbol{z}^{\{k-1\}}]. \tag{5.2.4}$$

Writing T_r^* as $T_r^* = I y_r^d + \mathrm{red}(T_r^*)$ and multiplying both sides of (5.2.4) by y_r^d, one gets

$$QT_r^* = Ry_r^d + Q \cdot \mathrm{red}(T_r^*) - PT_k y_r^d, \tag{5.2.5}$$

where $d = \mathrm{ldeg}(T_r^*)$. Set $\hat{T}_r = Ry_r^d + Q \cdot \mathrm{red}(T_r^*)$. Evidently, $\mathrm{lv}(\hat{T}_r) = \mathrm{lv}(T_r^*) = y_r$. This implies that $\hat{\mathbb{T}} = [T_1, \ldots, T_{r-1}, \hat{T}_r]$ is a triangular set. We want to show that

$$\mathrm{Zero}(\tilde{\mathbb{T}}) \subset \mathrm{Zero}(\hat{\mathbb{T}}), \quad \mathrm{Zero}(\hat{\mathbb{T}}/\mathrm{ini}(\hat{\mathbb{T}})) \subset \mathrm{Zero}(\tilde{\mathbb{T}}/\mathrm{ini}(\tilde{\mathbb{T}})).$$

Since $\hat{T}_r$ can be written as a linear combination of T_k and T_r^* with polynomial coefficients, the first relation holds obviously. Note that $\mathrm{ini}(\hat{T}_r) = R$. Hence, for any $\bar{\boldsymbol{z}} \in \mathrm{Zero}(\hat{\mathbb{T}}/\mathrm{ini}(\hat{\mathbb{T}}))$ one has

$$\begin{aligned} &T_j(\bar{\boldsymbol{z}}) = 0, \quad I_j(\bar{\boldsymbol{z}}) \neq 0, \quad 1 \le j \le r - 1, \\ &I(\bar{\boldsymbol{z}}) \neq 0, \quad R(\bar{\boldsymbol{z}}) \neq 0. \end{aligned}$$

From (5.2.5) and the determination of $\hat{T}_r$, one sees that $Q(\bar{\boldsymbol{z}})T_r^*(\bar{\boldsymbol{z}}) = 0$. On the other hand, $Q(\bar{\boldsymbol{z}})I(\bar{\boldsymbol{z}}) \neq 0$ by (5.2.4). It follows that

$$T_r^*(\bar{\boldsymbol{z}}) = 0, \quad I(\bar{\boldsymbol{z}}) \neq 0.$$

Therefore, $\bar{\boldsymbol{z}} \in \mathrm{Zero}(\tilde{\mathbb{T}}/\mathrm{ini}(\tilde{\mathbb{T}}))$ and the second zero relation is proved.

Case 2. T_k and I are not relatively prime with respect to y_k. In this case, they have a common divisor whose leading variable is y_k. Let us simply remove all possible factors of R, the GCD of T_k and I with respect to y_k, from T_k as done by the subalgorithm Remo and denote the obtained polynomial by D. If $\mathrm{cls}(D) < \mathrm{cls}(T_k)$, then the algorithm terminates with $\mathbb{T}^* = \mathtt{Fail}$ returned. Otherwise,

$$\mathbb{T}' = [T_1, \ldots, T_{k-1}, D, T_{k+1}, \ldots, T_{r-1}, T_r^*]$$

is a triangular set. Thus,

$$\mathrm{Zero}(\tilde{\mathbb{T}}/R) \subset \mathrm{Zero}(\mathbb{T}'), \quad \mathrm{Zero}(\tilde{\mathbb{T}}/\mathrm{ini}(\tilde{\mathbb{T}})) = \mathrm{Zero}(\mathbb{T}'/\mathrm{ini}(\mathbb{T}')).$$

As D and I now are relatively prime with respect to y_k, the problem is reduced, by regarding $\mathbb{T}'$ as $\tilde{\mathbb{T}}$, to case 1. Therefore, one can determine $\hat{\mathbb{T}}$ and $\hat{\mathbb{F}}$ such that

$$\begin{aligned} &\mathrm{Zero}(\tilde{\mathbb{T}}/\hat{\mathbb{F}}) \subset \mathrm{Zero}(\mathbb{T}') \subset \mathrm{Zero}(\hat{\mathbb{T}}), \\ &\mathrm{Zero}(\hat{\mathbb{T}}/\mathrm{ini}(\hat{\mathbb{T}})) \subset \mathrm{Zero}(\mathbb{T}'/\mathrm{ini}(\mathbb{T}')) = \mathrm{Zero}(\tilde{\mathbb{T}}/\mathrm{ini}(\tilde{\mathbb{T}})). \end{aligned}$$

Hence, in any case the iteration step N2 either fails with $\mathbb{T}^* = \mathtt{Fail}$ or produces a sequence of triangular sets $\mathbb{T} = \mathbb{T}_r, \ldots, \mathbb{T}_1$ and polynomial sets $\mathbb{F}_{r-1}, \ldots, \mathbb{F}_1$ satisfying

$$\begin{aligned} &\mathrm{Zero}(\mathbb{T}_r/\mathbb{F}_{r-1}) \subset \mathrm{Zero}(\mathbb{T}_{r-1}), \ldots, \mathrm{Zero}(\mathbb{T}_2/\mathbb{F}_1) \subset \mathrm{Zero}(\mathbb{T}_1), \\ &\mathrm{Zero}(\mathbb{T}_1/\mathrm{ini}(\mathbb{T}_1)) \subset \cdots \subset \mathrm{Zero}(\mathbb{T}_{r-1}/\mathrm{ini}(\mathbb{T}_{r-1})) \\ &\quad \subset \mathrm{Zero}(\mathbb{T}_r/\mathrm{ini}(\mathbb{T}_r)). \end{aligned}$$

Setting $\bar{\mathbb{F}} = \mathbb{F}_{r-1} \cup \cdots \cup \mathbb{F}_1$, we have

$$\begin{aligned} &\mathrm{Zero}(\mathbb{T}/\bar{\mathbb{F}}) = \mathrm{Zero}(\mathbb{T}_r/\bar{\mathbb{F}}) \subset \mathrm{Zero}(\mathbb{T}_1), \\ &\mathrm{Zero}(\mathbb{T}_1/\mathrm{ini}(\mathbb{T}_1)) \subset \mathrm{Zero}(\mathbb{T}_r/\mathrm{ini}(\mathbb{T}_r)) = \mathrm{Zero}(\mathbb{T}/\mathrm{ini}(\mathbb{T})). \end{aligned}$$

Let

$$\mathbb{T}_1 = [T_1', \ldots, T_r'], \quad \mathbb{T}_1' = [T_1', \ldots, T_{r-1}'].$$

Observe that $\mathrm{ini}(T_r') \in \boldsymbol{K}[\boldsymbol{u}]$. Since $\mathbb{T}_1'$ contains $r-1$ polynomials, one can compute, if not failing, a fine normal triangular set $\mathbb{T}^\star$ and a polynomial set $\mathbb{F}^\star$ by induction as in step N3 such that

$$\mathrm{Zero}(\mathbb{T}_1'/\mathbb{F}^\star) \subset \mathrm{Zero}(\mathbb{T}^\star), \quad \mathrm{Zero}(\mathbb{T}^\star/\mathrm{ini}(\mathbb{T}^\star)) \subset \mathrm{Zero}(\mathbb{T}_1'/\mathrm{ini}(\mathbb{T}_1')).$$

Now, let $\mathbb{T}^* = \mathbb{T}^\star \cup [T_r']$ and $\mathbb{F} = \bar{\mathbb{F}} \cup \mathbb{F}^\star$. Then the zero relations in (5.2.3) hold. As we wanted, all the initials of the polynomials in $\mathbb{T}^*$ are now in $\boldsymbol{K}[\boldsymbol{u}]$; therefore, they are all reduced with respect to $\mathbb{T}^*$. In other words, $\mathbb{T}^*$ is a fine normal triangular set, and the correctness of the algorithm is proved. □

Remark 5.2.2. For the normal triangular set $\mathbb{T}^*$ computed from any triangular set $\mathbb{T}$ by algorithm Norm or NormG, there is no guarantee that

$$\mathrm{Zero}(\mathbb{T}/\mathrm{ini}(\mathbb{T})) = \mathrm{Zero}(\mathbb{T}^*/\mathrm{ini}(\mathbb{T}^*)),$$

even if $\mathbb{T}$ is simple. This is why the additional polynomial set $\mathbb{F}$ needs to be computed by algorithm Norm. Consider, for example,

$$\mathbb{T} = [x_2^2 + x_1, (x_3 - x_2)x_4 + 1].$$

It is a simple set with respect to $x_1 \prec \cdots \prec x_4$ because $\mathfrak{S} = [\mathbb{T}, [x_1, x_3 - x_2]]$ is a simple system. $\mathbb{T}$ is also irreducible. Normalization of $\mathbb{T}$ yields

$$\mathbb{T}^* = [x_2^2 + x_1, (x_3^2 + x_1)x_4 + x_3 + x_2].$$

Now

$$\begin{aligned}\mathrm{Zero}(\mathbb{T}/\mathrm{ini}(\mathbb{T})) &= \mathrm{Zero}(\mathbb{T}/(x_3 - x_2)) \neq \mathrm{Zero}(\mathbb{T}^*/(x_3^2 + x_1)) \\ &= \mathrm{Zero}(\mathbb{T}^*/\mathrm{ini}(\mathbb{T}^*)).\end{aligned}$$

This may be seen by verifying that

$$\begin{aligned}&(-1, 1, -1, 1/2) \\ &\quad \in \mathrm{Zero}(\mathbb{T}/(x_3 - x_2)), \quad \text{but} \quad \notin \mathrm{Zero}(\mathbb{T}^*/(x_3^2 + x_1)).\end{aligned}$$

In fact, $\mathbb{T}$ may be decomposed into two normal simple sets $\mathbb{T}^*$ and

$$\mathbb{T}' = [x_2^2 + x_1, x_3 + x_2, 2x_1x_4 + x_2]$$

such that

$$\mathrm{Zero}(\mathbb{T}/(x_3 - x_2)) = \mathrm{Zero}(\mathbb{T}^*/(x_3^2 + x_1)) \cup \mathrm{Zero}(\mathbb{T}'/x_1).$$

Also, one cannot get a normal simple system $\mathfrak{S}^*$ from $\mathfrak{S}$ such that

$$\mathrm{Zero}(\mathfrak{S}) = \mathrm{Zero}(\mathfrak{S}^*).$$

$\mathfrak{S}$ may decompose into two normal simple systems

$$\mathfrak{S}^* = [\mathbb{T}^*, [x_1, x_3^2 + x_1]], \quad \mathfrak{S}' = [\mathbb{T}', [x_1]]$$

such that

$$\mathrm{Zero}(\mathfrak{S}) = \mathrm{Zero}(\mathfrak{S}^*) \cup \mathrm{Zero}(\mathfrak{S}').$$

However, if $\mathbb{T}$ is regular, simple or irreducible, then $\mathbb{T}$ and $\mathbb{T}^*$ have the same set of regular or generic zeros. This can be easily proved by using the fact that the resultant R computed in N2.1 of Norm does not vanish at any regular zero of $\mathbb{T}$.

A polynomial P is *monic* if $\mathrm{lc}(T) = 1$. A polynomial set $\mathbb{P}$ is said to be *monic* if every $P \in \mathbb{P}$ is monic.

Definition 5.2.2. A triangular set $\mathbb{T} \subset \boldsymbol{K}[\boldsymbol{x}]$ is said to be *canonical* if it is normal, simple, reduced, primitive and monic.

The definition of a canonical triangular set here is similar to but slightly stronger than that of a triangular set given in Lazard (1991). For example,

$$[x_1^2 - 1, (x_2 - x_1)x_3 + 1]$$

with $x_1 \prec x_2 \prec x_3$ is a triangular set according to Lazard's definition, but it is not canonical by Definition 5.2.2.

Now consider any polynomial set $\mathbb{P}$. One knows how to compute simple systems $[\mathbb{T}_1, \tilde{\mathbb{T}}_1], \ldots, [\mathbb{T}_t, \tilde{\mathbb{T}}_t]$ from $\mathbb{P}$ by algorithm SimSer such that

$$\mathrm{Zero}(\mathbb{P}) = \bigcup_{i=1}^{t} \mathrm{Zero}(\mathbb{T}_i/\tilde{\mathbb{T}}_i).$$

By algorithm Norm and Lemma 5.2.1, one can compute, from each simple set $\mathbb{T}_i$, a reduced, normal, and primitive simple set $\mathbb{T}_i^*$ and a polynomial set $\mathbb{F}_i$ such that

$$\mathrm{Zero}(\mathbb{T}_i/\tilde{\mathbb{T}}_i) = \mathrm{Zero}(\mathbb{T}_i^*/\tilde{\mathbb{T}}_i \cup \mathbb{F}) \cup \bigcup_{F \in \mathbb{F}_i} \mathrm{Zero}(\mathbb{T}_i \cup \{F\}/\tilde{\mathbb{T}}_i).$$

Applying SimSer to each polynomial system $[\mathbb{T}_i \cup \{F\}, \tilde{\mathbb{T}}_i]$, one may obtain other reduced, normal, and primitive simple sets and the corresponding zero decompositions. Since each $F \in \mathbb{F}_i$ does not involve the dependents of $\mathbb{T}_i$, the first triangular set in any simple system from a simple series of $[\mathbb{T}_i \cup \{F\}, \tilde{\mathbb{T}}_i]$ should contain more polynomials than $\mathbb{T}$. Hence, the recursive process must terminate. Finally one should reach a zero decomposition of the form

$$\mathrm{Zero}(\mathbb{P}) = \bigcup_{i=1}^{e} \mathrm{Zero}(\mathbb{T}_i/\tilde{\mathbb{T}}_i), \tag{5.2.6}$$

where each triangular set $\mathbb{T}_i$ is normal, simple, reduced, and primitive. According to Remark 5.2.1, $\mathrm{prem}(P, \mathbb{T}_i) = 0$ for any $P \in \mathbb{P}$. A simple reasoning similar to the proof of Theorem 3.4.6 shows that each $\tilde{\mathbb{T}}_i$ in (5.2.6) can be replaced by $\mathrm{ini}(\mathbb{T}_i)$. For every $T \in \mathbb{T}_i$, it is trivial to make T monic: one divides T by $\mathrm{lc}(T)$. The following theorem is therefore established.

Theorem 5.2.2. There is an algorithm which computes, from any polynomial set $\mathbb{P} \subset \boldsymbol{K}[\boldsymbol{x}]$, a finite number of canonical triangular sets $\mathbb{T}_1, \ldots, \mathbb{T}_e$ such that

$$\mathrm{Zero}(\mathbb{P}) = \bigcup_{i=1}^{e} \mathrm{Zero}(\mathbb{T}_i/\mathrm{ini}(\mathbb{T}_i)).$$

The above zero decomposition is not necessarily *minimal*. Some redundant zero sets may be removed by using Corollary 3.4.5.

Example 5.2.1. Refer to the polynomial set $\mathbb{P}$ in Example 2.4.1 and its simple series in Example 3.3.4. The simple sets $\mathbb{T}_i$ are normal only for $i = 2, 4, 5$ but not for the others. Let us first consider $\mathbb{T}_1 = [T_1, T_2, T_3]$, where

$$\begin{aligned} T_1 &= z^3 - z^2 + r^2 - 1, \\ T_2 &= x^4 + z^2x^2 - r^2x^2 + z^4 - 2z^2 + 1, \\ T_3 &= xy + z^2 - 1. \end{aligned}$$

One sees that $\text{ini}(T_1) = \text{ini}(T_2) = 1$ and $\text{ini}(T_3) = x$. It is easy to verify that

$$R = \text{res}(x, [T_1, T_2]) = (r^2 - 1)^2(r^2 - 3)^2 = xQ + Q_1T_1 + Q_2T_2,$$

where

$$\begin{aligned} Q = & -x(x^2 + z^2 - r^2)(r^4z^2 - 2r^2z^2 + 2z^2 - 2r^4z \\ & + 3r^2z - z + 3r^4 - 7r^2 + 4). \end{aligned}$$

All irreducible factors of R are divisors of the only polynomial in $\tilde{\mathbb{T}}_1$ (see Example 3.3.4), so R is not needed. Hence, the output $\mathbb{F}$ from $\text{Norm}(\mathbb{T}_1)$ is empty, and $\mathbb{T}_1$ is normalized to $\mathbb{T}_1^* = [T_1, T_2, T_3^*]$ with $T_3^* = Ry + Q(z^2 - 1)$ such that

$$\text{Zero}(\mathbb{T}_1/\tilde{\mathbb{T}}_1) = \text{Zero}(\mathbb{T}_1^*/\mathbb{U}_1).$$

Reducing T_3^* by T_2 and T_1 and taking the primitive part of the remainder, we have

$$\begin{aligned} \hat{T}_3 &= \text{pp}(\text{prem}(T_3^*, [T_1, T_2]), y) \\ &= (r^4 - 4r^2 + 3)y - z^2x^3 + r^2zx^3 - zx^3 - r^2x^3 + x^3 \\ &\quad + r^2z^2x - z^2x - r^4zx + 2r^2zx - zx + 2r^2x - 2x. \end{aligned}$$

$\hat{T}_3$ is monic, so $\hat{\mathbb{T}}_1 = [T_1, T_2, \hat{T}_3]$ is a canonical triangular set.

Observe that for the other abnormal simple sets, the corresponding resultants R_i are all constants. This is because $|\mathbb{T}_i| = 4$, the number of variables, for $i > 1$. Therefore, one can obtain a canonical triangular set $\hat{\mathbb{T}}_i$ from each $\mathbb{T}_i$ for $i = 3, 6, \ldots, 9$. The polynomials in these canonical triangular sets should all have constant initials. In particular, $\hat{\mathbb{T}}_i = \mathbb{T}_i$ for $i = 2, 3, 5$. Thus, we have

$$\text{Zero}(\mathbb{P}) = \text{Zero}(\hat{\mathbb{T}}_1/(r^2 - 1)(r^2 - 3)) \cup \bigcup_{i=2}^{9} \text{Zero}(\hat{\mathbb{T}}_i).$$

This decomposition is not minimal: $\text{Zero}(\hat{\mathbb{T}}_i)$ can be removed for $i = 3, 4, 6, \ldots, 9$. In other words, the summation index i ranges only for 2 and 5, viz.,

$$\text{Zero}(\mathbb{P}) = \text{Zero}(\hat{\mathbb{T}}_1/(r^2 - 1)(r^2 - 3)) \cup \text{Zero}(\mathbb{T}_2) \cup \text{Zero}(\mathbb{T}_5).$$

In the above example, a number of redundant simple sets are computed, normalized, and finally removed in order to arrive at a canonical zero decomposition. A crucial question is how to avoid computing such redundant simple sets or systems. A complete answer to this question is not easy, but in practice one must develop effective strategies to detect the redundant components as early as possible. When efficiency is of concern, one is advised to compute irreducible triangular series rather than simple series. A canonical zero decomposition can be obtained more easily via the former than via the latter. As we have mentioned early, simple series is of value more theoretically than practically.

The normalization process may also be incorporated into SimSer and other decomposition algorithms. Moreover, resultant computation can be substituted by subresultant computation; the latter has been used in several algorithms including SimSer and RegSer. Actually, one can design an algorithm that computes, from any polynomial set, a simple or regular series with all simple or regular systems therein normal. For each normal simple or regular system $[\mathbb{T}, \tilde{\mathbb{T}}]$, one can also require that every polynomial $P \in \mathbb{T} \cup \tilde{\mathbb{T}}$ does not involve the dependents of $\mathbb{T} \setminus [P]$. We do not go any further in this direction.

Another algorithm to decompose polynomial sets into canonical triangular sets is presented in Lazard (1991). It makes use of incremental computations over field extensions and is rather involved. A technical description of the algorithm is provided without formal proof in the cited reference.

5.3 Gröbner bases

The method of Gröbner bases introduced by Buchberger (1965) provides another powerful device for polynomial elimination. It has been well studied and described in great detail in several books including Adams and Loustaunau (1994), Becker and Weispfenning (1993), Cox et al. (1992, chap. 2), and Mishra (1993, chaps. 2 and 3), so we have no intention to give another comprehensive exposition. We shall be satisfied by only giving a brief review of the method with emphasis on its elimination aspects.

With a fixed variable ordering, one may introduce different *admissible* term orderings. Two commonly used examples of them are the *total degree* and *purely lexicographical* orderings. For our purpose of variable elimination, we shall use the purely lexicographical term ordering which has been explained in Sect. 1.1. Some of the notations used below are also given there. All the polynomials mentioned in this section are assumed to be in $\boldsymbol{K}[\boldsymbol{x}]$.

Buchberger's algorithm

Definition 5.3.1. Let $\mathbb{P}$ be a polynomial set and G any polynomial in $\boldsymbol{K}[\boldsymbol{x}]$. G is said to be *reducible* with respect to $\mathbb{P}$ if there exist a polynomial $P \in \mathbb{P}$ and a term λ such that $\mathrm{coef}(G, \lambda \cdot \mathrm{lt}(P)) \neq 0$. If no such P and λ exist, G is said to be *reduced* or in *normal form* with respect to $\mathbb{P}$.

If G is reducible with respect to $\mathbb{P}$, then one can find a polynomial $P \in \mathbb{P}$ with the term $\lambda \cdot \mathrm{lt}(P)$ maximal (with respect to the term ordering) such that

$$G = b \cdot \lambda \cdot P + H,$$

where

$$b = \frac{\mathrm{coef}(G, \lambda \cdot \mathrm{lt}(P))}{\mathrm{lc}(P)}.$$

This is a one-step reduction of G to H so that one term of G is eliminated. In other words, the term $\lambda \cdot \mathrm{lt}(P)$ does not appear in H.

If H is reducible with respect to $\mathbb{P}$, then one can reduce H to another polynomial in the same way by choosing P, b, and λ. As the reduction is a Noetherian relation, such a process will terminate. That is, after a finite number of reduction steps, the obtained polynomial R will be reduced with respect to $\mathbb{P}$. In this case, one gets a *remainder formula* of the form

$$G = \sum_{j=1}^{s} Q_j P_j + R, \tag{5.3.1}$$

in which $P_j \in \mathbb{P}$, Q_j, $R \in \boldsymbol{K}[\boldsymbol{x}]$ and R is reduced with respect to $\mathbb{P}$. The polynomial R is called the *remainder* or *normal form* of G with respect to $\mathbb{P}$ and denoted $\mathrm{rem}(G, \mathbb{P})$. The procedure for getting R from G is called a *reduction* of G with respect to $\mathbb{P}$. As usual, for any $\mathbb{Q} \subset \boldsymbol{K}[\boldsymbol{x}]$

$$\mathrm{rem}(\mathbb{Q}, \mathbb{P}) \triangleq \{\mathrm{rem}(Q, \mathbb{P}):\ Q \in \mathbb{Q}\}.$$

Example 5.3.1. Consider the following polynomials

$$\begin{aligned} P_1 &= x_1x_4 + x_3 - x_1x_2, \\ P_2 &= 2x_4^2 - 2x_3x_4 + 5x_1x_2x_4 - 5x_1x_2x_3, \\ G &= x_1x_4^2 + x_4^2 - x_1x_2x_4 - x_2x_4 + x_1x_2 + 3x_2. \end{aligned}$$

The monomials in P_1, P_2, and G are ordered according to the purely lexicographical ordering. In symbols, we have

$$\mathrm{lt}(P_1) = x_1x_4,\ \ \mathrm{lt}(P_2) = x_4^2,\ \ \mathrm{lt}(G) = x_1x_4^2$$

and

$$\mathrm{lc}(P_1) = \mathrm{lc}(G) = 1, \quad \mathrm{lc}(P_2) = 2.$$

Set $\mathbb{P} = \{P_1, P_2\}$. G is clearly reducible with respect to $\mathbb{P}$. For example, we have

$$G = b \cdot \lambda \cdot \mathrm{lt}(P_1) + H$$

with

$$\begin{aligned} b &= -1,\ \ \lambda = x_2, \\ H &= x_1x_4^2 + x_4^2 - x_2x_4 + x_2x_3 - x_1x_2^2 + x_1x_2 + 3x_2. \end{aligned}$$

Here, the term $x_1x_2x_4$ does not appear in H. In the above reduction, the term is not maximal with respect to the term ordering. To select the maximal term, one has to reduce the leading monomial $x_1x_4^2$ in G first. The following is a reduction of G to its remainder with respect to $\mathbb{P}$:

$$G = x_4P_2 + H_1,\ \ H_1 = \tfrac{1}{2}P_2 + H_2,\ \ H_2 = -\tfrac{5}{2}P_1 + H_3,$$

where

$$H_1 = x_4^2 - x_3x_4 - x_2x_4 + x_1x_2 + 3x_2,$$
$$H_2 = -\tfrac{5}{2}x_1x_2x_4 - x_2x_4 + \tfrac{5}{2}x_1x_2x_3 + x_1x_2 + 3x_2,$$
$$H_3 = -x_2x_4 + \tfrac{5}{2}x_1x_2x_3 + \tfrac{5}{2}x_2x_3 - \tfrac{5}{2}x_1x_2^2 + x_1x_2 + 3x_2.$$

Now H_3 is reduced with respect to $\mathbb{P}$, so no further reduction is possible. Therefore,

$$R = \mathrm{rem}(G, \mathbb{P}) = H_3 = G + \tfrac{5}{2}P_1 - (x_4 + \tfrac{1}{2})P_2.$$

In general the remainder R is not unique; that is, different choices of P_j from $\mathbb{P}$ in (5.3.1) may produce different remainders. Those polynomial sets, with respect to which the remainders of any polynomial are always the same, are of special significance.

Definition 5.3.2. A polynomial set $\mathbb{G} \subset \boldsymbol{K}[\boldsymbol{x}]$ is called a *Gröbner basis* if the remainder $\mathrm{rem}(G, \mathbb{G})$ is unique for all $G \in \boldsymbol{K}[\boldsymbol{x}]$.

$\mathbb{G}$ is called a *Gröbner basis* of a polynomial set $\mathbb{P} \subset \boldsymbol{K}[\boldsymbol{x}]$ or for Ideal($\mathbb{P}$) if $\mathbb{G}$ is a Gröbner basis and $\mathrm{Ideal}(\mathbb{P}) = \mathrm{Ideal}(\mathbb{G})$.

Definition 5.3.3. The *S-polynomial* of two nonzero polynomials F and G in $\boldsymbol{K}[\boldsymbol{x}]$ is defined to be

$$\mathrm{spol}(F, G) \triangleq \mu \cdot F - \frac{\mathrm{lc}(F)}{\mathrm{lc}(G)} \cdot \nu \cdot G,$$

where μ and ν are terms such that

$$\mathrm{lt}(F) \cdot \mu = \mathrm{lt}(G) \cdot \nu = \mathrm{lcm}(\mathrm{lt}(F), \mathrm{lt}(G)).$$

Example 5.3.2. For the polynomials P_1 and P_2 in Example 5.3.1, we have

$$\begin{aligned}\mathrm{spol}(P_1, P_2) &= \mu_1 \cdot P_1 - \frac{\mathrm{lc}(P_1)}{\mathrm{lc}(P_2)} \cdot \mu_2 \cdot P_2\\ &= x_1x_3x_4 + x_3x_4 - \tfrac{5}{2}x_1^2x_2x_4 - x_1x_2x_4 + \tfrac{5}{2}x_1^2x_2x_3,\end{aligned}$$

where $\mu_1 = x_4$ and $\mu_2 = x_1$.

Theorem 5.3.1. A polynomial set $\mathbb{G} \subset \boldsymbol{K}[\boldsymbol{x}]$ is a Gröbner basis if and only if

$$\mathrm{rem}(\mathrm{spol}(F, G), \mathbb{G}) = 0 \quad \text{for any } F, G \in \mathbb{G}.$$

This theorem provides an algorithmic characterization of Gröbner bases. Whether a polynomial set $\mathbb{P}$ is a Gröbner basis can be tested by considering only finitely many pairs of polynomials in $\mathbb{P}$. On the basis of Theorem 5.3.1 we are ready to describe the following algorithm due to Buchberger (1965, 1985).

Algorithm GroBas: $\mathbb{G} \leftarrow \text{GroBas}(\mathbb{P})$. Given a nonempty polynomial set $\mathbb{P} \subset \boldsymbol{K}[\boldsymbol{x}]$, this algorithm computes a Gröbner basis $\mathbb{G}$ of $\mathbb{P}$.
G1. Set $\mathbb{G} \leftarrow \mathbb{P}$, $\Theta \leftarrow \{\{F, G\}\colon\ F \neq G,\ F, G \in \mathbb{P}\}$.
G2. While $\Theta \neq \emptyset$, do:
  G2.1. Let $\{F, G\}$ be an element of Θ and set $\Theta \leftarrow \Theta \setminus \{\{F, G\}\}$.
  G2.2. Compute $R \leftarrow \text{rem}(\text{spol}(F, G), \mathbb{G})$.
  G2.3. If $R \neq 0$, then set

$$\Theta \leftarrow \Theta \cup \{\{R, G\}\colon\ G \in \mathbb{G}\},\ \mathbb{G} \leftarrow \mathbb{G} \cup \{R\}.$$

The above algorithm for computing Gröbner bases may be sketched as follows:

$$\begin{array}{ccccccc} \mathbb{P} = \mathbb{G}_1 & \subset & \mathbb{G}_2 & \subset \cdots \subset & \mathbb{G}_m = \mathbb{G} \\ \Theta_1 & & \Theta_2 & \cdots & \Theta_m \\ \mathbb{R}_1 & & \mathbb{R}_2 & \cdots & \mathbb{R}_m = \emptyset \end{array} \tag{5.3.2}$$

where

$$\Theta_1 = \{\{F, G\}\colon\ F \neq G,\ F, G \in \mathbb{P}\}$$

and

$$\begin{aligned} &\mathbb{R}_i = \text{rem}(\bar{\Theta}_i, \mathbb{G}_i)\setminus\{0\} \ \text{ with } \ |\mathbb{R}_i| = 1 \ \text{ for some } \ \bar{\Theta}_i \subset \Theta_i, \\ &\Theta_{i+1} = \Theta_i \setminus \bar{\Theta}_i \cup \{\text{spol}(R, G)\colon\ R \in \mathbb{R}_i,\ G \in \mathbb{G}_i\}, \\ &\mathbb{G}_{i+1} = \mathbb{G}_i \cup \mathbb{R}_i \end{aligned}$$

for $1 \leq i \leq m - 1$. The algorithm terminates at the mth step with

$$\mathbb{R}_m = \text{rem}(\Theta_m, \mathbb{G}_m)\setminus\{0\} = \emptyset.$$

The correctness that $\mathbb{G} = \mathbb{G}_m$ is a Gröbner basis of $\mathbb{P}$ follows from Theorem 5.3.1. To see the termination, one considers the sequence of ideals

$$\text{Ideal}(\mathbb{F}_1) \subset \text{Ideal}(\mathbb{F}_2) \subset \cdots \subset \text{Ideal}(\mathbb{F}_i) \subset \cdots,$$

where $\mathbb{F}_i$ is the set of leading terms of the polynomials in $\mathbb{G}_i$ and $\mathbb{G}_i$ is enlarged from $\mathbb{P}$ for the ith time. The inclusions in the above sequence are proper, so by Hilbert's theorem on ascending chains of ideals in $\boldsymbol{K}[\boldsymbol{x}]$ the sequence must be finite. See Buchberger (1985), Adams and Loustaunau (1994, pp. 42 f), and Becker and Weispfenning (1993, pp. 213–215) for more details.

A polynomial set $\mathbb{P}$ is said to be *reduced* if every polynomial $P \in \mathbb{P}$ is monic and reduced with respect to $\mathbb{P} \setminus \{P\}$. The following algorithm computes, from any Gröbner basis, the unique *reduced Gröbner basis* (see Theorem 5.3.3).

Algorithm RedGroBas: $\mathbb{G}^* \leftarrow \text{RedGroBas}(\mathbb{G})$. Given a Gröbner basis $\mathbb{G} \subset \boldsymbol{K}[\boldsymbol{x}]$, this algorithm computes the reduced Gröbner basis $\mathbb{G}^*$ of $\mathbb{G}$.
R1. Set $\mathbb{P} \leftarrow \mathbb{G}$, $\mathbb{G}^* \leftarrow \emptyset$.
R2. While $\mathbb{P} \neq \emptyset$, do:

R2.1. Select a polynomial $G \in \mathbb{P}$ and set $\mathbb{P} \leftarrow \mathbb{P} \setminus \{G\}$.
R2.2. If $\operatorname{lt}(P) \nmid \operatorname{lt}(G)$ for all $P \in \mathbb{P} \cup \mathbb{G}^*$, then set $\mathbb{G}^* \leftarrow \mathbb{G}^* \cup \{G\}$.
R3. While $\mathbb{G}^*$ is not reduced, do:
R3.1. Select a $G \in \mathbb{G}^*$ which is reducible with respect to $\mathbb{G}^* \setminus \{G\}$ and set $\mathbb{G}^* \leftarrow \mathbb{G}^* \setminus \{G\}$.
R3.2. Compute $R \leftarrow \operatorname{rem}(G, \mathbb{G}^*)$. If $R \neq 0$, then set $\mathbb{G}^* \leftarrow \mathbb{G}^* \cup \{R\}$.
R4. Set $\mathbb{G}^* \leftarrow \{G/\operatorname{lc}(G)\colon\ G \in \mathbb{G}^*\}$.

We refer to Becker and Weispfenning (1993, pp. 203 f and 216 f) for the proof of this algorithm.

Example 5.3.3. Recall the polynomials in Example 5.3.1 and let

$$P_3 = x_3x_4 - 2x_2^2 - x_1x_2 - 1.$$

The reduced Gröbner basis of $\{P_1, G, P_3\}$ with respect to the purely lexicographical term ordering determined by $x_1 \prec \cdots \prec x_4$ is

$$\mathbb{G} = \begin{bmatrix} x_1x_2^2 + x_2^2 - x_1x_2 + \frac{1}{2}x_1 + \frac{1}{2}, \\ x_3^2 - x_1x_2x_3 - 2x_2^2 + x_1^2x_2 + 2x_1x_2 - 1, \\ x_1x_4 + x_3 - x_1x_2, \\ x_2^2x_4 + \frac{1}{2}x_4 - x_2^2x_3 + x_2x_3 - \frac{1}{2}x_3 - x_2^3 - \frac{1}{2}x_2, \\ x_3x_4 - 2x_2^2 - x_1x_2 - 1, \\ x_4^2 - x_2x_4 - 2x_2^2 + 3x_2 - 1 \end{bmatrix}.$$

The reader may compare this Gröbner basis with the characteristic set in Example 2.2.3.

With the same variable and term ordering, a Gröbner basis of $\{P_1, P_2, P_3\}$ consists of 9 polynomials. These polynomials are quite large and are not listed here.

Algorithm GroBas is not optimized and thus not practically efficient. Several improved versions of the algorithm exist. Such improved algorithms take into account criteria for optimal selection of pairs for the S-polynomial formation, additional reduction, and detection of unnecessary S-polynomials before they are produced. Moreover, some alternative algorithms have also been developed for Gröbner bases computation. We do not pursue any further on these developments and refer to the previously cited books on the theory and method of Gröbner bases.

Properties

Gröbner bases are very well behaved in terms of properties and structure. A Gröbner basis $\mathbb{G}$ not containing any constant can be written as

$$\mathbb{G} = \left[\begin{array}{l} G_1(x_1, \ldots, x_{p_1}), \\ \quad \ldots \\ G_{q_1}(x_1, \ldots, x_{p_1}), \\ G_{q_1+1}(x_1, \ldots, x_{p_1}, \ldots, x_{p_2}), \\ \quad \ldots \\ G_{q_2}(x_1, \ldots, x_{p_1}, \ldots, x_{p_2}), \\ \quad \ldots\ldots \\ G_{q_{r-1}+1}(x_1, \ldots, x_{p_1}, \ldots, x_{p_2}, \ldots, x_{p_r}), \\ \quad \ldots \\ G_{q_r}(x_1, \ldots, x_{p_1}, \ldots, x_{p_2}, \ldots, x_{p_r}) \end{array}\right],$$

where

$$\begin{aligned} 0 < p_1 < p_2 < \cdots < p_r \leq n, \\ p_i = \mathrm{cls}(G_{q_{i-1}+1}) = \cdots = \mathrm{cls}(G_{q_i}), \\ x_{p_i} = \mathrm{lv}(G_{q_{i-1}+1}) = \cdots = \mathrm{lv}(G_{q_i}) \end{aligned}$$

with $q_0 = 0$ and $q_{i-1} < q_i$ for $1 \leq i \leq r$. The above form compares readily with (2.1.1).

In what follows we list some of the nice properties of Gröbner bases, which have closer relevance with polynomial elimination, the theme of this book. The reader may refer to the previously mentioned works for elaborations of many other properties.

Theorem 5.3.2. The following properties are equivalent.

a. $\mathbb{G}$ is a Gröbner basis in $\boldsymbol{K}[\boldsymbol{x}]$,
b. For all F and G in $\boldsymbol{K}[\boldsymbol{x}]$,

$$F - G \in \mathrm{Ideal}(\mathbb{G}) \iff \mathrm{rem}(F, \mathbb{G}) = \mathrm{rem}(G, \mathbb{G}).$$

c. Every nonzero polynomial $F \in \mathrm{Ideal}(\mathbb{G})$ is reducible with respect to $\mathbb{G}$.
d. For every nonzero polynomial $F \in \mathrm{Ideal}(\mathbb{G})$, there exists a polynomial $G \in \mathbb{G}$ such that $\mathrm{lt}(G) \mid \mathrm{lt}(F)$.
e. For all $F \in \boldsymbol{K}[\boldsymbol{x}]$,

$$F \in \mathrm{Ideal}(\mathbb{G}) \iff F = \sum_{G \in \mathbb{G}} H_G G$$
$$\text{with } \mathrm{lt}(F) = \max_{G \in \mathbb{G}} \mathrm{lt}(H_G) \cdot \mathrm{lt}(G).$$

f. $\mathrm{Ideal}(\{\mathrm{lm}(G)\colon\ G \in \mathbb{G}\}) = \mathrm{Ideal}(\{\mathrm{lm}(G)\colon\ G \in \mathrm{Ideal}(\mathbb{G})\})$.

Proof. Theorem 6.1 in Buchberger (1985), theorem 1.6.2 in Adams and Loustaunau (1994, pp. 32 f), and proposition 5.38 in Becker and Weispfenning (1993, pp. 207 f). □

The significance of introducing reduced Gröbner bases lies partially on the fact that for any polynomial ideal, its reduced Gröbner basis is unique. In other words, we have the following theorem.

Theorem 5.3.3. Let $\mathbb{G}_1$ and $\mathbb{G}_2$ be reduced Gröbner bases of two polynomial sets $\mathbb{P}_1$ and $\mathbb{P}_2$ in $\boldsymbol{K}[\boldsymbol{x}]$, respectively. If $\text{Ideal}(\mathbb{P}_1) = \text{Ideal}(\mathbb{P}_2)$, then $\mathbb{G}_1 = \mathbb{G}_2$.

Proof. Theorem 6.3 in Buchberger (1985), theorem 1.8.7 in Adams and Loustaunau (1994, pp. 48 f), or theorem 5.43 in Becker and Weispfenning (1993, p. 209). □

For any polynomial set $\mathbb{P} \subset \boldsymbol{K}[\boldsymbol{x}]$, let $\text{GB}(\mathbb{P})$ denote the unique *reduced Gröbner basis* of $\mathbb{P}$.

Corollary 5.3.4. Let $\mathbb{P}$ be any polynomial set in $\boldsymbol{K}[\boldsymbol{x}]$. Then

$$\text{Zero}(\mathbb{P}) = \emptyset \iff \text{GB}(\mathbb{P}) = [1].$$

Proof. If $\text{Zero}(\mathbb{P}) = \emptyset$, then $1 \in \text{Ideal}(\mathbb{P})$ according to Theorem 1.6.2. It follows that $\text{Ideal}(\mathbb{P}) = \text{Ideal}(\{1\})$. Hence, by Theorem 5.3.3

$$\text{GB}(\mathbb{P}) = \text{GB}(\{1\}) = [1].$$

On the other hand, $\text{GB}(\mathbb{P}) = [1]$ implies that $\text{Zero}(\mathbb{P}) = \text{Zero}([1]) = \emptyset$. □

The following elimination property of Gröbner bases, observed first by W. Trinks, can be easily proved. It is of particular importance for successive zero determination and will also play a crucial role in the following chapter.

Theorem 5.3.5. Let $\mathbb{G}$ be a Gröbner basis over $\boldsymbol{K}$ with respect to the purely lexicographical term ordering determined by $x_1 \prec \cdots \prec x_n$. Then for any $1 \leq i \leq n$

$$\text{Ideal}(\mathbb{G}) \cap \boldsymbol{K}[\boldsymbol{x}^{\{i\}}] = \text{Ideal}(\mathbb{G} \cap \boldsymbol{K}[\boldsymbol{x}^{\{i\}}]), \tag{5.3.3}$$

where the ideal on the right-hand side is formed in $\boldsymbol{K}[\boldsymbol{x}^{\{i\}}]$.

Proof. The right-hand side is obviously contained in the left-hand side of (5.3.3). To show the other direction, let $G \in \text{Ideal}(\mathbb{G}) \cap \boldsymbol{K}[\boldsymbol{x}^{\{i\}}]$; then $\text{rem}(G, \mathbb{G}) = 0$. Note that in the reduction of G to 0 all the polynomials involve only the variables $\boldsymbol{x}^{\{i\}}$. Thus, in the corresponding remainder formula (5.3.1) we have

$$R = 0, \quad P_j \in \mathbb{G} \cap \boldsymbol{K}[\boldsymbol{x}^{\{i\}}], \quad Q_j \in \boldsymbol{K}[\boldsymbol{x}^{\{i\}}].$$

Hence G belongs to the right-hand side of (5.3.3). □

Gröbner series

Let $G \in \mathbb{G}$ be a polynomial reducible over $\boldsymbol{K}$ and with a factorization $G = G_1 G_2$. Let $\mathbb{P}_i = \mathbb{G} \cup \{G_i\}$ and $\mathbb{G}_i$ be a Gröbner basis of $\mathbb{P}_i$ for $i = 1, 2$. Then the following zero decomposition holds

$$\text{Zero}(\mathbb{G}) = \text{Zero}(\mathbb{G}_1) \cup \text{Zero}(\mathbb{G}_2).$$

Regarding each $\mathbb{G}_i$ as $\mathbb{G}$ and continuing in this way, one shall finally get a decomposition of the form

$$\text{Zero}(\mathbb{P}) = \bigcup_{i=1}^{e} \text{Zero}(\mathbb{G}_i), \tag{5.3.4}$$

where $\mathbb{G}_i$ is a Gröbner basis and all the polynomials in $\mathbb{G}_i$ are irreducible over $\boldsymbol{K}$ for each i.

Definition 5.3.4. A finite set or sequence Ψ of Gröbner bases $\mathbb{G}_1, \ldots, \mathbb{G}_e$ is called a *Gröbner series* of a polynomial set $\mathbb{P}$ in $\boldsymbol{K}[\boldsymbol{x}]$ if the zero decomposition (5.3.4) holds.

A finite set or sequence Ψ of polynomial systems $[\mathbb{G}_1, \mathbb{D}_1], \ldots, [\mathbb{G}_e, \mathbb{D}_e]$ is called a *Gröbner series* of a polynomial system $\mathfrak{P}$ in $\boldsymbol{K}[\boldsymbol{x}]$ if

$$\operatorname{Zero}(\mathfrak{P}) = \bigcup_{i=1}^{e} \operatorname{Zero}(\mathbb{G}_i/\mathbb{D}_i)$$

and each $\mathbb{G}_i$ is a Gröbner basis. Of course, one may assume that $0 \notin \operatorname{rem}(\mathbb{D}_i, \mathbb{G}_i)$ for each i.

Ψ is said to be *quasi-irreducible* if all the polynomials in $\mathbb{G}_i$ are irreducible over $\boldsymbol{K}$ for $1 \le i \le e$.

Example 5.3.4. The last polynomial in the Gröbner basis $\mathbb{G}$ in Example 5.3.3 is reducible over $\mathbf{Q}$. Splitting $\mathbb{G}$ according to the factorization of this polynomial, one may get two Gröbner bases

$$\begin{aligned}
\mathbb{G}_1 &= [2x_2^2 + 2x_1x_2^2 - 2x_1x_2 + x_1 + 1, x_3 - 2x_1x_2 + x_1, x_4 + x_2 - 1], \\
\mathbb{G}_2 &= [2x_2^2 + 2x_1x_2^2 - 2x_1x_2 + x_1 + 1, x_3 + x_1x_2 - x_1, x_4 - 2x_2 + 1]
\end{aligned}$$

such that

$$\operatorname{Zero}(\{P_1, G, P_3\}) = \operatorname{Zero}(\mathbb{G}_1) \cup \operatorname{Zero}(\mathbb{G}_2).$$

Refer to Examples 5.3.1 and 5.3.3 for P_1, P_2, P_3, and G. A Gröbner series of $\{P_1, P_2, P_3\}$ consists of the following two Gröbner bases

$$\begin{bmatrix}
x_1^2x_2^2 + 4x_1x_2^2 + 2x_2^2 + x_1^3x_2 + 2x_1^2x_2 + x_1x_2 + x_1^2 + 2x_1 + 1, \\
x_1x_3 + x_3 - x_1x_2, x_2x_3 + x_1x_2^2 + 2x_2^2 + x_1^2x_2 + x_1x_2 + x_1 + 1, \\
x_3^2 - 2x_2^2 - x_1x_2 - 1, x_4 - x_3
\end{bmatrix},$$

$$\left[25x_1^3x_2^2 + 10x_1^2x_2^2 + 8x_2^2 + 4x_1x_2 + 4, 2x_3 - 5x_1^2x_2 - 2x_1x_2, 2x_4 + 5x_1x_2\right].$$

5.4 Resultant elimination

This section summarizes the main (classical) elimination techniques using resultants. Our presentation is based on the materials in Chionh and Goldman (1995), Kapur and Lakshman (1992), and van der Waerden (1950, chap. XI).

Resultants revisited

The Sylvester resultant has been introduced in Sect. 1.3. Another formulation of univariate resultants due to É. Bézout and A. Cayley, with its extension to the bivariate case by Dixon (1908), is described below.

Bézout–Cayley resultant
Consider two univariate polynomials $F, G \in \boldsymbol{R}[x]$ of respective degrees m and l in x with $m \geq l > 0$ as in Sect. 1.3. Let α be a new indeterminate. The determinant

$$\Delta(x, \alpha) = \begin{vmatrix} F(x) & G(x) \\ F(\alpha) & G(\alpha) \end{vmatrix}$$

is a polynomial in x and α and is equal to 0 when $x = \alpha$. So $x - \alpha$ is a divisor of Δ. The polynomial

$$\Lambda(x, \alpha) = \frac{\Delta(x, \alpha)}{x - \alpha}$$

has degree $m-1$ in α and is symmetric with respect to both x and α. As $\Lambda(\bar{x}, \alpha) = 0$ for any $\bar{x} \in \mathrm{Zero}(\{F, G\})$ no matter what value α has, all the coefficients of Λ as a polynomial in α, $B_i(x) = \mathrm{coef}(\Lambda, \alpha^i)$, are 0 at $x = \bar{x}$. Consider the following m polynomial equations in x:

$$B_0(x) = 0, \ldots, B_{m-1}(x) = 0; \tag{5.4.1}$$

the maximum degree of the B_i in x is $m - 1$. Any common zero of F and G is a solution of (5.4.1), and the equations in (5.4.1) have a common solution if the determinant R of the B_i's coefficient matrix is 0.

The determinant R of the $m \times m$ matrix is called the Bézout–Cayley resultant of F and G with respect to x. It is identical to the Sylvester resultant defined in Sect. 1.3 when $m = l$ and has an extraneous factor $\mathrm{lc}(F, x)^{m-l}$ when $m > l$. Note that the Sylvester resultant of F and G with respect to x was formulated as the determinant of an $(l + m) \times (l + m)$ matrix.

Example 5.4.1. Consider the univariate quartic polynomial

$$F = x^4 + x_1 x^3 + x_2 x^2 + x_3 x + x_4.$$

We want to compute the discriminant of F with respect to x, which is defined to be the resultant of F and its derivative

$$G = dF/dx = 4x^3 + 3x_1 x^2 + 2x_2 x + x_3.$$

Following the above method, we first compute

$$\Lambda = \frac{1}{x - \alpha} \begin{vmatrix} F(x) & G(x) \\ F(\alpha) & G(\alpha) \end{vmatrix} = G\alpha^3 + B_2\alpha^2 + B_1\alpha + B_0,$$

where

$$\begin{aligned}
B_2 &= 3x_1x^3 - (2x_2 - 3x_1^2)x^2 - (3x_3 - 2x_1x_2)x - 4x_4 + x_1x_3, \\
B_1 &= 2x_2x^3 - (3x_3 - 2x_1x_2)x^2 - (4x_4 + 2x_1x_3 - 2x_2^2)x - 3x_1x_4 + x_2x_3, \\
B_0 &= x_3x^3 - (4x_4 - x_1x_3)x^2 - (3x_1x_4 - x_2x_3)x - 2x_2x_4 + x_3^2.
\end{aligned}$$

By equating the coefficients of the terms of α in Λ to 0, one gets four equations

$$G = 0, \quad B_2 = 0, \quad B_1 = 0, \quad B_0 = 0.$$

Considered as homogeneous linear equations in the unknowns x^3, x^2, x^1, x^0, they have a common solution if and only if the determinant of the coefficient matrix is 0, viz.,

$$\begin{aligned} R &= \begin{vmatrix} 4 & 3x_1 & 2x_2 & x_3 \\ 3x_1 & -2x_2 + 3x_1^2 & -3x_3 + 2x_1x_2 & -4x_4 + x_1x_3 \\ 2x_2 & -3x_3 + 2x_1x_2 & -4x_4 - 2x_1x_3 + 2x_2^2 & -3x_1x_4 + x_2x_3 \\ x_3 & -4x_4 + x_1x_3 & -3x_1x_4 + x_2x_3 & -2x_2x_4 + x_3^2 \end{vmatrix} \\ &= 256x_4^3 - 192x_1x_3x_4^2 - 128x_2^2x_4^2 + 144x_1^2x_2x_4^2 - 27x_1^4x_4^2 + 144x_2x_3^2x_4 \\ &\quad - 6x_1^2x_3^2x_4 - 80x_1x_2^2x_3x_4 + 18x_1^3x_2x_3x_4 + 16x_2^4x_4 - 4x_1^2x_2^3x_4 \\ &\quad - 27x_3^4 + 18x_1x_2x_3^3 - 4x_1^3x_3^3 - 4x_2^3x_3^2 + x_1^2x_2^2x_3^2 \\ &= 0. \end{aligned}$$

The above determinant which is the discriminant of F will be used in Example 7.4.4.

Dixon bidegree resultant

The formulation of Bézout–Cayley resultants may be extended to three polynomials F, G, and H of bidegree (l, m) in two variables x and y and other restricted cases. This was shown by Dixon (1908). Here, *bidegree* means that the polynomials $F, G, H \in \boldsymbol{R}[x, y]$ have total degree $l + m$ in x and y but only degree l in x and m in y. Let us consider this case. The determinant

$$\Delta(x, y, \alpha, \beta) = \begin{vmatrix} F(x, y) & G(x, y) & H(x, y) \\ F(\alpha, y) & G(\alpha, y) & H(\alpha, y) \\ F(\alpha, \beta) & G(\alpha, \beta) & H(\alpha, \beta) \end{vmatrix}$$

vanishes when one replaces α by x, or β by y. It follows that $(x - \alpha)(y - \beta) \mid \Delta$. Hence

$$\Lambda(x, y, \alpha, \beta) = \frac{\Delta(x, y, \alpha, \beta)}{(x - \alpha)(y - \beta)}$$

is a polynomial in x, y, α, β with

$$\begin{aligned} \deg(\Lambda, \alpha) &= 2l - 1, \quad & \deg(\Lambda, x) &= l - 1, \\ \deg(\Lambda, \beta) &= m - 1, \quad & \deg(\Lambda, y) &= 2m - 1. \end{aligned}$$

Since $\Lambda(\bar{x}, \bar{y}, \alpha, \beta) = 0$ for any $(\bar{x}, \bar{y}) \in \text{Zero}(\{F, G, H\})$ no matter what α and β are, the coefficients $D_{ij} = \text{coef}(\Lambda, \alpha^i\beta^j)$ for $0 \le i \le 2l - 1$ and $0 \le j \le m - 1$ have common zeros for x and y, which contain $\text{Zero}(\{F, G, H\})$. Consider

$$D_{ij}(x, y) = 0 \quad (0 \le i \le l - 1, \ 0 \le j \le 2m - 1)$$

as $2lm$ homogeneous linear equations in the $2lm$ terms

$$x^i y^j \quad (0 \le i \le l-1,\ 0 \le j \le 2m-1).$$

In matrix form, we have

$$\Lambda(x, y, \alpha, \beta) = (x^{l-1}y^{2m-1} \dots y^{2m-1} \dots x^{l-1} \dots 1)\,\mathbf{D}\begin{pmatrix} \alpha^{2l-1}\beta^{m-1} \\ \vdots \\ \beta^{m-1} \\ \vdots \\ \alpha^{2l-1} \\ \vdots \\ 1 \end{pmatrix},$$

where $\mathbf{D}$ is the coefficient matrix of the D_{ij}. The matrix $\mathbf{D}$ and the determinant R of $\mathbf{D}$ are called the *Dixon matrix* and the *Dixon resultant* of $\{F, G, H\}$ with respect to x and y, respectively.

For arbitrary three polynomials $F, G, H \in \boldsymbol{R}[x, y]$, one can also construct the corresponding Dixon matrix $\mathbf{D}$ in a similar way. In this case, $\mathbf{D}$ is not necessarily square; or even if it is square, it may be singular, i.e., $\det(\mathbf{D}) = 0$. So the method does not work in general. However, as far as the Dixon matrix $\mathbf{D}$ is square and nonsingular, the determinant of $\mathbf{D}$ differs only by a constant factor from the usual resultant, and is called the Dixon resultant of $\{F, G, H\}$ with respect to x and y. The following example is provided as an illustration.

Example 5.4.2. Consider the binary cubic polynomial

$$F(x, y) = y^2 + a_1xy + a_3y - x^3 - a_2x^2 - a_4x - a_6.$$

The resultant R of

$$\mathbb{P} = \{F, \partial F/\partial x, \partial F/\partial y\}$$

with respect to x and y is also called the *discriminant* of F; $R = 0$ gives a necessary and sufficient condition for the cubic curve $F(x, y) = 0$ to have singularities (see Sect. 7.4). If $R \neq 0$, then $F(x, y) = 0$ is an elliptic curve.

To obtain R, one first computes the polynomial $\Lambda(x, y, \alpha, \beta)$ which consists of 45 terms and can be written as

$$(xy\ \ y\ \ x^2\ \ x\ \ 1)\begin{pmatrix} 0 & 6 & 0 & 3a_1 & 3a_3 \\ 6 & a_1^2 + 4a_2 & 6a_1 & d_{24} & d_{25} \\ 0 & 0 & -6 & d_{34} & d_{35} \\ 3a_1 & 3a_3 & 2a_1^2 - 4a_2 & d_{44} & d_{45} \\ 3a_3 & 2a_2a_3 - a_1a_4 & 2a_1a_3 - 2a_4 & d_{54} & d_{55} \end{pmatrix}\begin{pmatrix} \alpha\beta \\ \beta \\ \alpha^2 \\ \alpha \\ 1 \end{pmatrix},$$

where

$$\begin{aligned}
d_{24} &= a_1^3 + 4a_1a_2 + 3a_3, \\
d_{25} &= a_1^2a_3 + 2a_2a_3 + a_1a_4, \\
d_{34} &= -a_1^2 - 4a_2, \\
d_{35} &= -a_1a_3 - 2a_4, \\
d_{44} &= -a_1^2a_2 - 4a_2^2 + 5a_1a_3 + 4a_4, \\
d_{45} &= -a_1a_2a_3 + 3a_3^2 - 2a_2a_4 + 6a_6, \\
d_{54} &= a_1a_2a_3 + 3a_3^2 - a_1^2a_4 - 2a_2a_4 + 6a_6, \\
d_{55} &= 2a_2a_3^2 - 2a_1a_3a_4 - 2a_4^2 + a_1^2a_6 + 4a_2a_6.
\end{aligned}$$

The determinant of the 5×5 matrix

$$\begin{aligned}
R = 18(&72a_2a_3^2a_4 + 288a_2a_4a_6 + 72a_1^2a_4a_6 - 8a_1^2a_2^2a_3^2 - 12a_1^4a_2a_6 \\
&+ 8a_1^2a_2a_4^2 + 36a_1a_2a_3^3 - 30a_1^2a_3^2a_4 + 36a_1^3a_3a_6 - 96a_1a_3a_4^2 \\
&- 48a_1^2a_2^2a_6 - a_1^4a_2a_3^2 + a_1^5a_3a_4 + a_1^4a_4^2 - a_1^6a_6 + a_1^3a_3^3 \\
&+ 16a_1a_2^2a_3a_4 + 144a_1a_2a_3a_6 + 8a_1^3a_2a_3a_4 - 64a_4^3 - 27a_3^4 \\
&+ 16a_2^2a_4^2 - 216a_3^2a_6 - 432a_6^2 - 64a_2^3a_6 - 16a_2^3a_3^2)
\end{aligned}$$

consists of 26 terms and is the Dixon resultant of $\mathbb{P}$ with respect to x and y. It can be written as

$$R = 18(-b_2^2b_8 - 8b_4^3 - 27b_6^2 + 9b_2b_4b_6),$$

where

$$\begin{aligned}
&b_2 = a_1^2 + 4a_2, \quad b_4 = a_1a_3 + 2a_4, \quad b_6 = a_3^2 + 4a_6, \\
&b_8 = a_1^2a_6 + 4a_2a_6 - a_1a_3a_4 + a_2a_3^2 - a_4^2.
\end{aligned}$$

These are familiar expressions in the arithmetic of elliptic curves.

We do not go further with Dixon's method for three equal-degree polynomials and other cases, nor its recent generalizations. The interested reader may refer to Dixon (1908), Chionh and Goldman (1995), Kapur and Lakshman (1992), Kapur and Saxena (1995), and references therein for more information and technical discussions.

Multivariate resultants

In this section we explain Macaulay's method that constructs a resultant from any n homogeneous polynomials in n variables; so several variables are eliminated at once. This is clearly a generalization of univariate and bivariate resultants. Again, we proceed to form a system of m linear equations in m terms which may be considered as unknowns. This will be done by the dialytic method which takes certain terms as multipliers for the polynomials.

Macaulay matrix

Consider a set of n homogeneous polynomials, $\mathbb{P} = \{P_1, \ldots, P_n\}$, in n variables $\boldsymbol{x} = (x_1, \ldots, x_n)$ with indeterminante coefficients and $d_i = \operatorname{tdeg}(P_i)$. Let

$$d = 1 + \sum_{i=1}^{n}(d_i - 1)$$

and

$$\mathcal{M} = \{x_1^{i_1} \ldots x_n^{i_n}\colon\ i_1 + \cdots + i_n = d\}.$$

Then

$$m = |\mathcal{M}| = \binom{d+n-1}{n-1}.$$

We want to multiply each polynomial P_i by appropriate terms to generate m equations in m terms of degree d. For this purpose, let

$$\begin{aligned}
\mathcal{M}_1 &= \{\mu/x_1^{d_1}\colon\ x_1^{d_1} \mid \mu, \mu \in \mathcal{M}\},\\
\mathcal{M}_i &= \{\mu/x_i^{d_i}\colon\ x_i^{d_i} \mid \mu, \mu \in \mathcal{M} \setminus \{x_j^{d_j}\nu_j\colon\ \nu_j \in \mathcal{M}_j, 1 \le j \le i-1\}\},\\
&\qquad 2 \le i \le n.
\end{aligned}$$

Set $m_i = |\mathcal{M}_i|$ for $1 \le i \le n$. Macaulay (1964, pp. 7f) showed that

$$m_1 + \cdots + m_n = m.$$

In fact,

$$\mathcal{M} = \{x_i^{d_i}\mu_i\colon\ \mu_i \in \mathcal{M}_i, 1 \le i \le n\}.$$

Now, we form a square matrix $\mathbf{M}$ of dimension $m \times m$ as follows. Let the columns of $\mathbf{M}$ be labeled by the terms in $\mathcal{M}$. And, let the first m_1 rows be labeled by the terms in $\mathcal{M}_1$, the next m_2 rows be labeled by the terms in $\mathcal{M}_2$, and so forth. In each row of $\mathcal{M}$ labeled by the term $\mu \in \mathcal{M}_i$, fill in the coefficient $\operatorname{coef}(\mu P_i, \nu)$ under the column labeled by ν for all $\nu \in \mathcal{M}$ (observing that $\operatorname{tdeg}(\mu P_i) = d$). The matrix $\mathbf{M}$ so constructed is called the *Macaulay matrix* of $P_1, \ldots, P_n$, or of $\mathbb{P}$, with respect to $\boldsymbol{x}$.

Macaulay resultant

Let $\mathcal{N}_i$ be the set of those terms in $\mathcal{M}_i$ which are divisible by $x_j^{d_j}$ for at least one j, where $2 \le i+1 \le j \le n$. If all the $\mathcal{N}_i$ are empty, then set $\mathbf{N}$ to be the trivial matrix (1) of dimension 1×1. Otherwise, let $\mathbf{N}$ be the minor of $\mathbf{M}$ whose columns are labeled by the terms in

$$\{x_i^{d_i}\mu_i\colon\ \mu_i \in \mathcal{N}_i, 1 \le i \le n-1\},$$

and whose rows are labeled by the terms in

$$\mathcal{N}_1 \cup \cdots \cup \mathcal{N}_{n-1}.$$

The determinant of $\mathbf{M}$ is a polynomial homogeneous in the coefficients of each P_i. Assume that the determinant of $\mathbf{N}$ is nonzero (see Remark 5.4.2). The quotient

$$R = \det(\mathbf{M})/\det(\mathbf{N})$$

is defined to be the *Macaulay resultant* of $P_1, \ldots, P_n$ or of $\mathbb{P}$ with respect to $\boldsymbol{x}$.

The above discussions are recapitulated in the form of the following algorithm.

Algorithm MacRes: $R \leftarrow \operatorname{MacRes}(\mathbb{P})$. Given a set $\mathbb{P} = \{P_1, \ldots, P_n\}$ of n homogeneous polynomials in n variables $\boldsymbol{x}$ (with indeterminate coefficients) over $\boldsymbol{K}$, this algorithm computes the Macaulay resultant R of $\mathbb{P}$ with respect to $\boldsymbol{x}$.

M1. Set

$$\begin{aligned} & d_i \leftarrow \operatorname{tdeg}(P_i), \quad i = 1, \ldots, n, \\ & d \leftarrow 1 + \textstyle\sum_{i=1}^{n} (d_i - 1), \\ & \mathcal{M} \leftarrow \{x_1^{i_1} \ldots x_n^{i_n} \colon i_1 + \cdots + i_n = d\}, \\ & \mathcal{T} \leftarrow \mathcal{M}, \\ & \mathbb{M} \leftarrow \emptyset. \end{aligned}$$

M2. For $i = 1, \ldots, n$ do:

M2.1. Set

$$\begin{aligned} & \mathcal{S} \leftarrow \{\mu \in \mathcal{T} \colon x_i^{d_i} \mid \mu\}, \\ & \mathcal{M}_i \leftarrow \{\mu / x_i^{d_i} \colon \mu \in \mathcal{S}\}, \\ & \mathcal{T} \leftarrow \mathcal{T} \setminus \mathcal{S}. \end{aligned}$$

M2.2. Compute $\mathbb{M} \leftarrow \mathbb{M} \cup \{\mu P_i \colon \mu \in \mathcal{M}_i\}$.

M3. For $i = 1, \ldots, n-1$ do:

$$\mathcal{N}_i \leftarrow \{\mu \in \mathcal{M}_i \colon \exists j, i+1 \le j \le n, \text{ such that } x_j^{d_j} \mid \mu\}.$$

M4. Let $\mathbf{M}$ be the coefficient matrix of the polynomials in $\mathbb{M}$ with the terms in $\mathcal{M}$ as unknowns and set

$$\mathcal{N} \leftarrow \mathcal{N}_1 \cup \cdots \cup \mathcal{N}_{n-1}.$$

If $\mathcal{N} = \emptyset$, then set $\mathbf{N} \leftarrow (1)$; else let $\mathbf{N}$ be the minor of $\mathbf{M}$ whose rows are labeled by the terms in $\mathcal{N}$ and whose columns are labeled by the terms in

$$\{x_i^{d_i} \mu_i \colon \mu_i \in \mathcal{N}_i, 1 \le i \le n-1\}.$$

Return $R \leftarrow \det(\mathbf{M}) / \det(\mathbf{N})$.

Example 5.4.3. Consider the following set $\mathbb{P}$ of three polynomials in three variables with indeterminate coefficients

$$\begin{aligned} P_1 &= a_{11}x_1^2 + a_{12}x_1x_2 + a_{13}x_1x_3 + a_{22}x_2^2 + a_{23}x_2x_3 + a_{33}x_3^2, \\ P_2 &= b_{11}x_1^2 + b_{12}x_1x_2 + b_{13}x_1x_3 + b_{22}x_2^2 + b_{23}x_2x_3 + b_{33}x_3^2, \\ P_3 &= c_1x_1 + c_2x_2 + c_3x_3. \end{aligned}$$

Using the above notations, we have

$$d_1 = d_2 = 2, \quad d_3 = 1, \quad d = 3, \quad m = 10.$$

The Macaulay matrix **M** of dimension 10×10 together with the labeled terms is shown below

$$\begin{array}{c} \\ x_1 \\ x_2 \\ x_3 \\ x_1 \\ x_2 \\ x_3 \\ x_1x_2 \\ x_1x_3 \\ x_2x_3 \\ x_3^2 \end{array}
\begin{array}{c} \begin{array}{cccccccccc} x_1^3 & x_1^2x_2 & x_1^2x_3 & x_1x_2^2 & x_1x_2x_3 & x_1x_3^2 & x_2^3 & x_2^2x_3 & x_2x_3^2 & x_3^3 \end{array} \\
\left(\begin{array}{cccccccccc}
a_{11} & a_{12} & a_{13} & a_{22} & a_{23} & a_{33} & 0 & 0 & 0 & 0 \\
0 & a_{11} & 0 & a_{12} & a_{13} & 0 & a_{22} & a_{23} & a_{33} & 0 \\
0 & 0 & a_{11} & 0 & a_{12} & a_{13} & 0 & a_{22} & a_{23} & a_{33} \\
b_{11} & b_{12} & b_{13} & b_{22} & b_{23} & b_{33} & 0 & 0 & 0 & 0 \\
0 & b_{11} & 0 & b_{12} & b_{13} & 0 & b_{22} & b_{23} & b_{33} & 0 \\
0 & 0 & b_{11} & 0 & b_{12} & b_{13} & 0 & b_{22} & b_{23} & b_{33} \\
0 & c_1 & 0 & c_2 & c_3 & 0 & 0 & 0 & 0 & 0 \\
0 & 0 & c_1 & 0 & c_2 & c_3 & 0 & 0 & 0 & 0 \\
0 & 0 & 0 & 0 & c_1 & 0 & 0 & c_2 & c_3 & 0 \\
0 & 0 & 0 & 0 & 0 & c_1 & 0 & 0 & c_2 & c_3
\end{array} \right) \end{array}.$$

It is constructed as follows.

As the terms labeled on the first three columns of **M** are divisible by x_1^2, we have $\mathcal{M}_1 = \{x_1, x_2, x_3\}$. Multiplying P_1 by the x_i in $\mathcal{M}_1$ respectively and filling in the corresponding coefficients, one obtains the first three rows of **M**. The terms labeled on the fourth, the seventh, and the eighth column of **M** are divisible by x_2^2, so $\mathcal{M}_2 = \{x_1, x_2, x_3\}$. Thus, the next three rows are obtained by filling in the coefficients of $x_1 P_2, x_2 P_2, x_3 P_2$ respectively. Dividing the remaining four terms labeled on the columns by x_3 yields

$$\mathcal{M}_3 = \{x_1x_2, x_1x_3, x_2x_3, x_3^2\}.$$

Accordingly, the last four rows are obtained by filling in the coefficients of μP_3 for $\mu \in \mathcal{M}_3$.

The determinant of **M** is a polynomial consisting of 432 terms in a_{ij}, b_{ij} and c_k. To see the corresponding minor **N** of **M**, one may find that

$$\mathcal{N}_1 = \mathcal{N}_2 = \{x_3\}.$$

Taking the third and the eighth columns, and the third and the sixth rows of **M**, produces **N** as follows

$$\begin{array}{c} \\ x_3 \\ x_3 \end{array}
\begin{array}{c} \begin{array}{cc} x_1^2x_3 & x_2^2x_3 \end{array} \\
\left(\begin{array}{cc} a_{11} & a_{22} \\ b_{11} & b_{22} \end{array} \right) \end{array}.$$

The Macaulay resultant of $\mathbb{P}$, a polynomial consisting of 234 terms in a_{ij}, b_{ij} and c_k, is finally obtained by taking the quotient $\det(\mathbf{M}) / \det(\mathbf{N})$.

The following theorem lists some important properties about Macaulay resultants.

Theorem 5.4.1. Let $\mathbb{P} = \{P_1, \dots, P_n\}$ be a set of n homogeneous polynomials in $\boldsymbol{x}$ with indeterminate coefficients over $\boldsymbol{K}$, R the Macaulay resultant of $\mathbb{P}$ (with respect to $\boldsymbol{x}$), and $\mathbf{0} = (0, \dots, 0)$. Then

a. $R = 0$ if and only if $\mathrm{Zero}(\mathbb{P}) \supsetneq \{\mathbf{0}\}$;
b. R is irreducible over any algebraic closure of $\boldsymbol{K}$ and invariant under linear coordinate transformations – thus $R = 0$ is the smallest necessary condition for $\mathrm{Zero}(\mathbb{P}) \supsetneq \{\mathbf{0}\}$;
c. R is homogeneous and has degree $\prod_{1 \le j \le n, j \ne i} d_j$ in the coefficients of each P_i, where $d_i = \mathrm{tdeg}(P_i)$ for $1 \le i \le n$;
d. if $P_i = FG$ for some $1 \le i \le n$ and specialized coefficients, then R is the product of the Macaulay resultants R_1 of $\mathbb{P} \setminus \{P_i\} \cup \{F\}$ and R_2 of $\mathbb{P} \setminus \{P_i\} \cup \{G\}$ with respect to $\boldsymbol{x}$.

Proof. Sects. 7–11 in Macaulay (1964, pp. 8–15). □

Remark 5.4.1. Macaulay (1921) gave an improved algorithm for constructing the resultant of $\mathbb{P}$ when all the P_i have the same degree, i.e., $d_1 = \cdots = d_n$. In this case, the dimensions of the corresponding matrices are made smaller; see Chionh and Goldman (1995). Macaulay's methods mainly deal with sets of homogeneous polynomials and their zeros in projective space $\mathbf{P}^n$. For nonhomogeneous polynomial sets, one has to homogenize the polynomials before applying the methods. Zeros at infinity may be included and have to be handled separately if one is only interested in affine zeros.

Remark 5.4.2. The Macaulay resultant as a quotient of two determinants is defined if the submatrix $\mathbf{N}$ is nonsingular. The condition is satisfied "in general," or when the polynomials have indeterminate coefficients. For specialized polynomials, the theoretical approach is to compute the Macaulay resultant R of the polynomials with indeterminate coefficients and then evaluate R by specializing the coefficient values. However, this is not practically feasible because of the large size of R even for polynomials of small degree. To compute R with specialized coefficients, one may encounter the situation in which $\mathbf{N}$ is singular. To deal with this in practice, more advanced techniques such as perturbation are required (see Lazard 1981 and the end of this section).

Resultant systems and u-resultants

Resultant system

Write $\boldsymbol{x}^{\{i\}}$ for $x_1, \dots, x_i$ with $\boldsymbol{x} = \boldsymbol{x}^{\{n\}}$ as before and let

$$\mathbb{P} = \{P_1, \dots, P_s\}$$

be a finite set of s (≥ 2) polynomials in $\boldsymbol{K}[\boldsymbol{x}]$. We want to determine another polynomial set $\mathbb{R} = \{R_1, \dots, R_r\} \subset \boldsymbol{K}[\boldsymbol{x}^{\{n-1\}}]$ (with the variable x_n eliminated) and establish some zero relation between $\mathbb{P}$ and $\mathbb{R}$.

For this purpose, let

$$d_i = \deg(P_i, x_n), \quad 1 \le i \le s, \quad \text{and } d = \max_{1 \le i \le s} d_i$$

and construct a new polynomial set $\mathbb{F} = \{F_1, \ldots, F_t\}$ from $\mathbb{P}$ by replacing those P_i for which $d_i < d$ with $x_n^{d-d_i} P_i$ and $(x_n - 1)^{d-d_i} P_i$ so that the polynomials in $\mathbb{F}$ have the same degree d in x_n and $\mathrm{Zero}(\mathbb{F}) = \mathrm{Zero}(\mathbb{P})$. With respect to x_n, we form the resultant R of the two polynomials

$$F_1 u_1 + \cdots + F_t u_t, \quad F_1 v_1 + \cdots + F_t v_t,$$

where $\boldsymbol{u} = (u_1, \ldots, u_t)$ and $\boldsymbol{v} = (v_1, \ldots, v_t)$ are new indeterminates. Clearly, R is a polynomial in $\boldsymbol{x}^{\{n-1\}}$ and $\boldsymbol{u}, \boldsymbol{v}$. Consider R as polynomial in $\boldsymbol{u}$ and $\boldsymbol{v}$ only and let its nonzero coefficients be $R_1, \ldots, R_e$. The polynomial set $\mathbb{R} = \{R_1, \ldots, R_e\} \subset \boldsymbol{K}[\boldsymbol{x}^{\{n-1\}}]$ is called a *resultant system* of $\mathbb{P}$ with respect to x_n. It is empty when $R \equiv 0$. According to van der Waerden (1950, p. 1), the above method of constructing resultant systems is due to L. Kronecker.

Theorem 5.4.2. Let $\mathbb{R}$ be a resultant system of any polynomial set $\mathbb{P} \subset \boldsymbol{K}[\boldsymbol{x}]$ with respect to x_n, and $\bar{\boldsymbol{x}}^{\{n-1\}} \in \tilde{\boldsymbol{K}}^{n-1}$. Then, $\bar{\boldsymbol{x}}^{\{n-1\}} \in \mathrm{Zero}(\mathbb{R})$ if and only if either $\mathrm{Zero}(\mathbb{P}^{\langle \bar{x}, n-1 \rangle}) \neq \emptyset$, or $\bar{\boldsymbol{x}}^{\{n-1\}} \in \mathrm{Zero}(\{\mathrm{lc}(P, x_n) \colon P \in \mathbb{P}\})$.

Proof. Let $F_u = F_1 u_1 + \cdots + F_t u_t$ and $F_v = F_1 v_1 + \cdots + F_t v_t$ and $\mathbb{F}$ as above. Since F_u is independent of $\boldsymbol{v}$ and so is F_v of $\boldsymbol{u}$, every common divisor of F_u and F_v must be independent of $\boldsymbol{u}$ and $\boldsymbol{v}$ and thus divides $F_1, \ldots, F_t$. Conversely, any common divisor of $F_1, \ldots, F_t$ also divides F_u and F_v. Therefore,

$$\mathrm{Zero}(\mathbb{F}) \neq \emptyset \iff \mathrm{Zero}(\{F_u, F_v\}) \neq \emptyset.$$

Let $R = \mathrm{res}(F_u, F_v, x_n)$ and $\bar{\boldsymbol{x}}^{\{n-1\}} \in \tilde{\boldsymbol{K}}^{n-1}$. By Theorem 1.3.2, $R(\boldsymbol{u}, \boldsymbol{v}, \bar{\boldsymbol{x}}^{\{n-1\}}) = 0$ if and only if either $F_u(\boldsymbol{u}, \bar{\boldsymbol{x}}^{\{n-1\}})$ and $F_v(\boldsymbol{v}, \bar{\boldsymbol{x}}^{\{n-1\}})$ have a common zero for x_n, or

$$\mathrm{lc}(F_u, x_n)(\boldsymbol{u}, \bar{\boldsymbol{x}}^{\{n-1\}}) = \mathrm{lc}(F_v, x_n)(\boldsymbol{v}, \bar{\boldsymbol{x}}^{\{n-1\}}) = 0;$$

and thus if and only if

$$\mathrm{Zero}(\mathbb{P}^{\langle \bar{x}, n-1 \rangle}) = \mathrm{Zero}(\{F_1, \ldots, F_t\}|_{\boldsymbol{x}^{\{n-1\}} = \bar{\boldsymbol{x}}^{\{n-1\}}}) \neq \emptyset,$$

or

$$\bar{\boldsymbol{x}}^{\{n-1\}} \in \mathrm{Zero}(\{\mathrm{lc}(P, x_n) \colon P \in \mathbb{P}\}).$$

As $\boldsymbol{u}$ and $\boldsymbol{v}$ are indeterminates, $R(\boldsymbol{u}, \boldsymbol{v}, \bar{\boldsymbol{x}}^{\{n-1\}}) = 0$ if and only if all the coefficients of R considered as a polynomial in $\boldsymbol{u}$ and $\boldsymbol{v}$ vanish at $\boldsymbol{x}^{\{n-1\}} = \bar{\boldsymbol{x}}^{\{n-1\}}$, i.e., $\bar{\boldsymbol{x}}^{\{n-1\}} \in \mathrm{Zero}(\mathbb{R})$. □

Example 5.4.4. Let $\mathbb{P} = \{P_1, P_2, P_3\}$ with

$$P_1 = x - rt, \quad P_2 = y - rt^2, \quad P_3 = z - r^2$$

and $x \prec y \prec z \prec t \prec r$. These polynomials will appear again in Example 7.4.1. To compute a resultant system of $\mathbb{P}$ with respect to r, we first form the following polynomials

$$G_1 = rP_1, \quad G_2 = (r-1)P_1, \quad G_3 = rP_2,$$
$$G_4 = (r-1)P_2, \quad G_5 = P_3.$$

The resultant R of

$$G_1u_1 + \cdots + G_5u_5 \text{ and } G_1v_1 + \cdots + G_5v_5$$

with respect to r is a polynomial consisting of 710 terms in x, y, z, t and the indeterminates u_i, v_j. By collecting all the coefficients of R in u_i and v_j, one gets a resultant system of $\mathbb{P}$, which contains 76 polynomials in x, y, z, and t.

As remarked in van der Waerden (1950, p. 2), if one of the formal leading coefficients of P_i, say $\mathrm{lc}(P_1, x_n)$, does not vanish, then the construction of $\mathbb{F}$ is not needed and the resultant system may be obtained simply by forming the resultant of P_1 and $v_2P_2 + \cdots + v_nP_n$ instead.

For Example 5.4.4, $\mathrm{lc}(P_3, r) = -1 \neq 0$, so we only need to compute

$$\begin{aligned} R &= \mathrm{res}(P_3, v_1P_1 + v_2P_2, r) \\ &= -x^2v_1^2 - 2xyv_1v_2 - y^2v_2^2 + zt^2v_1^2 + 2zt^3v_1v_2 + zt^4v_2^2. \end{aligned}$$

Collecting the coefficients of R as a polynomial in v_1 and v_2, one obtains a much simpler resultant system of $\mathbb{P}$ as follows:

$$\mathbb{R} = \{zt^2 - x^2, zt^4 - y^2, 2zt^3 - 2xy\}. \tag{5.4.2}$$

Zero determination

Now we explain how to determine all zeros of an arbitrary polynomial set $\mathbb{P} = \{P_1, \ldots, P_s\}$ by using resultant systems. Following van der Waerden (1950, p. 3), one can assume that $\mathbb{P}$ contains one polynomial with nonvanishing leading coefficient with respect to x_n. If the assumption does not hold, it may be brought about as follows. Leaving out the trivial case in which all P_i vanish identically, we assume, without loss of generality, that P_n does not vanish identically. Under this hypothesis, introduce the following variable transformation

$$\begin{aligned} x_1 &= z_1 + u_1z_n, \\ &\cdots\cdots \\ x_{n-1} &= z_{n-1} + u_{n-1}z_n, \\ x_n &= u_nz_n, \end{aligned}$$

where $\boldsymbol{u} = (u_1, \ldots, u_n)$ are indeterminates or some special values to be determined later. This transformation maps P_n to a polynomial whose leading coefficient with respect to x_n is a nonvanishing polynomial in $\boldsymbol{u}$. One can take any values from $\boldsymbol{K}$ or some extension field of $\boldsymbol{K}$ for $\boldsymbol{u}$ as far as the leading coefficient does not vanish.

Let $\mathbb{R}_n = \mathbb{P}$ and assume that $\mathbb{R}_n$ contains one polynomial having nonvanishing leading coefficient with respect to x_n. Compute a resultant system $\mathbb{R}_{n-1} \subset \boldsymbol{K}[\boldsymbol{x}^{\{n-1\}}]$ of $\mathbb{R}_n$. Then, $\mathrm{Zero}(\mathbb{R}_n^{\langle\bar{x},n-1\rangle}) \neq \emptyset$ for any $\bar{\boldsymbol{x}}^{\{n-1\}} \in \mathrm{Zero}(\mathbb{R}_{n-1})$. In fact, all the zeros can be obtained from the GCD of the polynomials in $\mathbb{R}_n^{\langle\bar{x},n-1\rangle}$ with respect to x_n.

Therefore, the problem is reduced to determining the zeros of $\mathbb{R}_{n-1}$. Again, we can assume that $\mathbb{R}_{n-1}$ contains one polynomial whose leading coefficient with respect to x_{n-1} does not vanish and compute a resultant system $\mathbb{R}_{n-2} \subset \boldsymbol{K}[\boldsymbol{x}^{\{n-2\}}]$ of $\mathbb{R}_{n-1}$, and so on. In this way, two cases may happen: the process either stops at the ith step with $i \leq n$ and $\mathbb{R}_{n-i} = \{0\}$ or continues until $\mathbb{R}_0$ is computed and it contains a nonzero constant. In the latter case, $\mathrm{Zero}(\mathbb{P}) = \emptyset$. For the former, one can determine successively the zeros for $x_{n-i+1}, \ldots, x_n$ from the resultant systems $\mathbb{R}_{n-i+1}, \ldots, \mathbb{R}_n$ by replacing $x_1, \ldots, x_{n-i}$ with arbitrary values. The number of zeros is finite if and only if $i = n$. If some linear variable transformations have been made in the process of elimination, the zeros of the original polynomial set may be recovered by transforming back to the original variables.

In view of the complexity of computing resultant systems, the above-described method is however not practically applicable. The successive elimination is rather straightforward, but the variable transformations necessary for making the hypothesis satisfied complicate the process. We do not go further to give an algorithmic presentation of the method. Instead, the previous example is recalled for illustration.

Example 5.4.5. Refer to Example 5.4.4. For $\mathbb{R}$ in (5.4.2), we take a simple variable transformation $z = w + t$. Then the three polynomials in $\mathbb{R}$ are mapped to

$$\begin{aligned} Q_1 &= (w+t)t^2 - x^2 = t^3 + wt^2 - x^2, \\ Q_2 &= (w+t)t^4 - y^2 = t^5 + wt^4 - y^2, \\ Q_3 &= 2(w+t)t^3 - 2xy = 2t^4 + 2wt^3 - 2xy, \end{aligned}$$

whose leading coefficients with respect to t are all constants. The resultant of Q_1 and $v_2Q_2 + v_3Q_3$ with respect to t is R_1R_2 with

$$\begin{aligned} R_1 &= x^5 - y^3 - xy^2w, \\ R_2 &= y^3v_2^3 + 6xy^2v_2^2v_3 - xy^2wv_2^3 - 4x^2ywv_2^2v_3 + 12x^2yv_2v_3^2 \\ &\quad - 4x^3wv_2v_3^2 + 8x^3v_3^3 + x^5v_2^3, \end{aligned}$$

from which the following resultant system of $\{Q_1, Q_2, Q_3\}$ with respect to t is obtained:

$$\begin{aligned} \mathbb{R}_1 = \{&(x^5 + y^3 - xy^2w)R_1, 4x^2(3y - xw)R_1, \\ &2xy(3y - 2xw)R_1, 8x^3R_1\}. \end{aligned}$$

Since all the polynomials in $\mathbb{R}_1$ have a common divisor, any resultant system of $\mathbb{R}_1$ with respect to any of the variables x, y, w should be equal to $\{0\}$.

For any given values of x and y, the zeros for w, t, and r can be successively computed from $\mathbb{R}_1$, $\mathbb{R}$, and $\mathbb{P}$ respectively. The zeros for z are obtained as the corresponding $w + t$. In the generic case, x and y are regarded as indeterminates, and thus $xy \neq 0$. The GCD of the four polynomials in $\mathbb{R}_1$ is R_1. Solving $R_1 = 0$ for w, one gets $w = (x^5 - y^3)/(xy^2)$. Substituting this solution into Q_1, Q_2, Q_3 and computing their GCD, one finds the only solution for t: $t = y/x$. Now the zero for z can be recovered: $z = w + t = x^4/y^2$. Substituting the solution for z and t into the original polynomials in $\mathbb{P}$ and computing their GCD, one finally obtains the only solution for r: $r = x^2/y$. Therefore, the only zero of $\mathbb{P}$ for z, t, r in terms of generic x and y is determined as $(x^4/y^2, y/x, x^2/y)$.

Solvability criteria

Using the Macaulay resultant, we have established solvability criteria for n homogeneous polynomials in n variables. In what follows, an algebraic criterion is derived for the solvability of an arbitrary set of homogeneous polynomial equations by using resultant systems.

In the rest of this section, $\boldsymbol{x}$ stands for $n+1$ variables $x_0, x_1, \ldots, x_n$ with $\boldsymbol{x}^{\{i\}} = (x_0, x_1, \ldots, x_i)$; similar abbreviations are used with $\bar{\boldsymbol{x}}$, $\boldsymbol{u}$, $\boldsymbol{\lambda}$, etc. Let $P_1, \ldots, P_s$ be homogeneous nonconstant polynomials in $\boldsymbol{x}$ with indeterminate coefficients over $\boldsymbol{K}$. They always have the "trivial" zero $\mathbf{0} = (0, \ldots, 0)$ at least. So the criterion should be for the existence of nontrivial zeros of $\mathbb{P} = \{P_1, \ldots, P_s\}$. The following approach based on Kronecker's method of successive elimination is due to H. Kapferer (see van der Waerden 1950, p. 7).

Form the resultant system $\mathbb{R}$ of $\mathbb{P}$ with respect to x_n according to the method explained above without the linear variable transformation. We now show that

$$\mathrm{Zero}(\mathbb{P}) \supsetneqq \{\mathbf{0}\} \iff \mathrm{Zero}(\mathbb{R}) \supsetneqq \{\mathbf{0}\} \tag{5.4.3}$$

in some extension field of $\boldsymbol{K}$.

Let $d_i = \mathrm{tdeg}(P_i)$ for $1 \leq i \leq s$. Consider first the case in which the coefficients $\mathrm{coef}(P_i, x_n^{d_i})$ do not all vanish. Then by Theorem 5.4.2, for every nontrivial zero $\bar{\boldsymbol{x}}^{\{n-1\}}$ of $\mathbb{R}$, $\mathbb{P}^{\langle \bar{x}, n-1\rangle}$ has at least one zero $\bar{x}_n$ for x_n. The zero $\bar{\boldsymbol{x}}$ of course cannot be trivial. Conversely, every nontrivial zero $\bar{\boldsymbol{x}}$ of $\mathbb{P}$ gives rise to a zero $\bar{\boldsymbol{x}}^{\{n-1\}}$ of $\mathbb{R}$, which cannot be trivial either since $\bar{\boldsymbol{x}}^{\{n-1\}} = 0$ would lead immediately to $\bar{x}_n = 0$ (noting that each P_i is homogeneous).

If $\mathrm{coef}(P_i, x_n^{d_i})$ vanishes for all i, then $\mathbb{R} = \emptyset$ according to Theorem 5.4.2. Hence, $\mathbb{R}$ has a nontrivial zero, say $(1, \ldots, 1)$. In this case, $(0, \ldots, 0, 1)$ is a nontrivial zero of $\mathbb{P}$ as the terms of P_i with the highest power of x_n are all omitted. This proves (5.4.3).

Now the polynomials in $\mathbb{R}$, if any, are homogeneous in $\boldsymbol{x}^{\{n-1\}}$ and one can form a resultant system of $\mathbb{R}$ with respect to x_{n-1}. Let this elimination process continue for $x_{n-1}, \ldots, x_1$. Finally, a finite set of homogeneous polynomials in x_0

$$R_1 x_0^{k_1}, \ldots, R_t x_0^{k_t} \tag{5.4.4}$$

will be obtained. These polynomials have a nontrivial zero if and only if $R_1 = \cdots = R_t = 0$.

Clearly, $R_1, \ldots, R_t$ are polynomials in the coefficients of the P_i. From their construction, it is easy to show that they are homogeneous in the coefficients of every individual P_i (see van der Waerden 1950, p. 8). The set of polynomials $R_1, \ldots, R_t$ is also called a *resultant system* of $P_1, \ldots, P_s$ or of $\mathbb{P}$ with respect to $\boldsymbol{x}$. It may be empty: in this case $t = 0$.

Summing up the above discussions, we have the following.

Theorem 5.4.3. From any set $\mathbb{P}$ of homogeneous polynomials in $\boldsymbol{x}$ with indeterminate coefficients $\boldsymbol{u}$ over $\boldsymbol{K}$, one can determine a finite set $\mathbb{R}$ of polynomials in $\boldsymbol{K}[\boldsymbol{u}]$ such that for any special values $\bar{\boldsymbol{u}}$ of $\boldsymbol{u}$ in an arbitrary extension field of $\boldsymbol{K}$

$$\bar{\boldsymbol{u}} \in \mathrm{Zero}(\mathbb{R}) \iff \mathrm{Zero}(\mathbb{P}|_{\boldsymbol{u}=\bar{\boldsymbol{u}}}) \supsetneqq \{\mathbf{0}\}.$$

The polynomials in $\mathbb{R}$ are homogeneous in the coefficients of every individual polynomial in $\mathbb{P}$.

The resultant system $\mathbb{R}$ of $\mathbb{P}$ may contain numerous polynomials. Theorem 5.4.1 implies that, when $|\mathbb{P}| = s = n + 1$ (the number of variables), the single Macaulay resultant is sufficient. In general no condition for solvability is necessary if $s < n + 1$.

u-Resultant

Consider a set of n homogeneous polynomials $\mathbb{P} = \{P_1, \ldots, P_n\} \subset \boldsymbol{K}[\boldsymbol{x}]$. Let $d_i = \mathrm{tdeg}(P_i)$ for $1 \le i \le n$ and

$$P_u = x_0 u_0 + x_1 u_1 + \cdots + x_n u_n,$$

where $\boldsymbol{u} = (u_0, u_1, \ldots, u_n)$ are $n + 1$ new indeterminates.

Definition 5.4.1. The Macaulay resultant R_u of the $n+1$ homogeneous polynomials $P_1, \ldots, P_n, P_u$ with respect to the $n + 1$ variables $\boldsymbol{x}$ is called the *u-resultant* of $P_1, \ldots, P_n$ or of $\mathbb{P}$ with respect to $\boldsymbol{x}$.

The u-resultant may also be defined for an arbitrary set of s (not necessarily n) homogeneous polynomials in $\boldsymbol{x}$ that has only finitely many zeros (van der Waerden 1950, pp. 15 f). For $n = 2$, it can be constructed alternatively by using the bivariate resultant (Chionh and Goldman 1995).

Let R_u be the *u-resultant* of $\mathbb{P}$, a set of n homogeneous polynomials in $\boldsymbol{K}[\boldsymbol{x}]$, with respect to $\boldsymbol{x}$. If $R_u \equiv 0$, then $\mathrm{Zero}(\mathbb{P})$ is infinite. Otherwise, R_u is a polynomial homogeneous in $\boldsymbol{u}$ of degree $D = d_1 \cdots d_n$ by Theorem 5.4.1 c. In this case, R_u can be factorized into linear factors:

$$R_u = \prod_{j=1}^{D} (\lambda_{0j} u_0 + \lambda_{1j} u_1 + \cdots + \lambda_{nj} u_n)$$

over some algebraic-extension field of $\boldsymbol{K}$. Thus,

$$(\lambda_{0j}, \lambda_{1j}, \ldots, \lambda_{nj}) \in \mathrm{Zero}(\mathbb{P}) \tag{5.4.5}$$

for any $1 \le j \le D$. On the contrary, if (5.4.5) holds, then

$$\lambda_{0j} u_0 + \lambda_{1j} u_1 + \cdots + \lambda_{nj} u_n$$

must be a factor of R_u. This gives a method for the exact determination of $\mathrm{Zero}(\mathbb{P})$ as well as the multiplicity of each zero (as the degree of the corresponding linear factor) (cf. Lazard 1981).

To see the correctness of the method, consider any

$$\bar{\boldsymbol{x}} = (\bar{x}_0, \bar{x}_1, \ldots, \bar{x}_n) \in \mathrm{Zero}(\mathbb{P}).$$

For any $\bar{\boldsymbol{u}} = (\bar{u}_0, \bar{u}_1, , \ldots, \bar{u}_n)$ satisfying

$$\bar{x}_0 \bar{u}_0 + \bar{x}_1 \bar{u}_1 + \cdots + \bar{x}_n \bar{u}_n = 0, \tag{5.4.6}$$

the linear equation $P_{\bar{u}} = 0$ represents a hyperplane passing through the point $\bar{\boldsymbol{x}}$. It follows that

$$\bar{\boldsymbol{x}} \in \text{Zero}(\mathbb{P} \cup \{P_{\bar{u}}\}).$$

Hence, $R_{\bar{u}} = 0$ by Theorem 5.4.1 a. As this is true for any $\bar{\boldsymbol{u}}$ satisfying (5.4.6),

$$\bar{x}_0 u_0 + \bar{x}_1 u_1 + \cdots + \bar{x}_n u_n$$

is a factor of R_u by the divisibility of polynomials.

For any linear factor

$$L = \lambda_0 u_0 + \lambda_1 u_1 + \cdots + \lambda_n u_n$$

of R_u, we call the number of all those linear factors (including L itself) of R_u which differ from L only by constant factors (in some algebraic extension of $\boldsymbol{K}$) the *multiplicity* of

$$(\lambda_0, \lambda_1, \ldots, \lambda_n) \in \text{Zero}(\mathbb{P}).$$

As a consequence, we have the following constructive version of Bézout's theorem.

Theorem 5.4.4. Let $\mathbb{P}$ be a set of n homogeneous polynomials in $\boldsymbol{K}[\boldsymbol{x}]$. Then either Zero($\mathbb{P}$) is infinite, or the sum of the multiplicities of all $\bar{\boldsymbol{x}} \in \text{Zero}(\mathbb{P})$ is equal to $\prod_{P \in \mathbb{P}} \text{tdeg}(P)$.

If the given polynomials P_i are nonhomogeneous but ordinary ones in n variables $x_1, \ldots, x_n$, one can introduce a new variable x_0 to homogenize them. Let the obtained set of homogeneous polynomials be

$$\tilde{\mathbb{P}} = \{\tilde{P}_1, \ldots, \tilde{P}_n\}.$$

Unlikely to cause confusion, the u-resultant R_u of $\tilde{\mathbb{P}}$ is also said to be the *u-resultant* of $\mathbb{P}$. R_u may be used to determine Zero($\mathbb{P}$) as well. This is illustrated by the following example.

Example 5.4.6. Find the intersection of the circle and ellipse given respectively by

$$\begin{aligned} P_1 &= x_1^2 + x_2^2 - 2 = 0, \\ P_2 &= x_1^2 + 6x_2^2 - 3 = 0. \end{aligned}$$

We do so by computing the u-resultant R of $\{P_1, P_2\}$ with respect to x_1 and x_2. By definition, R is the Macaulay resultant of

$$\begin{aligned} \tilde{P}_1 &= x_1^2 + x_2^2 - 2x_0^2, \\ \tilde{P}_2 &= x_1^2 + 6x_2^2 - 3x_0, \\ P_u &= u_0 x_0 + u_1 x_1 + u_2 x_2, \end{aligned}$$

where x_0 is introduced to homogenize P_1 and P_2. R may be obtained from the Macaulay resultant computed in Example 5.4.3 with $x_3 = x_0$ by substituting

a_{ij}, b_{ij} with the corresponding numerical coefficients of $\tilde{P}_1$, $\tilde{P}_2$ and c_i with u_i (of course $u_3 = u_0$). One can find that

$$R = 25u_0^4 - 90u_0^2u_1^2 - 10u_0^2u_2^2 + 81u_1^4 - 18u_1^2u_2^2 + u_2^4,$$

which can be factorized to

$$(\sqrt{5}u_0 + 3u_1 + u_2)(\sqrt{5}u_0 + 3u_1 - u_2)$$
$$(\sqrt{5}u_0 - 3u_1 + u_2)(\sqrt{5}u_0 - 3u_1 - u_2).$$

From the linear factors, one gets the four points of intersection

$$\left(\frac{3}{\sqrt{5}}, \frac{1}{\sqrt{5}}\right), \left(\frac{3}{\sqrt{5}}, -\frac{1}{\sqrt{5}}\right),$$
$$\left(-\frac{3}{\sqrt{5}}, \frac{1}{\sqrt{5}}\right), \left(-\frac{3}{\sqrt{5}}, -\frac{1}{\sqrt{5}}\right).$$

The above method of determining Zero($\mathbb{P}$) on the basis of the computation of the u-resultant R_u of $\mathbb{P}$ is applicable only if $R_u \not\equiv 0$, i.e., Zero($\tilde{\mathbb{P}}$) is finite. It may happen that Zero($\mathbb{P}$) is finite, but Zero($\tilde{\mathbb{P}}$) is not. In other words, $\mathbb{P}$ may have infinitely many zeros at infinity. Thus, R_u may be identically 0 even if Zero($\mathbb{P}$) is finite. When this happens, Zero($\mathbb{P}$) is said to have *excess components* at infinity. For example, let

$$\mathbb{P} = \{x_1(x_1 + \cdots + x_n) - 1, \ldots, x_n(x_1 + \cdots + x_n) - 1\};$$

Zero($\mathbb{P}$) consists of two (affine) zeros

$$\left(\frac{1}{\sqrt{n}}, \ldots, \frac{1}{\sqrt{n}}\right), \left(-\frac{1}{\sqrt{n}}, \ldots, -\frac{1}{\sqrt{n}}\right)$$

and has an excess component at infinity given by $x_1 + \cdots + x_n = 0$ for $n \geq 2$. The u-resultant R_u of $\mathbb{P}$ is zero when $n \geq 3$. In the case $n = 2$, R_u is nonzero because the homogenized polynomial set $\tilde{\mathbb{P}}$ has only finitely many zeros.

To deal with such sets of nonhomogeneous polynomials which have finitely many affine zeros with excess components at infinity, one may employ a modified version of the method which permits us to find all the affine zeros. The modification explained below is due to J. F. Canny, A. L. Chistov, and D. Yu. Grigor'ev according to Kapur and Lakshman (1992).

Consider an arbitrary set of n polynomials, $\mathbb{P} = \{P_1, \ldots, P_n\} \subset \boldsymbol{K}[x_1, \ldots, x_n]$. Let $\tilde{P}_i$ be the homogenization of P_i by x_0 and

$$F_i = \tilde{P}_i + v x_i^{d_i}$$

for $1 \leq i \leq n$, and let

$$F_u = (u_0 + v)x_0 + u_1x_1 + \cdots + u_nx_n,$$

where v is a new variable. Compute the Macaulay resultant $R_u = R_u(v, \boldsymbol{u})$ of $F_1, \ldots, F_n, F_u$, regarded as homogeneous polynomials in $x_0, x_1, \ldots, x_n$; R_u is called the *generalized characteristic polynomial* of $\mathbb{P}$ with respect to $x_1, \ldots, x_n$. Now consider R_u as a polynomial in v, written in the following form

$$R_u = v^q + R_{q-1}v^{q-1} + \cdots + R_k v^k,$$

where $k \geq 0$ and the R_i are polynomials in $\boldsymbol{K}[\boldsymbol{u}]$. If $k = 0$, then R_k is the same as the u-resultant R_u of $\mathbb{P}$. However, if $\mathbb{P}$ has excess components at infinity, then $k > 0$. In this case, the trailing coefficient R_k shares a nice property with R_u: R_k may be factorized into linear factors

$$R_k = \prod_j (\lambda_{0j}u_0 + \lambda_{1j}u_1 + \cdots + \lambda_{nj}u_n)$$

over some algebraic-extension field of $\boldsymbol{K}$ and thus

$$(\lambda_{0j}, \lambda_{1j}, \ldots, \lambda_{nj}) \in \mathrm{Zero}(\tilde{\mathbb{P}})$$

for each j. On the contrary, if $(\bar{x}_1, \ldots, \bar{x}_n) \in \mathrm{Zero}(\mathbb{P})$, then

$$u_0 + \bar{x}_1 u_1 + \cdots + \bar{x}_n u_n$$

is a divisor of R_k. This provides a way to recover all the affine zeros of $\mathbb{P}$ even in the presence of excess components at infinity.

Remark 5.4.3. Computing full u-resultants and thus complete generalized characteristic polynomials is almost impossible for polynomial sets of moderate size. For practical computation of zeros, one may construct the u-resultant for specialized values of some of the indeterminates u_i, so that the zeros for some of the variables are determined first. Techniques of this type come from recent research. For more details, the interested reader may consult relevant publications by J. F. Canny, Y. N. Lakshman, and their co-workers.

6 Computational algebraic geometry and polynomial-ideal theory

Among the fundamental objects studied in algebraic geometry are algebraic varieties which are aggregates of common zeros of polynomial sets, viewed as points in an affine space. In contrast, ideals generated by polynomial sets are typical examples dealt with in commutative algebra. Elimination algorithms provide powerful constructive tools for many problems in these two related areas. In this chapter, we investigate some computational aspects of a few such problems.

6.1 Dimension

As in the previous chapters, all considered polynomials are in n variables $\boldsymbol{x}$ with coefficients in a fixed field $\boldsymbol{K}$ of characteristic 0 unless stated otherwise.

Definition 6.1.1. The dimension of a perfect triangular set $\mathbb{T} \subset \boldsymbol{K}[\boldsymbol{x}]$ is defined to be

$$\dim(\mathbb{T}) \triangleq n - |\mathbb{T}|.$$

It is also called the dimension of any perfect triangular system $[\mathbb{T}, \mathbb{U}]$ in $\boldsymbol{K}[\boldsymbol{x}]$.

Lemma 6.1.1. One can compute an irreducible triangular series Ψ of any perfect triangular system $\mathfrak{T}$ in $\boldsymbol{K}[\boldsymbol{x}]$ such that $\dim(\mathfrak{T}) = \max_{\mathfrak{T}^* \in \Psi} \dim(\mathfrak{T}^*)$.

Proof. Applying algorithm Decom to $\mathfrak{T} = [\mathbb{T}, \mathbb{U}]$, one can obtain $[\mathbb{T}_1, \mathbb{U}_1], \ldots, [\mathbb{T}_e, \mathbb{U}_e]$ and $[\mathbb{P}_1, \mathbb{Q}_1, \mathbb{T}_1^*], \ldots, [\mathbb{P}_h, \mathbb{Q}_h, \mathbb{T}_h^*]$ such that (4.2.3) holds and each irreducible triangular set $\mathbb{T}_i$ has the same set of parameters as $\mathbb{T}$ and thus $\dim(\mathbb{T}_i) = \dim(\mathbb{T})$. We assume that in all the algebraic factorization of T in D2.2.2 of Decom the polynomial D is so chosen that it does not involve the dependents of $\mathbb{T}'$. Then each $\mathbb{P}_j$ in (4.2.3) is obtained actually from a triangular set $\mathbb{T}_j^-$ by adjoining a single polynomial D_j. Moreover, $\mathbb{T}_j^-$ has the same set of parameters as $\mathbb{T}$ and D_j involves only these parameters. Let

$$[\mathbb{T}_{j1}, \mathbb{U}_{j1}], \ldots, [\mathbb{T}_{jt_j}, \mathbb{U}_{jt_j}]$$

be a triangular series of $\{D_j\}$ and $\mathbb{T}_{jl}^* = \mathbb{T}_{jl} \cup \mathbb{T}_j^- \cup \mathbb{T}_j^*$ for $l = 1, \ldots, t_j$. Then

$$\mathrm{Zero}(\mathbb{P}_j \cup \mathbb{T}_j^*/\mathbb{Q}_j) = \bigcup_{l=1}^{t_j} \mathrm{Zero}(\mathbb{T}_{jl}^*/\mathbb{Q}_j \cup \mathbb{U}_{jl}),$$

each $\mathbb{T}^*_{jl}$ can be ordered as a triangular set and $\mathfrak{T}_{jl} = [\mathbb{T}^*_{jl}, \mathbb{Q}_j \cup \mathbb{U}_{jl}]$ is a triangular system. If $\mathfrak{T}_{jl}$ is perfect, then $\dim(\mathfrak{T}_{jl}) < \dim(\mathbb{T})$. Now consider each of the perfect triangular systems $\mathfrak{T}_{jl}$ as $[\mathbb{T}, \mathbb{U}]$ and proceed as above recursively. The procedure will terminate finally to give an irreducible triangular series Ψ of $\mathfrak{T}$. This proves that $\dim(\mathfrak{T}) \geq \max_{\mathfrak{T}^* \in \Psi} \dim(\mathfrak{T}^*)$.

It remains to be shown that $e \neq 0$. By Lemma 5.1.3, $\mathfrak{T}$ has a regular zero $\boldsymbol{\xi}$. If $e = 0$, then the number of parameters of $\mathbb{T}^*$ is smaller than that of $\mathbb{T}$ for any $[\mathbb{T}^*, \mathbb{U}^*] \in \Psi$. Hence, $\boldsymbol{\xi}$ cannot be a zero of any such triangular system $[\mathbb{T}^*, \mathbb{U}^*]$. This derives a contradiction, so $e > 0$ and the lemma is proved. □

Corollary 6.1.2. For any irreducible triangular series Ψ of a perfect triangular system $\mathfrak{T}$ in $\boldsymbol{K}[\boldsymbol{x}]$, $\dim(\mathfrak{T}) = \max_{\mathfrak{T}^* \in \Psi} \dim(\mathfrak{T}^*)$.

Proof. Compute an irreducible triangular series $\bar{\Psi}$ of $\mathfrak{T}$ according to Lemma 6.1.1 such that $\dim(\mathfrak{T}) = \max_{\bar{\mathfrak{T}} \in \bar{\Psi}} \dim(\bar{\mathfrak{T}})$. Clearly,

$$\bigcup_{\bar{\mathfrak{T}} \in \bar{\Psi}} \mathrm{Zero}(\bar{\mathfrak{T}}) = \bigcup_{\mathfrak{T}^* \in \Psi} \mathrm{Zero}(\mathfrak{T}^*) \tag{6.1.1}$$

holds. If $\max_{\bar{\mathfrak{T}} \in \bar{\Psi}} \dim(\bar{\mathfrak{T}}) > \max_{\mathfrak{T}^* \in \Psi} \dim(\mathfrak{T}^*)$, then there exists a $\bar{\mathfrak{T}} \in \bar{\Psi}$ such that $\dim(\bar{\mathfrak{T}}) > \dim(\mathfrak{T}^*)$ for all $\mathfrak{T}^* \in \Psi$. Let $\boldsymbol{\xi} \in \mathrm{RegZero}(\bar{\mathfrak{T}})$. It follows that $\boldsymbol{\xi}$ cannot be a zero of any $\mathfrak{T}^* \in \Psi$. This contradicts with (6.1.1). For the same reason, $\max_{\bar{\mathfrak{T}} \in \bar{\Psi}} \dim(\bar{\mathfrak{T}})$ cannot be smaller than $\max_{\mathfrak{T}^* \in \Psi} \dim(\mathfrak{T}^*)$. Therefore, $\dim(\mathfrak{T}) = \max_{\bar{\mathfrak{T}} \in \bar{\Psi}} \dim(\bar{\mathfrak{T}}) = \max_{\mathfrak{T}^* \in \Psi} \dim(\mathfrak{T}^*)$, and the proof is complete. □

Lemma 6.1.3. Any perfect triangular system in $\boldsymbol{K}[\boldsymbol{x}]$ is also perfect over an algebraic closure of $\boldsymbol{K}$.

Proof. Let $\mathfrak{T}$ be a perfect triangular system and Ψ an irreducible triangular series of $\mathfrak{T}$; then $\Psi \neq \emptyset$. Let $\mathfrak{T}^* \in \Psi$. By Theorem 4.3.3 $\mathfrak{T}^*$ has a zero in the algebraic closure $\bar{\boldsymbol{K}}$ of $\boldsymbol{K}$. It is also a zero of $\mathfrak{T}$. Hence $\mathfrak{T}$ is perfect over $\bar{\boldsymbol{K}}$. □

Corollary 6.1.4. Any triangular system in $\boldsymbol{K}[\boldsymbol{x}]$ is perfect if and only if it is perfect over an algebraic closure of $\boldsymbol{K}$.

Theorem 5.1.12 can also be considered as a corollary of Lemma 6.1.3.

A new notation: $\mathrm{ITS}(\mathfrak{P})$ stands for an *irreducible triangular series* of any polynomial set or system $\mathfrak{P}$ in $\boldsymbol{K}[\boldsymbol{x}]$.

Lemma 6.1.5. Let Ψ_1 and Ψ_2 be two triangular series in $\boldsymbol{K}[\boldsymbol{x}]$, with all triangular systems in Ψ_1 and Ψ_2 perfect, such that

$$\bigcup_{\mathfrak{T}_1 \in \Psi_1} \mathrm{Zero}(\mathfrak{T}_1) = \bigcup_{\mathfrak{T}_2 \in \Psi_2} \mathrm{Zero}(\mathfrak{T}_2).$$

Then $\max_{\mathfrak{T}_1 \in \Psi_1} \dim(\mathfrak{T}_1) = \max_{\mathfrak{T}_2 \in \Psi_2} \dim(\mathfrak{T}_2)$.

Proof. Note that $\Psi_i^* = \bigcup_{\mathfrak{T}_i \in \Psi_i} \mathrm{ITS}(\mathfrak{T}_i)$, for $i = 1, 2$, are two irreducible triangular series such that

$$\bigcup_{\mathfrak{T}_1 \in \Psi_1^*} \mathrm{Zero}(\mathfrak{T}_1) = \bigcup_{\mathfrak{T}_2 \in \Psi_2^*} \mathrm{Zero}(\mathfrak{T}_2).$$

By Corollary 6.1.2 we have

$$\max_{\mathfrak{T}_i \in \Psi_i} \dim(\mathfrak{T}_i) = \max_{\mathfrak{T}_i \in \Psi_i} \max_{\mathfrak{T}_i^* \in \mathrm{ITS}(\mathfrak{T}_i)} \dim(\mathfrak{T}_i^*) = \max_{\mathfrak{T} \in \Psi_i^*} \dim(\mathfrak{T}),$$

for $i = 1, 2$. Repeating the reasoning in the proof of Corollary 6.1.2 shows that

$$\max_{\mathfrak{T}_1 \in \Psi_1^*} \dim(\mathfrak{T}_1) = \max_{\mathfrak{T}_2 \in \Psi_2^*} \dim(\mathfrak{T}_2).$$

This implies that $\max_{\mathfrak{T}_1 \in \Psi_1} \dim(\mathfrak{T}_1) = \max_{\mathfrak{T}_2 \in \Psi_2} \dim(\mathfrak{T}_2)$. □

As a consequence of this lemma, we have the following.

Corollary 6.1.6. Let Ψ be any triangular series of a perfect triangular system $\mathfrak{T}$ in $\boldsymbol{K}[\boldsymbol{x}]$, with all triangular systems in Ψ perfect. Then

$$\dim(\mathfrak{T}) = \max_{\mathfrak{T}^* \in \Psi} \dim(\mathfrak{T}^*).$$

By Lemma 6.1.5, the following definition is proper.

Definition 6.1.2. Let $\mathfrak{P}$ be a polynomial system in $\boldsymbol{K}[\boldsymbol{x}]$ with $\mathrm{Zero}(\mathfrak{P}) \neq \emptyset$, and Ψ any triangular series of $\mathfrak{P}$, with all triangular systems in Ψ perfect. The *dimension* of $\mathfrak{P}$ is defined to be

$$\mathrm{Dim}(\mathfrak{P}) \triangleq \max_{\mathfrak{T} \in \Psi} \dim(\mathfrak{T}).$$

$\mathrm{Dim}([\mathbb{P}, \emptyset])$ is also called the dimension of $\mathbb{P}$.

Remark 6.1.1. The notation Dim is used to distinguish the dimension of a polynomial set or system from that of a triangular set or system. Consider, for example,

$$\mathbb{T} = [x(x-1), xy + u, xz - u]$$

in 4-dimensional space with $u \prec x \prec y \prec z$. As a polynomial set, $\mathbb{T}$ is clearly of dimension 2. However, $\mathbb{T}$ as a triangular set is perfect of dimension $4 - |\mathbb{T}| = 1$. Hence

$$\mathrm{Dim}(\mathbb{T}) = 2 \neq 1 = \dim(\mathbb{T}).$$

Now we introduce a few concepts related to *algebraic varieties* or *manifolds* which are geometric objects defined by sets of algebraic equations in an n-dimensional space.

Definition 6.1.3. Let $\mathcal{V}$ be a collection of points in an n-dimensional affine space $\mathbf{A}^n_{\tilde{\boldsymbol{K}}}$ with coordinates $\boldsymbol{x}$ over some extension field $\tilde{\boldsymbol{K}}$ of $\boldsymbol{K}$. $\mathcal{V}$ is called an (affine) *algebraic variety*, or simply a *variety*, if there is a polynomial set $\mathbb{P} \subset \boldsymbol{K}[\boldsymbol{x}]$ such that $\mathcal{V} = \mathrm{Zero}(\mathbb{P})$. We call $\mathbb{P}$ the *defining set* and $\mathbb{P} = 0$ the *defining equations* of $\mathcal{V}$.

A variety $\mathcal{V}_1$ is called a *subvariety* of another variety $\mathcal{V}_2$, which is denoted as $\mathcal{V}_1 \subset \mathcal{V}_2$, if any point in $\mathcal{V}_1$ is also in $\mathcal{V}_2$. A variety $\mathcal{V}_1$ is called a *true* subvariety of $\mathcal{V}_2$ if $\mathcal{V}_1 \subset \mathcal{V}_2$ and $\mathcal{V}_1 \neq \mathcal{V}_2$.

Definition 6.1.4. A variety $\mathcal{V} \subset \mathbf{A}^n_{\tilde{K}}$ is said to be *irreducible* if it cannot be expressed as the union of two true subvarieties $\mathcal{V}_1$ and $\mathcal{V}_2$ of $\mathcal{V}$. In this case, the defining set of $\mathcal{V}$ is also said to be irreducible.

Any point $\boldsymbol{\xi}$ of an algebraic variety $\mathcal{V}$ over some extension of $\boldsymbol{K}$, which is such that every polynomial in $\boldsymbol{K}[\boldsymbol{x}]$ annulled by $\boldsymbol{\xi}$ vanishes on $\mathcal{V}$, is called a *generic point* of $\mathcal{V}$.

Definition 6.1.5. Let an algebraic variety $\mathcal{V} \subset \mathbf{A}^n_{\tilde{K}}$ be defined by the polynomial set $\mathbb{P} \subset \boldsymbol{K}[\boldsymbol{x}]$ and $\mathcal{V} \neq \emptyset$. The dimension of $\mathbb{P}$ is also called the *dimension* of $\mathcal{V}$ or $\text{Zero}(\mathbb{P})$. Symbolically,

$$\text{Dim}(\mathcal{V}) = \text{Dim}(\text{Zero}(\mathbb{P})) = \text{Dim}(\mathbb{P}).$$

The dimension of a nonempty algebraic variety is one of the fundamental invariants that characterize the variety. The definition given here is equivalent to those in standard books of algebraic geometry. This can be seen from the following fact which will be proved in the next section. From each irreducible triangular set $\mathbb{T}$ in an irreducible triangular series Ψ of $\mathbb{P}$, one can construct an irreducible algebraic variety $\mathcal{V}_{\mathbb{T}} \subset \mathcal{V} = \text{Zero}(\mathbb{P})$ such that any generic zero of $\mathbb{T}$ is a generic point of $\mathcal{V}_{\mathbb{T}}$ and $\mathcal{V} = \bigcup_{\mathbb{T}\in\Psi} \mathcal{V}_{\mathbb{T}}$. Therefore, $\text{Dim}(\mathcal{V}_{\mathbb{T}}) = \dim(\mathbb{T})$ coincides with the dimension of $\mathcal{V}_{\mathbb{T}}$ defined in algebraic geometry, and so does $\text{Dim}(\mathcal{V}) = \text{Dim}(\mathbb{P})$.

Definition 6.1.6. An *irreducible component* of an algebraic variety $\mathcal{V} \subset \mathbf{A}^n_{\tilde{K}}$ is an irreducible subvariety $\mathcal{W}$ of $\mathcal{V}$. Any defining polynomial set of $\mathcal{W}$ is also called an *irreducible component* of the defining set $\mathbb{P} \subset \boldsymbol{K}[\boldsymbol{x}]$ of $\mathcal{V}$. $\mathcal{W}$ is said to be *irredundant* if it is not contained in another irreducible subvariety of $\mathcal{V}$.

Note that an irreducible component referred to in algebraic geometry usually means an irredundant irreducible component. In what follows we recall several results on dimension from algebraic geometry (see, e.g., Hartshorne 1977, pp. 7f and 48). Some of them can be easily proved by using triangular series. We omit the proofs; they may be worked out as exercises.

Proposition 6.1.7. An irreducible polynomial set $\mathbb{P} \subset \boldsymbol{K}[\boldsymbol{x}]$ has dimension $n-1$ if and only if $\text{Zero}(\mathbb{P}) = \text{Zero}(P)$, where P is a nonconstant polynomial irreducible over $\boldsymbol{K}$.

Proposition 6.1.8. Let $\mathbb{P}$ be an irreducible polynomial set and P any polynomial in $\boldsymbol{K}[\boldsymbol{x}]$ with $\text{Zero}(\mathbb{P}) \not\subset \text{Zero}(P)$. If $\text{Zero}(\mathbb{P} \cup \{P\}) \neq \emptyset$, then all the irredundant irreducible components of $\mathbb{P} \cup \{P\}$ have the same dimension $\text{Dim}(\mathbb{P}) - 1$, and so does $\mathbb{P} \cup \{P\}$ itself.

See Wu (1994, pp. 186f), for a proof of the above proposition in weak form: $\text{Dim}(\mathbb{P} \cup \{P\}) < \text{Dim}(\mathbb{P})$.

Proposition 6.1.9. Let $\mathbb{P} \subset \boldsymbol{K}[\boldsymbol{x}]$ be any polynomial set with $\text{Zero}(\mathbb{P}) \neq \emptyset$. Then every irredundant irreducible component of $\mathbb{P}$ has dimension $\geq n - |\mathbb{P}|$. In particular, $\text{Dim}(\mathbb{P}) \geq n - |\mathbb{P}|$.

Proposition 6.1.10 (Affine dimension theorem). Let $\mathbb{P}_1, \mathbb{P}_2 \subset \boldsymbol{K}[\boldsymbol{x}]$ be two irreducible polynomial sets of dimensions s_1, s_2 respectively. Then every irredundant irreducible component of $\mathbb{P}_1 \cup \mathbb{P}_2$ has dimension $\geq s_1 + s_2 - n$, and so does $\mathbb{P}_1 \cup \mathbb{P}_2$ itself.

Theorem 6.1.11. Let $\mathbb{T}$ be a regular set and P any polynomial in $\boldsymbol{K}[\boldsymbol{x}]$ such that $P(\boldsymbol{\xi}) \neq 0$ for any $\boldsymbol{\xi} \in \mathrm{RegZero}(\mathbb{T})$, and Ψ a triangular series of $[\mathbb{T} \cup \{P\}, \mathrm{ini}(\mathbb{T})]$. Then either $\mathfrak{T}$ is not perfect or $\dim(\mathfrak{T}) < \dim(\mathbb{T})$ for each $\mathfrak{T} \in \Psi$.

Proof. By Proposition 5.1.5, $R = \mathrm{res}(P, \mathbb{T})$ is a nonzero polynomial not involving the dependents of $\mathbb{T}$. Let $\mathbb{T}_1, \ldots, \mathbb{T}_e$ be a characteristic series of $\{R\}$. Then each $\mathbb{T} \cup \mathbb{T}_i$ can be ordered as a triangular set $\mathbb{T}_i^*$. Either $\mathbb{T}_i^*$ is not perfect or $\dim(\mathbb{T}_i^*) \leq \dim(\mathbb{T}) - 1 < \dim(\mathbb{T})$. On the other hand,

$$\mathrm{Zero}(\mathbb{T} \cup \{P\}/\mathrm{ini}(\mathbb{T})) \subset \mathrm{Zero}(\mathbb{T} \cup \{R\}/\mathrm{ini}(\mathbb{T})) = \bigcup_{i=1}^{e} \mathrm{Zero}(\mathbb{T}_i^*/\mathrm{ini}(\mathbb{T}_i^*)).$$

If $e = 0$ (when $R \in \boldsymbol{K}$) or all the $\mathbb{T}_i^*$ are not perfect, then $\mathrm{Zero}(\mathbb{T} \cup \{P\}/\mathrm{ini}(\mathbb{T})) = \emptyset$ and every $\mathfrak{T} \in \Psi$ is not perfect. Hence, for each $\mathfrak{T} \in \Psi$ either $\mathfrak{T}$ is not perfect or $\dim(\mathfrak{T}) < \dim(\mathbb{T})$. The theorem is proved. □

This theorem holds true when $\mathbb{T}$ is irreducible and $\mathrm{prem}(P, \mathbb{T}) \neq 0$. For any irreducible triangular set $\mathbb{T}$ is regular, and $P(\boldsymbol{\xi}) \neq 0$ for any generic zero $\boldsymbol{\xi}$ of $\mathbb{T}$ if and only if $\mathrm{prem}(P, \mathbb{T}) \neq 0$ (see Lemma 4.3.1). The theorem is also valid if Ψ is a triangular series of $[\mathbb{T} \cup \{P\}, \mathbb{U}]$, where $\mathbb{U}$ is any polynomial set such that $[\mathbb{T}, \mathbb{U}]$ constitutes a triangular system.

6.2 Decomposition of algebraic varieties

Decomposing given algebraic varieties into irreducible or equidimensional components is a fundamental task in classical algebraic geometry and has various applications in modern geometry engineering. Among such applications we can mention two: one in computer-aided geometric design where the considered geometric objects are desired to be decomposed into *simpler* subobjects and the other in automated geometry theorem proving where the configuration of the geometric hypotheses needs to be decomposed in order to determine on which components the geometric theorem holds true.

In view of the relationship between varieties and ideals, a decomposition of an algebraic variety will lead to one of the radicals of the corresponding ideal, and vice versa. So the two kinds of decomposition are presented and mixed together in this section.

Ideal saturation for triangular sets

Definition 6.2.1. Let $\mathfrak{I}$ be an ideal and F a polynomial in $\boldsymbol{K}[\boldsymbol{x}]$. The *saturation* of $\mathfrak{I}$ with respect to F is the infinite set

$$\mathfrak{I} : F^\infty \triangleq \{P \in \boldsymbol{K}[\boldsymbol{x}]:\ F^q P \in \mathfrak{I} \text{ for some integer } q > 0\}.$$

It is easy to verify by definition that $\mathfrak{I} : F^\infty$ is an ideal. This can also be seen from the following lemma.

Lemma 6.2.1. Let $\mathbb{P}$ be a polynomial set and F a polynomial in $\boldsymbol{K}[\boldsymbol{x}]$, and $\mathbb{P}^* = \mathbb{P} \cup \{zF - 1\}$, where z is a new variable. Then $P \in \text{Ideal}(\mathbb{P}^*) \cap \boldsymbol{K}[\boldsymbol{x}]$ if and only if there exists an integer $q > 0$ such that $F^q P \in \text{Ideal}(\mathbb{P})$.

Proof. Let $P \in \text{Ideal}(\mathbb{P}^*) \cap \boldsymbol{K}[\boldsymbol{x}]$; then there are polynomials $Q_i, Q \in \boldsymbol{K}[\boldsymbol{x}, z]$ such that $P = \sum_{P_i \in \mathbb{P}} Q_i P_i + Q(zF - 1)$. In this equality, z is arbitrary, so we can substitute z by $1/F$. Cleaning the denominators of the substituted equality, one gets an expression of the form

$$F^s P = \sum_{P_i \in \mathbb{P}} Q_i^* P_i$$

for some integer $s \geq 0$ and polynomials $Q_i^* \in \boldsymbol{K}[\boldsymbol{x}]$. It follows that $F^q P \in \text{Ideal}(\mathbb{P})$, where $q = \max(s, 1) > 0$.

On the other hand, if $F^q P \in \text{Ideal}(\mathbb{P})$ for some integer $q > 0$, then

$$(zF)^q P \in \text{Ideal}(\mathbb{P}^*) \subset \boldsymbol{K}[\boldsymbol{x}, z].$$

Hence

$$\begin{aligned} P &= (zF)^q P - [(zF)^q - 1]P \\ &= (zF)^q P - (zF - 1)[(zF)^{q-1} + \cdots + 1]P \in \text{Ideal}(\mathbb{P}^*). \end{aligned}$$ □

The following lemma and Lemma 6.2.1 are parallel, and so are their proofs.

Lemma 6.2.2. Let $\mathbb{P}$ be a polynomial set and $F_1, \ldots, F_t$ be t polynomials in $\boldsymbol{K}[\boldsymbol{x}]$, and

$$\mathbb{P}^* = \mathbb{P} \cup \{z_i F_i - 1 : \ 1 \leq i \leq t\},$$

where $z_1, \ldots, z_t$ are new variables. Then $P \in \text{Ideal}(\mathbb{P}^*) \cap \boldsymbol{K}[\boldsymbol{x}]$ if and only if there exist integers $q_1 > 0, \ldots, q_t > 0$ such that $F_1^{q_1} \cdots F_t^{q_t} P \in \text{Ideal}(\mathbb{P})$.

Proof. Let $P \in \text{Ideal}(\mathbb{P}^*) \cap \boldsymbol{K}[\boldsymbol{x}]$; then there are polynomials $Q_i, H_j \in \boldsymbol{K}[\boldsymbol{x}, z_1, \ldots, z_t]$ such that

$$P = \sum_{P_i \in \mathbb{P}} Q_i P_i + \sum_{j=1}^{t} H_j (z_j F_j - 1).$$

This equality holds for arbitrary $z_1, \ldots, z_t$, wherefore one can substitute z_j by $1/F_j$ for each j. Cleaning the denominators of the obtained expression (and multiplying the result by F_i when necessary), we have

$$F_1^{q_1} \cdots F_t^{q_t} P = \sum_{P_i \in \mathbb{P}} Q_i^* P_i \in \text{Ideal}(\mathbb{P}),$$

in which $q_1 > 0, \ldots, q_t > 0$ and $Q_i^* \in \boldsymbol{K}[\boldsymbol{x}]$.

Conversely, let $F_1^{q_1} \cdots F_t^{q_t} P \in \text{Ideal}(\mathbb{P})$ for some integers $q_1 > 0, \ldots, q_t > 0$. Then

$$(z_1 F_1)^{q_1} \cdots (z_t F_t)^{q_t} P \in \text{Ideal}(\mathbb{P}^*) \subset \boldsymbol{K}[\boldsymbol{x}, z_1, \ldots, z_t].$$

The left-hand side of this expression can be written as

$$[(z_1F_1-1)+1]^{q_1}\cdots[(z_tF_t-1)+1]^{q_t}\,P=\sum_{i=1}^{t}R_i(z_iF_i-1)+P,$$

where $R_i \in \boldsymbol{K}[\boldsymbol{x}, z_1, \ldots, z_t]$. This implies that $P \in \text{Ideal}(\mathbb{P}^\star) \cap \boldsymbol{K}[\boldsymbol{x}]$, and the lemma is proved. □

Lemma 6.2.3. Let $\mathfrak{J}$ be an ideal generated by $\mathbb{P}$ and F a polynomial in $\boldsymbol{K}[\boldsymbol{x}]$; $F_1, \ldots, F_t$ be t factors of F such that $F_1 \cdots F_t \neq 0 \iff F \neq 0$;

$$\mathbb{P}^* = \mathbb{P} \cup \{zF-1\},\quad \mathbb{P}^\star = \mathbb{P} \cup \{z_iF_i-1:\ 1 \le i \le t\},$$

where $z, z_1, \ldots, z_t$ are new variables; and $\mathbb{G}^*$, $\mathbb{G}^\star$ are the Gröbner bases of $\mathbb{P}^*$ in $\boldsymbol{K}[\boldsymbol{x}, z]$ and of $\mathbb{P}^\star$ in $\boldsymbol{K}[\boldsymbol{x}, z_1, \ldots, z_t]$ with respect to the purely lexicographical ordering determined with $x_l \prec z$ and $x_l \prec z_j$, respectively. Then

$$\begin{aligned}\mathfrak{J} : F^\infty &= \text{Ideal}(\mathbb{P}^*) \cap \boldsymbol{K}[\boldsymbol{x}] = \text{Ideal}(\mathbb{G}^* \cap \boldsymbol{K}[\boldsymbol{x}])\\ &= \text{Ideal}(\mathbb{P}^\star) \cap \boldsymbol{K}[\boldsymbol{x}] = \text{Ideal}(\mathbb{G}^\star \cap \boldsymbol{K}[\boldsymbol{x}]).\end{aligned}$$

Proof. The first equality is a corollary of Lemma 6.2.1. The two equalities on the right-hand side follow from the elimination property of Gröbner bases (see Theorem 5.3.5). So we only need to show that

$$\text{Ideal}(\mathbb{P}^*) \cap \boldsymbol{K}[\boldsymbol{x}] = \text{Ideal}(\mathbb{P}^\star) \cap \boldsymbol{K}[\boldsymbol{x}].$$

This is proved if, for any $P \in \boldsymbol{K}[\boldsymbol{x}]$, there exists an integer $q > 0$ such that $F^q P \in \mathfrak{J}$ if and only if there exist integers $q_1 > 0, \ldots, q_t > 0$ such that $F_1^{q_1} \cdots F_t^{q_t} P \in \mathfrak{J}$. This is obvious because each F_i is a factor of F and $F_1 \cdots F_t \neq 0 \iff F \neq 0$. □

In fact, for the Gröbner bases computation any compatible ordering in which $x_1^{i_1} \ldots x_n^{i_n} \prec z$ does. The above technique of computing saturation bases was introduced independently by several researchers, for example, Gianni et al. (1988), Chou et al. (1990), and Wang (1989).

There is another method for determining a finite basis for any $\mathfrak{J} : F^\infty$ that may be more efficient in practice. The method proceeds by computing the bases for the ideal quotients $\mathfrak{J} : F^k$ with k increasing from 1. A basis for $\mathfrak{J} : F^\infty$ is obtained when $\mathfrak{J} : F^k = \mathfrak{J} : F^{k+1}$ for some k; in this case $\mathfrak{J} : F^k = \mathfrak{J} : F^\infty$. See Definition 6.4.2, Lemma 6.4.1, and Cox et al. (1992).

Definition 6.2.2. Let $\mathbb{T}$ be any triangular set in $\boldsymbol{K}[\boldsymbol{x}]$. The *saturation* of $\mathbb{T}$ is the ideal

$$\text{sat}(\mathbb{T}) \triangleq \text{Ideal}(\mathbb{T}) : J^\infty,$$

where $J = \prod_{T \in \mathbb{T}} \text{ini}(T)$.

Let $\mathbb{P}$ be a finite basis for $\text{sat}(\mathbb{T})$; the following relation is obvious:

$$\text{Ideal}(\mathbb{T}) \subset \text{sat}(\mathbb{T}) = \text{Ideal}(\mathbb{P}).$$

Definition 6.2.3. Let $\mathbb{T}$ be any triangular set in $\boldsymbol{K}[\boldsymbol{x}]$. The *p-saturation* of $\mathbb{T}$ is the infinite set

$$\text{p-sat}(\mathbb{T}) \triangleq \{P \in \boldsymbol{K}[\boldsymbol{x}]: \ \text{prem}(P, \mathbb{T}) = 0\}.$$

Theorem 6.2.4. For any regular set $\mathbb{T} \subset \boldsymbol{K}[\boldsymbol{x}]$, $\text{sat}(\mathbb{T}) = \text{p-sat}(\mathbb{T})$.

Proof. Let $P \in \text{p-sat}(\mathbb{T})$ and $J = \prod_{T \in \mathbb{T}} \text{ini}(T)$; then $\text{prem}(P, \mathbb{T}) = 0$. By the remainder formula (2.1.2), there is an exponent $q > 0$ such that $J^q P \in \text{Ideal}(\mathbb{T})$. It follows from Definitions 6.2.1 and 6.2.2 that $P \in \text{sat}(\mathbb{T})$.

To show the other direction, write $\mathbb{T}$ as

$$\mathbb{T} = [T_1, \ldots, T_r]$$

with $I_i = \text{ini}(T_i)$ and $J_i = I_1 \cdots I_i$ for $1 \leq i \leq r$. Then, for any $P \in \text{sat}(\mathbb{T})$ there exist an integer $q > 0$ and polynomials $Q_i \in \boldsymbol{K}[\boldsymbol{x}]$ such that

$$J_r^q P = Q_1 T_1 + \cdots + Q_r T_r. \tag{6.2.1}$$

We now prove the following assertion by induction on r.

If $P \in \text{sat}(\mathbb{T})$ is reduced with respect to $\mathbb{T}$, then $P \equiv 0$.

If $r = 1$, then (6.2.1) becomes $J_1^q P = Q_1 T_1$. This is possible only if $Q_1 \equiv 0$. For P is reduced with respect to T_1, and thus $\text{ldeg}(T_1) > \deg(P, \text{lv}(T_1))$. Therefore, $P \equiv 0$.

Suppose that the assertion holds for any regular set $\mathbb{T}$ of length $<r$. We proceed to prove it for $r = |\mathbb{T}| > 1$. Let

$$x_{p_r} = \text{lv}(T_r), \ \ d_r = \text{ldeg}(T_r), \ \ m = \deg(Q_r, x_{p_r}) \geq -1.$$

In case $Q_r \neq 0$, consider the coefficients

$$F_r = \text{lc}(Q_r, x_{p_r}), \quad F_i = \text{coef}(Q_i, x_{p_r}^{m+d_r}), \ \ 1 \leq i \leq r-1.$$

Since $T_1, \ldots, T_{r-1}$ do not involve x_{p_r} and P is reduced with respect to T_r,

$$\begin{aligned} \sum_{i=1}^{r-1} F_i T_i + F_r I_r &= \text{coef}\left(\sum_{i=1}^{r} Q_i T_i, x_{p_r}^{m+d_r}\right) \\ &= \text{coef}(J_r^q P, x_{p_r}^{m+d_r}) = 0. \end{aligned} \tag{6.2.2}$$

Multiplying (6.2.1) by I_r and using (6.2.2), we have

$$J_r^q I_r P = Q_1' T_1 + \cdots + Q_r' T_r, \tag{6.2.3}$$

where

$$Q_i' = I_r Q_i - T_r F_i x_{p_r}^m, \ \ 1 \leq i \leq r-1, \quad Q_r' = I_r \text{red}(Q_r, x_{p_r}).$$

The right-hand side of (6.2.3) has the same form as that of (6.2.1), while $\deg(Q_r', x_{p_r}) < m = \deg(Q_r, x_{p_r})$. If $Q_r' \neq 0$, then we proceed in the same way to get

$$J_r^q I_r^2 P = Q_1'' T_1 + \cdots + Q_r'' T_r$$

with $\deg(Q''_r, x_{p_r}) < \deg(Q'_r, x_{p_r})$. This process must terminate at some point, so that

$$J_{r-1}^q I_r^s P = Q_1^* T_1 + \cdots + Q_{r-1}^* T_{r-1} \tag{6.2.4}$$

holds for some integer $s \geq q$ and polynomials $Q_i^* \in \boldsymbol{K}[\boldsymbol{x}]$.

Since $\mathbb{T}$ is regular, by Lemma 4.3.2 and Proposition 5.1.5 there exist polynomials $H, H_i \in \boldsymbol{K}[\boldsymbol{x}]$ such that

$$H I_r^s + H_1 T_1 + \cdots + H_{r-1} T_{r-1} = S = \operatorname{res}(I_r^s, \mathbb{T}^{\{r-1\}}) \neq 0. \tag{6.2.5}$$

Multiplying (6.2.4) by H and using (6.2.5), we obtain

$$J_{r-1}^q S P = \bar{Q}_1 T_1 + \cdots + \bar{Q}_{r-1} T_{r-1},$$

where $\bar{Q}_i = H Q_i^* + J_{r-1}^q H_i P$ for $1 \leq i \leq r-1$. Therefore, $SP \in \operatorname{sat}(\mathbb{T}^{\{r-1\}})$. As S does not involve the dependents of $\mathbb{T}$, SP is reduced with respect to $\mathbb{T}^{\{r-1\}}$. By the induction hypothesis, $SP \equiv 0$; this implies that $P \equiv 0$. The assertion is proved.

To complete the proof of Theorem 6.2.4, consider any $P \in \operatorname{sat}(\mathbb{T})$ and let $R = \operatorname{prem}(P, \mathbb{T})$; R is reduced with respect to $\mathbb{T}$. As $T_i \in \operatorname{sat}(\mathbb{T})$ obviously for each i, from the pseudo-remainder formula we know that $R \in \operatorname{sat}(\mathbb{T})$. According to the above assertion, $R \equiv 0$. Hence $P \in \text{p-sat}(\mathbb{T})$. □

Moreover, it can be proved that $\operatorname{sat}(\mathbb{T}) = \text{p-sat}(\mathbb{T})$ is also a sufficient condition for any triangular set $\mathbb{T} \subset \boldsymbol{K}[\boldsymbol{x}]$ to be regular (see Aubry et al. 1999). The following is a direct consequence of Theorem 6.2.4.

Corollary 6.2.5. Let $\mathbb{T}$ be any regular set in $\boldsymbol{K}[\boldsymbol{x}]$ and $\mathbb{P}$ a finite basis for $\operatorname{sat}(\mathbb{T})$. Then $\mathbb{T}$ is a W-characteristic set of $\mathbb{P}$.

In fact, one can state a result stronger than Corollary 6.2.5: Any reduced regular set $\mathbb{T}$ is a characteristic set of the ideal $\operatorname{sat}(\mathbb{T})$ in Ritt's definition (see Mishra 1993, pp. 174–176; Ritt 1950, pp. 4f).

For any irreducible triangular set $\mathbb{T}$, Theorem 6.2.14 asserts that $\operatorname{sat}(\mathbb{T})$ is a prime ideal. For any $F \in \boldsymbol{K}[\boldsymbol{x}]$, if $\operatorname{prem}(F, \mathbb{T}) \neq 0$, then $F \notin \operatorname{sat}(\mathbb{T})$ according to Theorem 6.2.4 and thus $\operatorname{sat}(\mathbb{T}) : F^\infty = \operatorname{sat}(\mathbb{T})$ by definition. This result is generalized in the following lemma for regular sets.

Lemma 6.2.6. Let $\mathbb{T}$ be a regular set and F any polynomial in $\boldsymbol{K}[\boldsymbol{x}]$. If $\operatorname{res}(F, \mathbb{T}) \neq 0$, then $\operatorname{sat}(\mathbb{T}) : F^\infty = \operatorname{sat}(\mathbb{T})$.

Proof. Obviously, $\operatorname{sat}(\mathbb{T}) \subset \operatorname{sat}(\mathbb{T}) : F^\infty$. To show the opposite direction, let $R = \operatorname{res}(F, \mathbb{T})$ and $\mathbb{T}$ be written in the form (5.1.1). Then $R \neq 0$ and $R \in \boldsymbol{K}[\boldsymbol{u}]$. By Lemma 4.3.2, there exists a polynomial $Q \in \boldsymbol{K}[\boldsymbol{u}, y_1, \ldots, y_r]$ such that $QF - R \in \operatorname{Ideal}(\mathbb{T}) \subset \operatorname{sat}(\mathbb{T})$. Now consider any $P \in \operatorname{sat}(\mathbb{T}) : F^\infty$. By definition, there exists an integer $q > 0$ such that $F^q P \in \operatorname{sat}(\mathbb{T})$. It follows that

$$R^q P = Q^q F^q P - (QF - R)[(QF)^{q-1} + \cdots + R^{q-1}]P \in \operatorname{sat}(\mathbb{T}).$$

Let $H = \operatorname{prem}(P, \mathbb{T})$; it is then easy to see from the pseudo-remainder formula that $R^q H \in \operatorname{sat}(\mathbb{T})$. By Theorem 6.2.4, $R^q H \in \text{p-sat}(\mathbb{T})$ and thus $\operatorname{prem}(R^q H, \mathbb{T}) = 0$.

Since $R \in \boldsymbol{K}[\boldsymbol{u}]$ does not involve the dependents of $\mathbb{T}$ and H is reduced with respect to $\mathbb{T}$, we have $R^q H = \text{prem}(R^q H, \mathbb{T}) = 0$. It follows that $\text{prem}(P, \mathbb{T}) = H = 0$, so $P \in \text{p-sat}(\mathbb{T}) = \text{sat}(\mathbb{T})$. The proof is complete. □

Proposition 6.2.7. Let $[\mathbb{T}, \mathbb{U}]$ be a regular system in $\boldsymbol{K}[\boldsymbol{x}]$ and $V = \prod_{U \in \mathbb{U}} U$. Then

$$\text{Ideal}(\mathbb{T}) : V^{\infty} = \text{sat}(\mathbb{T}). \tag{6.2.6}$$

Proof. Let $\mathfrak{I} = \text{Ideal}(\mathbb{T})$ and $J = \prod_{T \in \mathbb{T}} \text{ini}(T)$. Since $[\mathbb{T}, \mathbb{U}]$ is regular, $\text{res}(V, \mathbb{T}) \neq 0$. From Lemma 6.2.6 and Definition 6.2.1 one knows that

$$\text{sat}(\mathbb{T}) = \text{sat}(\mathbb{T}) : V^{\infty} = (\mathfrak{I} : J^{\infty}) : V^{\infty} = \mathfrak{I} : (JV)^{\infty}.$$

As $J(\bar{\boldsymbol{x}}) \neq 0$ for any $\bar{\boldsymbol{x}} \in \text{Zero}(\mathbb{T}/V)$, $\text{Zero}(\mathbb{T} \cup \{J\}) \subset \text{Zero}(V)$. By Hilbert's Nullstellensatz, there exists an exponent $s > 0$ and a polynomial $Q \in \boldsymbol{K}[\boldsymbol{x}]$ such that $V^s - QJ \in \mathfrak{I}$. Consider any $P \in \mathfrak{I} : (JV)^{\infty}$; then there exists an integer $q > 0$ such that $(JV)^q P \in \mathfrak{I}$. It follows that

$$V^{(s+1)q} P = V^q (V^s - QJ)[V^{s(q-1)} + \cdots + (QJ)^{q-1}]P + Q^q (JV)^q P \in \mathfrak{I}.$$

This implies that $P \in \mathfrak{I} : V^{\infty}$.

On the other hand, $\mathfrak{I} : V^{\infty} \subset \mathfrak{I} : (JV)^{\infty}$ by definition. It is thus proved that $\text{sat}(\mathbb{T}) = \mathfrak{I} : (JV)^{\infty} = \mathfrak{I} : V^{\infty}$. □

Proposition 6.2.7 may be used to give another simple proof of Theorem 5.1.9 (see Wang 2000). As a consequence of (6.2.6), we have

$$\text{Zero}(\text{Ideal}(\mathbb{T}) : V^{\infty}) = \text{Zero}(\text{sat}(\mathbb{T})).$$

Unmixed decomposition

Refer to the zero decomposition (2.2.7) which provides a representation of the variety $\mathcal{V}$ defined by $\mathbb{P}$ in terms of its subvarieties determined by $\mathbb{C}_i$. However, each $\text{Zero}(\mathbb{C}_i/\mathbb{I}_i)$ is not necessarily an algebraic variety; it is a *quasi-algebraic variety*. In what follows, we shall see how a corresponding variety decomposition may be obtained by determining, from each $\mathbb{C}_i$, a finite set of polynomials.

Theorem 6.2.8. Let $\mathbb{P}$ be a nonempty polynomial set in $\boldsymbol{K}[\boldsymbol{x}]$ and $\mathbb{T}_1, \ldots, \mathbb{T}_e$ a (weak-) characteristic series or a regular series of $\mathbb{P}$. Then

$$\text{Zero}(\mathbb{P}) = \bigcup_{i=1}^{e} \text{Zero}(\text{sat}(\mathbb{T}_i)). \tag{6.2.7}$$

Proof. If $\mathbb{T}_1, \ldots, \mathbb{T}_e$ is a (weak-) characteristic series of $\mathbb{P}$, then $\text{prem}(\mathbb{P}, \mathbb{T}_i) = \{0\}$ for each i; otherwise, by Theorem 5.1.11 a there exists an integer $d > 0$ such that $\text{prem}(P^d, \mathbb{T}_i) = 0$ for all $P \in \mathbb{P}$ and $1 \leq i \leq e$. In any case, it is easy to see from the pseudo-remainder formula that $\text{Zero}(\text{sat}(\mathbb{T}_i)) \subset \text{Zero}(\mathbb{P})$.

Now let $J_i = \prod_{T \in \mathbb{T}_i} \text{ini}(T)$ for each i. By definition and Theorem 5.1.11 c, we have $\text{Zero}(\mathbb{P}) = \bigcup_{i=1}^{e} \text{Zero}(\mathbb{T}_i/J_i)$. Hence, for any $\bar{\boldsymbol{x}} \in \text{Zero}(\mathbb{P})$ there exists an i such that $\bar{\boldsymbol{x}} \in \text{Zero}(\mathbb{T}_i/J_i)$. Let P be any polynomial in $\text{sat}(\mathbb{T}_i)$. Then there exists an integer $q > 0$ such that $J_i^q P \in \text{Ideal}(\mathbb{T}_i)$. It follows that $J_i(\bar{x})^q P(\bar{x}) = 0$. As $J_i(\bar{x}) \neq 0$, we have $P(\bar{x}) \neq 0$. This implies that $\bar{x} \in \text{Zero}(\text{sat}(\mathbb{T}_i))$. The theorem is proved. □

The following result used by Chou and Gao (1990b) provides a useful criterion for removing some redundant subvarieties in the decomposition (6.2.7) without computing their defining sets.

Lemma 6.2.9. Let $\mathbb{P}$ and $\mathbb{T}_i$ be as in Theorem 6.2.8. If $|\mathbb{T}_j| > |\mathbb{P}|$, then

$$\text{Zero}(\text{sat}(\mathbb{T}_j)) \subset \bigcup_{\substack{1 \le i \le e \\ i \neq j}} \text{Zero}(\text{sat}(\mathbb{T}_i));$$

thus $\text{Zero}(\text{sat}(\mathbb{T}_j))$ can be deleted from (6.2.7).

Proof. As $|\mathbb{T}_j| > |\mathbb{P}|$, $\dim(\mathbb{T}_j) < n - |\mathbb{P}|$. By Proposition 6.1.9 and Theorem 6.2.10, $\text{Zero}(\text{sat}(\mathbb{T}_j))$ is a redundant component of $\text{Zero}(\mathbb{P})$. □

Definition 6.2.4. An algebraic variety is said to be *unmixed* or *equidimensional* if all its irredundant irreducible components have the same dimension.

The following theorem is due to Gao and Chou (1993).

Theorem 6.2.10. Let $\mathbb{T}$ be any triangular set in $\boldsymbol{K}[\boldsymbol{x}]$. If $\mathbb{T}$ is not perfect, then $\text{sat}(\mathbb{T}) = \boldsymbol{K}[\boldsymbol{x}]$; if $\mathbb{T}$ is perfect, then $\text{Zero}(\text{sat}(\mathbb{T}))$ is an unmixed variety of dimension $n - |\mathbb{T}|$.

Proof. Let $J = \prod_{T \in \mathbb{T}} \text{ini}(T)$. If $\mathbb{T}$ is not perfect, then $\text{Zero}(\mathbb{T}) \subset \text{Zero}(J)$. By Theorem 1.6.3, there exists an integer $q > 0$ such that $J^q \in \text{Ideal}(\mathbb{T})$. Thus, $J^q P \in \text{Ideal}(\mathbb{T})$ for any $P \in \boldsymbol{K}[\boldsymbol{x}]$. It follows that any $P \in \boldsymbol{K}[\boldsymbol{x}]$ is contained in $\text{sat}(\mathbb{T})$, so $\text{sat}(\mathbb{T}) = \boldsymbol{K}[\boldsymbol{x}]$.

Now suppose that $\mathbb{T}$ is perfect and let $\mathbb{C}_1, \ldots, \mathbb{C}_e$ be an irreducible characteristic series of $\mathbb{T}$. Set

$$\Theta = \{i : |\mathbb{C}_i| \le |\mathbb{T}|, 1 \le i \le e\},$$
$$\Theta^* = \{i \in \Theta \colon \text{prem}(J, \mathbb{C}_i) \neq 0\}.$$

By Theorem 6.2.8 and Lemma 6.2.9, we have

$$\text{Zero}(\mathbb{T}) = \bigcup_{i \in \Theta} \text{Zero}(\text{sat}(\mathbb{C}_i)). \tag{6.2.8}$$

According to Corollary 6.1.2, $\max_{i \in \Theta^*} \dim(\mathbb{C}_i) = \dim(\mathbb{T}) = n - |\mathbb{T}|$. Whence, $\Theta^* \neq \emptyset$ and $\dim(\mathbb{C}_i) = \dim(\mathbb{T})$ for all $i \in \Theta^*$. From (6.2.8) one sees that

$$\text{Zero}(\mathbb{T}/J) = \bigcup_{i \in \Theta^*} \text{Zero}(\text{sat}(\mathbb{C}_i)/J).$$

This implies that

$$\text{Zero}(\text{sat}(\mathbb{T})) = \bigcup_{i \in \Theta^*} \text{Zero}(\text{sat}(\mathbb{C}_i) : J^\infty).$$

Let $i \in \Theta^*$ be fixed. Since $\mathbb{C}_i$ is irreducible and $\text{prem}(J, \mathbb{C}_i) \neq 0$, $\text{sat}(\mathbb{C}_i) : J^\infty = \text{sat}(\mathbb{C}_i)$ according to Lemma 6.2.6 or the remark before the Lemma. Note that $\text{Zero}(\text{sat}(\mathbb{C}_i))$ has dimension $n - |\mathbb{T}|$ for each $i \in \Theta^*$. It is thereby proved that $\text{Zero}(\text{sat}(\mathbb{T}))$ is unmixed of dimension $n - |\mathbb{T}|$. □

Recall that any regular, simple or irreducible triangular set $\mathbb{T}$ is perfect, so $\text{sat}(\mathbb{T}) = \text{p-sat}(\mathbb{T})$ and its variety is unmixed of dimension $n - |\mathbb{T}|$.

In (6.2.7), for each i let $\mathbb{P}_i$ be a finite basis for $\text{sat}(\mathbb{T}_i)$ which can be determined by computing a Gröbner basis according to Lemma 6.2.3. If $\text{sat}(\mathbb{T}_i) = \boldsymbol{K}[\boldsymbol{x}]$, then the constant 1 is contained in (the Gröbner basis of) $\mathbb{P}_i$. Let us assume that such $\mathbb{P}_i$ is simply removed. Thus, a variety decomposition of the following form is obtained:

$$\text{Zero}(\mathbb{P}) = \bigcup_{i=1}^{e} \text{Zero}(\mathbb{P}_i). \tag{6.2.9}$$

By Theorem 6.2.10, each $\mathbb{P}_i$ defines an unmixed algebraic variety.

Let $\mathcal{V}_i = \text{Zero}(\mathbb{P}_i)$; then the decomposition (6.2.9) can be rewritten as

$$\mathcal{V} = \mathcal{V}_1 \cup \cdots \cup \mathcal{V}_e. \tag{6.2.10}$$

This decomposition may be contractible; that is, some variety may be a subvariety of another. Some of the redundant subvarieties may be easily removed by using Lemma 6.2.9. The following lemma points out how to remove some other redundant components.

Lemma 6.2.11. Let $\mathbb{G}$ be a Gröbner basis and $\mathbb{P}$ an arbitrary polynomial set in $\boldsymbol{K}[\boldsymbol{x}]$. If every polynomial in $\mathbb{P}$ has remainder 0 with respect to $\mathbb{G}$, then $\text{Zero}(\mathbb{G}) \subset \text{Zero}(\mathbb{P})$.

Proof. Since every polynomial in $\mathbb{P}$ has remainder 0 with respect to $\mathbb{G}$, $\text{Ideal}(\mathbb{P}) \subset \text{Ideal}(\mathbb{G})$. It follows that $\text{Zero}(\mathbb{G}) \subset \text{Zero}(\mathbb{P})$. □

The method for decomposing an algebraic variety into unmixed components explained above can be described in the following algorithmic form.

Algorithm UnmVarDec: $\Psi \leftarrow \text{UnmVarDec}(\mathbb{P})$. Given a nonempty polynomial set $\mathbb{P} \subset \boldsymbol{K}[\boldsymbol{x}]$, this algorithm computes a finite set Ψ of polynomial sets $\mathbb{P}_1, \ldots, \mathbb{P}_e$ such that the decomposition (6.2.9) holds and each $\mathbb{P}_i$ defines an unmixed algebraic variety.

U1. Compute $\Phi \leftarrow \text{CharSer}(\mathbb{P})$ and set $\Psi \leftarrow \emptyset$.

U2. While $\Phi \neq \emptyset$, do:

  U2.1. Let $\mathbb{C}$ be an element of Φ and set $\Phi \leftarrow \Phi \setminus \{\mathbb{C}\}$. If $|\mathbb{C}| > |\mathbb{P}|$, then go to U2.

  U2.2. Compute a finite basis for $\text{sat}(\mathbb{C})$ according to Lemma 6.2.3, let it be given as a Gröbner basis $\mathbb{G}$, and set $\Psi \leftarrow \Psi \cup \{\mathbb{G}\}$.

U3. While $\exists \mathbb{G}, \mathbb{G}^* \in \Psi$ such that $\text{rem}(\mathbb{G}, \mathbb{G}^*) = \{0\}$, do:
  Set $\Psi \leftarrow \Psi \setminus \{\mathbb{G}^*\}$.

The termination of the algorithm is obvious. The variety decomposition (6.2.9) and the unmixture of each $\text{Zero}(\mathbb{P}_i)$ are guaranteed by Lemma 6.2.3 and Theorem 6.2.10.

The unmixed decomposition (6.2.9) computed by algorithm UnmVarDec is not necessarily irredundant. To remove *all* redundant components, extra computation is required. For an arbitrary regular set $\mathbb{T}$, sat($\mathbb{T}$) is not necessarily radical. It is so when $\mathbb{T}$ is a simple set.

Theorem 6.2.12. For any simple set $\mathbb{T} \subset \boldsymbol{K}[\boldsymbol{x}]$, the ideal p-sat($\mathbb{T}$) is radical.

Proof. Let $P^q \in$ p-sat($\mathbb{T}$); then $\text{Zero}(\mathbb{T}/\mathbb{I}) \subset \text{Zero}(P^q) = \text{Zero}(P)$; so by Corollary 3.4.5, we have $\text{prem}(P, \mathbb{T}) = 0$. Hence, $P \in$ p-sat($\mathbb{T}$) and p-sat($\mathbb{T}$) is radical. The theorem is proved. □

Therefore, if Φ in step U1 of UnmVarDec is a simple series of $\mathbb{P}$ computed by algorithm SimSer, then $\mathfrak{I}_i = \text{Ideal}(\mathbb{P}_i)$ is radical for each $\mathbb{P}_i \in \Psi$. This suggests the following ideal decomposition:

$$\sqrt{\mathfrak{I}} = \bigcap_{i=1}^{e} \mathfrak{I}_i,$$

where $\mathfrak{I} = \text{Ideal}(\mathbb{P})$ and each $\mathfrak{I}_i$ is radical and unmixed.

The removal of redundant subvarieties by examining the containment relations among the corresponding Gröbner bases has the drawback that one component can be removed only if the corresponding Gröbner basis has already been computed. The following lemma provides another criterion for removing redundant components.

Lemma 6.2.13. Let $\mathbb{T}$ be a regular set in $\boldsymbol{K}[\boldsymbol{x}]$ and $\mathbb{P}$ a finite basis for sat($\mathbb{T}$). If $\mathbb{P}^*$ is a polynomial set such that $\text{prem}(\mathbb{P}^*, \mathbb{T}) = \{0\}$, then $\text{Zero}(\mathbb{P}) \subset \text{Zero}(\mathbb{P}^*)$.

Proof. Since $\mathbb{T}$ is regular and $\text{prem}(\mathbb{P}^*, \mathbb{T}) = \{0\}$, $\mathbb{P}^* \subset \text{p-sat}(\mathbb{T}) = \text{sat}(\mathbb{T})$. It follows that

$$\text{Zero}(\mathbb{P}) = \text{Zero}(\text{sat}(\mathbb{T})) \subset \text{Zero}(\mathbb{P}^*).$$ □

Using Theorem 6.2.12 and Lemma 6.2.13, we can modify algorithm UnmVarDec as follows.

Algorithm UnmRadIdeDec: $\Psi \leftarrow$ UnmRadIdeDec($\mathbb{P}$). Given a nonempty polynomial set $\mathbb{P} \subset \boldsymbol{K}[\boldsymbol{x}]$, this algorithm computes a finite set Ψ of polynomial sets $\mathbb{P}_1, \ldots, \mathbb{P}_e$ such that the decomposition (6.2.9) holds and each $\mathbb{P}_i$ generates a radical and unmixed ideal.

U1. Compute $\Phi \leftarrow$ SimSer($\mathbb{P}$) and set

$$\Phi \leftarrow \{\mathbb{T}\colon\ |\mathbb{T}| \leq |\mathbb{P}|, [\mathbb{T}, \tilde{\mathbb{T}}] \in \Phi\}, \quad \Psi \leftarrow \emptyset.$$

U2. While $\Phi \neq \emptyset$, do:

- U2.1. Let $\mathbb{T}$ be an element of Φ of highest dimension and set $\Phi \leftarrow \Phi \setminus \{\mathbb{T}\}$.
- U2.2. Compute a finite basis for sat($\mathbb{T}$) according to Lemma 6.2.3, let it be given as a Gröbner basis $\mathbb{G}$, and set $\Psi \leftarrow \Psi \cup \{\mathbb{G}\}$.
- U2.3. While $\exists \mathbb{T}^* \in \Phi$ such that $\text{prem}(\mathbb{G}, \mathbb{T}^*) = \{0\}$, do:
  Set $\Phi \leftarrow \Phi \setminus \{\mathbb{T}^*\}$.

U3. While $\exists \mathbb{G}, \mathbb{G}^* \in \Psi$ such that $\operatorname{rem}(\mathbb{G}, \mathbb{G}^*) = \{0\}$ do:
Set $\Psi \leftarrow \Psi \setminus \{\mathbb{G}^*\}$.

Note that a variety $\mathcal{V}_1$ can be a true subvariety of another variety $\mathcal{V}_2$ only if $\operatorname{Dim}(\mathcal{V}_1) \leq \operatorname{Dim}(\mathcal{V}_2)$. The selection of $\mathbb{T}$ in step U2.1 and the detection in step U2.3 allows to remove some redundant components before their defining sets are computed. The last step U3 aims at removing those radical ideals which contain other ideals of the same dimension. Inspecting the algorithmic steps, one may see that for any simple series Φ computed by SimSer there should never exist $\mathbb{G}, \mathbb{G}' \in \Psi$ of the same dimension such that $\operatorname{rem}(\mathbb{G}, \mathbb{G}') = \{0\}$, i.e., $\operatorname{Ideal}(\mathbb{G}) \subset \operatorname{Ideal}(\mathbb{G}')$. However, the containment may happen for an arbitrary simple series Φ.

Together with ideal-intersection computation, algorithm UnmRadIdeDec provides a method for finding a generating set of $\sqrt{\mathfrak{I}}$ for any ideal $\mathfrak{I}$ with given generating set. The algorithms for computing simple series and Gröbner bases do not require polynomial factorization in theory, so neither does the algorithm for computing unmixed decompositions.

Irreducible decomposition

We come to decompose an arbitrary algebraic variety defined by a polynomial set into a family of irreducible subvarieties. This is done with an analogy to the unmixed decomposition of $\mathbb{P}$, requiring additionally that the characteristic series Φ is irreducible. Then any finite basis for $\operatorname{sat}(\mathbb{C})$, $\mathbb{C} \in \Phi$, will define an irreducible variety with any generic zero of $\mathbb{C}$ as its generic point.

Definition 6.2.5. An ideal $\mathfrak{I} \subset \boldsymbol{K}[\boldsymbol{x}]$ is said to be *prime* if, whenever $F, G \in \boldsymbol{K}[\boldsymbol{x}]$ and $FG \in \mathfrak{I}$, either $F \in \mathfrak{I}$ or $G \in \mathfrak{I}$.

Theorem 6.2.14. For any irreducible triangular set $\mathbb{T} \subset \boldsymbol{K}[\boldsymbol{x}]$, the ideal $\operatorname{p-sat}(\mathbb{T})$ is prime.

Proof. Let $\boldsymbol{\xi}$ be a generic zero of $\mathbb{T}$; then

$$\operatorname{prem}(P, \mathbb{T}) = 0 \iff P(\boldsymbol{\xi}) = 0,$$

for any $P \in \boldsymbol{K}[\boldsymbol{x}]$, by Lemma 4.3.1. Let $FG \in \operatorname{p-sat}(\mathbb{T})$. Then $\operatorname{prem}(FG, \mathbb{T}) = 0$, so $F(\boldsymbol{\xi})G(\boldsymbol{\xi}) = 0$. It follows that either $F(\boldsymbol{\xi}) = 0$ or $G(\boldsymbol{\xi}) = 0$; that is, either $\operatorname{prem}(F, \mathbb{T}) = 0$ or $\operatorname{prem}(G, \mathbb{T}) = 0$. In other words, either $F \in \operatorname{p-sat}(\mathbb{T})$ or $G \in \operatorname{p-sat}(\mathbb{T})$. Therefore, $\operatorname{p-sat}(\mathbb{T})$ is prime. □

When $\operatorname{sat}(\mathbb{T}) = \operatorname{p-sat}(\mathbb{T})$ is prime, its finite basis is called a *prime basis* of $\mathbb{T}$ and denoted by $\operatorname{PB}(\mathbb{T})$. Then the variety defined by $\operatorname{PB}(\mathbb{T})$ should have any generic zero of $\mathbb{T}$ as its generic point.

Proposition 6.2.15. Let $\mathbb{T}_1$ and $\mathbb{T}_2$ be two irreducible triangular sets in $\boldsymbol{K}[\boldsymbol{x}]$ which have the same set of generic zeros. Then $\operatorname{sat}(\mathbb{T}_1) = \operatorname{sat}(\mathbb{T}_2)$.

Proof. Since $\mathbb{T}_1$ and $\mathbb{T}_2$ are irreducible and have the same set of generic zeros, they have the same set of parameters and $\operatorname{prem}(\mathbb{T}_2, \mathbb{T}_1) = \operatorname{prem}(\mathbb{T}_1, \mathbb{T}_2) = \{0\}$ by Lemma 4.3.1. Thus

$$\operatorname{Ideal}(\mathbb{T}_2) \subset \operatorname{sat}(\mathbb{T}_1), \quad \operatorname{Ideal}(\mathbb{T}_1) \subset \operatorname{sat}(\mathbb{T}_2).$$

Consider any polynomial $P \in \boldsymbol{K}[\boldsymbol{x}]$. If $P \notin \operatorname{sat}(\mathbb{T}_2)$, then $\operatorname{prem}(P, \mathbb{T}_2) \neq 0$. According to Lemma 4.3.2, there exists a polynomial $Q \in \boldsymbol{K}[\boldsymbol{x}]$ such that

$$QP - R \in \operatorname{Ideal}(\mathbb{T}_2), \quad \text{where } R = \operatorname{res}(P, \mathbb{T}_2).$$

This implies that $QP - R \in \operatorname{sat}(\mathbb{T}_1)$. Since $\operatorname{prem}(R, \mathbb{T}_1) = R \neq 0$, $R \notin \operatorname{sat}(\mathbb{T}_1)$. Thus, P cannot be contained in $\operatorname{sat}(\mathbb{T}_1)$. This proves that $\operatorname{sat}(\mathbb{T}_1) \subset \operatorname{sat}(\mathbb{T}_2)$.

As $\mathbb{T}_1$ and $\mathbb{T}_2$ are symmetric, the same argument shows that $\operatorname{sat}(\mathbb{T}_2) \subset \operatorname{sat}(\mathbb{T}_1)$. The proof is complete. □

The conclusion in Proposition 6.2.15 still holds when $\mathbb{T}_1$ and $\mathbb{T}_2$ are simple sets having the same set of regular zeros. The proof of this needs a generalization of Corollary 3.4.5: for any simple set $\mathbb{T}$ and polynomial P in $\boldsymbol{K}[\boldsymbol{x}]$,

$$\operatorname{RegZero}(\mathbb{T}) \subset \operatorname{Zero}(P) \iff \operatorname{prem}(P, \mathbb{T}) = 0.$$

Proposition 6.2.16. Let $\mathbb{T}_1$ and $\mathbb{T}_2$ be two triangular sets in $\boldsymbol{K}[\boldsymbol{x}]$ which have the same set of parameters, and $\mathbb{T}_2$ be irreducible. If $\operatorname{prem}(\mathbb{T}_2, \mathbb{T}_1) = \{0\}$, then $\mathbb{T}_1$ is also irreducible and has the same set of generic zeros as $\mathbb{T}_2$; thus $\operatorname{sat}(\mathbb{T}_1) = \operatorname{sat}(\mathbb{T}_2)$.

Proof. Since $\mathbb{T}_1$ and $\mathbb{T}_2$ have the same set of parameters, they can be written as

$$\mathbb{T}_i = [T_{i1}(\boldsymbol{u}, y_1), \ldots, T_{ir}(\boldsymbol{u}, y_1, \ldots, y_r)], \quad i = 1, 2.$$

As $\operatorname{prem}(\mathbb{T}_2, \mathbb{T}_1) = \{0\}$, we have $\operatorname{prem}(T_{21}, T_{11}) = 0$. Thus, the irreducibility of T_{21} implies that T_{11} is also irreducible over $\boldsymbol{K}_0 = \boldsymbol{K}(\boldsymbol{u})$ and T_{11} differs from T_{21} only by a factor in $\boldsymbol{K}_0$. Similarly, $\operatorname{prem}(T_{22}, [T_{11}, T_{12}]) = 0$. Now T_{21} is irreducible over $\boldsymbol{K}_1 = \boldsymbol{K}_0(y_1)$ with adjoining polynomial T_{21} or T_{11} for y_1. From the pseudo-remainder formula, we know that T_{12} divides T_{22} over $\boldsymbol{K}_1$, so T_{12} differs from T_{22} only by a factor in $\boldsymbol{K}_1$.

Continuing with this argument, we shall see that T_{1k} and T_{2k} differ only by a factor in the algebraic-extension field $\boldsymbol{K}_{k-1} = \boldsymbol{K}_0(y_1, \ldots, y_{k-1})$ with adjoining triangular set $\mathbb{T}_1^{\{k-1\}}$ or $\mathbb{T}_2^{\{k-1\}}$ and thus have the same set of zeros for y_k in $\boldsymbol{K}_{k-1}$, $1 \leq k \leq r$. Hence, $\mathbb{T}_1$ is also irreducible and has the same set of generic zeros as $\mathbb{T}_2$. By Proposition 6.2.15, $\operatorname{sat}(\mathbb{T}_1) = \operatorname{sat}(\mathbb{T}_2)$. □

Proposition 6.2.16 generalizes a result in Chou and Gao (1990b); in the same paper the following is also proved.

Proposition 6.2.17. Let $\mathbb{T}_1$ and $\mathbb{T}_2$ be two triangular sets in $\boldsymbol{K}[\boldsymbol{x}]$, of which $\mathbb{T}_1$ is irreducible. If $\operatorname{prem}(\mathbb{T}_2, \mathbb{T}_1) = \{0\}$ and $0 \notin \operatorname{prem}(\operatorname{ini}(\mathbb{T}_2), \mathbb{T}_1)$, then $\operatorname{sat}(\mathbb{T}_2) \subset \operatorname{sat}(\mathbb{T}_1)$.

Proof. For any $P \in \operatorname{sat}(\mathbb{T}_2)$, by definition there exists an integer $q > 0$ such that $J_2^q P \in \operatorname{Ideal}(\mathbb{T}_2)$, where $J_2 = \prod_{T \in \mathbb{T}_2} \operatorname{ini}(T)$. As $\mathbb{T}_1$ is irreducible and $\operatorname{prem}(\mathbb{T}_2, \mathbb{T}_1) = \{0\}$, $\operatorname{Ideal}(\mathbb{T}_2) \subset \operatorname{sat}(\mathbb{T}_1)$. It follows that $J_2^q P \in \operatorname{sat}(\mathbb{T}_1)$. Since $\operatorname{sat}(\mathbb{T}_1)$ is prime and $0 \notin \operatorname{prem}(\operatorname{ini}(\mathbb{T}_2), \mathbb{T}_1)$ implies that $J_2^q \notin \operatorname{sat}(\mathbb{T}_1)$, we have $P \in \operatorname{sat}(\mathbb{T}_1)$. Therefore, $\operatorname{sat}(\mathbb{T}_2) \subset \operatorname{sat}(\mathbb{T}_1)$. □

By Theorem 6.2.14, to determine the prime basis of $\mathbb{T}$ one only needs to find the generators for $\operatorname{Ideal}(\mathbb{T}^\star) \cap \boldsymbol{K}[\boldsymbol{x}]$, by computing a Gröbner basis of $\mathbb{T}^\star$ according to Lemma 6.2.3.

Let each $\mathbb{T}_i$ in (6.2.7) be irreducible. Then we have the following zero decomposition

$$\mathrm{Zero}(\mathbb{P}) = \bigcup_{i=1}^{e} \mathrm{Zero}(\mathrm{PB}(\mathbb{T}_i)).$$

Now, each $\mathrm{PB}(\mathbb{T}_i)$ which can be exactly determined by using Gröbner bases defines an irreducible algebraic variety and we have thus accomplished an irreducible decomposition of the variety $\mathcal{V}$ defined by $\mathbb{P}$.

This decomposition is not necessarily minimal. The redundant subvarieties can be removed by using Proposition 6.2.17 and Lemma 6.2.13 or 6.2.11, so one can get a *minimal* irreducible decomposition.

Let us modify step U1 in algorithm UnmRadIdeDec as follows:

U1. Compute an irreducible characteristic series Φ of $\mathbb{P}$ by algorithm IrrCharSer, IrrCharSerE, or IrrTriSer and set $\Phi \leftarrow \{\mathbb{T} \in \Phi\colon\ |\mathbb{T}| \leq |\mathbb{P}|\}$, $\Psi \leftarrow \emptyset$.

Furthermore, delete from algorithm UnmRadIdeDec the detection step U3 (which is not needed when the ideals are prime). Let the resulting algorithm be named IrrVarDec; it has the following specification.

Algorithm IrrVarDec: $\Psi \leftarrow \mathrm{IrrVarDec}(\mathbb{P})$. Given a nonempty polynomial set $\mathbb{P} \subset \boldsymbol{K}[\boldsymbol{x}]$, this algorithm computes a finite set Ψ of polynomial sets $\mathbb{P}_1, \ldots, \mathbb{P}_e$ such that the decomposition (6.2.9) holds, it is minimal, and each $\mathbb{P}_i$ defines an irreducible algebraic variety.

Example 6.2.1. Let the algebraic variety $\mathcal{V}$ be defined by $\mathbb{P} = \{P_1, P_2, P_3\}$, where

$$\begin{aligned}
P_1 &= 3x_3x_4 - x_2^2 + 2x_1 - 2,\\
P_2 &= 3x_1^2x_4 + 4x_2x_3 + 6x_1x_3 - 2x_2^2 - 3x_1x_2,\\
P_3 &= 3x_3^2x_4 + x_1x_4 - x_2^2x_3 - x_2.
\end{aligned}$$

With $x_1 \prec \cdots \prec x_4$, $\mathbb{P}$ may be decomposed into two irreducible triangular sets $\mathbb{T}_1$ and $\mathbb{T}_2$ such that

$$\mathrm{Zero}(\mathbb{P}) = \mathrm{Zero}(\mathbb{T}_1/2x_2 + 3x_1^2) \cup \mathrm{Zero}(\mathbb{T}_2/x_2),$$

where

$$\begin{aligned}
\mathbb{T}_1 &= [T_1, T_2, 2x_2x_4 + 3x_1^2x_4 - 2x_2^2 - 3x_1x_2],\\
\mathbb{T}_2 &= [x_1, 2x_3 - x_2, 3x_2x_4 - 2x_2^2 - 4];\\
T_1 &= 2x_2^4 - 12x_1^2x_2^3 + 9x_1x_2^3 - 9x_1^4x_2^2 + 8x_1x_2^2 - 8x_2^2 + 24x_1^3x_2\\
&\quad - 24x_1^2x_2 + 18x_1^5 - 18x_1^4,\\
T_2 &= 2x_2x_3 + 3x_1^2x_3 - x_2^2.
\end{aligned}$$

To obtain an irreducible decomposition of $\mathcal{V}$, we determine the prime bases from $\mathbb{T}_1$ and $\mathbb{T}_2$ by computing the respective Gröbner bases $\mathbb{G}_1, \mathbb{G}_2$ of

$$\mathbb{T}_1 \cup \{z(2x_2 + 3x_1^2) - 1\}, \quad \mathbb{T}_2 \cup \{x_2z - 1\}$$

according to Lemma 6.2.3. The Gröbner bases may be found to consist of 8 and 4 polynomials respectively. Let $\mathbb{V}_i = \mathbb{G}_i \cap \mathbf{Q}[x_1, \ldots, x_4]$ and $\mathcal{V}_i = \mathrm{Zero}(\mathbb{V}_i)$, for $i = 1, 2$. We have

$$\mathbb{V}_1 = \begin{bmatrix} T_1, \\ 27x_1^4x_3 - 27x_1^3x_3 + 2x_2^3 - 15x_1^2x_2^2 + 9x_1x_2^2 + 8x_1x_2 \\ \quad - 8x_2 + 12x_1^3 - 12x_1^2, \\ T_2, \\ 12x_1x_3^2 - 12x_3^2 - 9x_1^2x_3 - 2x_1x_2^2 + 3x_2^2 + 4x_1^2 - 4x_1, \\ x_1x_4 - 2x_1x_3 + 2x_3 - x_2, \\ x_2x_4 + 3x_1^2x_3 - 3x_1x_3 - x_2^2, \\ P_1 \end{bmatrix}$$

and $\mathbb{V}_2 = \mathbb{T}_2$ such that $\mathcal{V} = \mathcal{V}_1 \cup \mathcal{V}_2$. One can check with ease that this decomposition is minimal.

Example 6.2.2. Consider the algebraic curve defined by

$$\mathbb{P} = \begin{Bmatrix} 3x^2 - 4y^2 + z^2 + 4xz - 8yz - 4x + 1, \\ x^2 + 2y^2 + xz + 2yz - 2x - y - 3z \end{Bmatrix},$$

which is the intersection of two algebraic surfaces in 3-dimensional space. With the variable ordering $z \prec y \prec x$, this curve may be decomposed into two irreducible components defined by

$$\mathbb{P}_1 = \{2y - 1, x + z\},$$

$$\mathbb{P}_2 = \begin{Bmatrix} 50y^3 + 140zy^2 - 5y^2 + 94z^2y - 58zy - 24y - 6z^3 \\ \quad - 74z^2 - 42z - 5, \\ zx + 2x - 10y^2 - 14zy + 3y + z^2 + 9z + 1, \\ 5yx - 13x + 70y^2 + 99zy - 29y - 6z^2 - 75z - 9, \\ x^2 - 4x + 12y^2 + 16zy - 4y - z^2 - 12z - 1 \end{Bmatrix};$$

the first is a line and the second is a twisted cubic. Except for points on the plane $z + 2 = 0$, the third and the fourth polynomial in $\mathbb{P}_2$ can be removed. The cubic contains 1 real and 2 complex points

$$(2, \tfrac{1}{2}, -2), \quad (2 \pm \tfrac{3}{5}\sqrt{-7}, \tfrac{13}{5}, -2)$$

on the plane $z + 2 = 0$. The real parts of the two curves for $-5 \leq x \leq 5$ are plotted in Fig. 4.

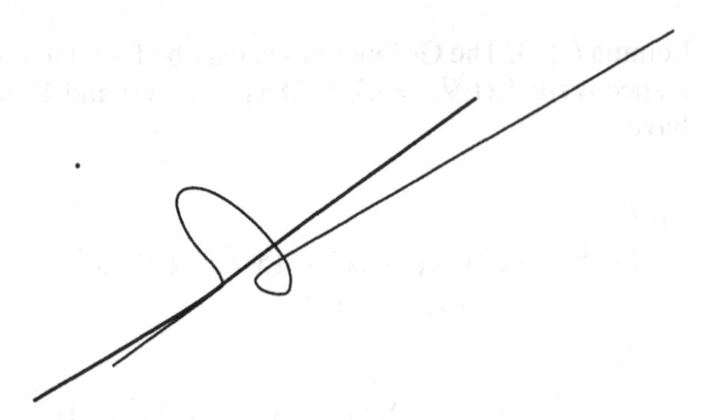

Fig. 4. Example 6.2.2

Example 6.2.3. As a more complicated example, consider the algebraic variety defined by the following five polynomials

$$\begin{aligned}
P_1 &= a_{20}a_{11} + a_{21} + a_{11}a_{02} + 3a_{03},\\
P_2 &= 54a_{20}a_{03} + 9a_{20}a_{11}a_{02} - 9a_{21}a_{02} - 9a_{11}a_{12} - 18a_{30}a_{11} - 2a_{11}^3,\\
P_3 &= 18a_{30}a_{03} - 9a_{20}^2a_{03} + 3a_{30}a_{11}a_{02} + 3a_{20}a_{02}a_{21} + 3a_{20}a_{12}a_{11}\\
&\quad - 3a_{21}a_{12} - 3a_{30}a_{21} - 2a_{11}^2a_{21},\\
P_4 &= 3a_{30}a_{21}a_{02} + 3a_{30}a_{11}a_{12} + 3a_{20}a_{21}a_{12} - 18a_{20}a_{30}a_{03} - 2a_{11}a_{21}^2,\\
P_5 &= 9a_{30}a_{21}a_{12} - 27a_{30}^2a_{03} - 2a_{21}^3.
\end{aligned}$$

Let $\mathbb{P} = \{P_1, \ldots, P_5\}$ and the variable ordering be $\omega_1:\ a_{21} \prec a_{11} \prec a_{30} \prec a_{20} \prec a_{03} \prec a_{02} \prec a_{12}$. Under ω_1, $\mathbb{P}$ can be decomposed into nine irreducible triangular sets $\mathbb{T}_i$ such that

$$\mathrm{Zero}(\mathbb{P}) = \bigcup_{i=1}^{9} \mathrm{Zero}(\mathbb{T}_i/\mathrm{ini}(\mathbb{T}_i)),$$

where

$$\begin{aligned}
\mathbb{T}_1 &= [9a_{11}^2a_{30}^3 + 2a_{21}^2a_{11}^2a_{30} + 2a_{21}^4,\ a_{21}a_{11}a_{20} - a_{11}^2a_{30} + a_{21}^2,\ P_1,\ P_2],\\
\mathbb{T}_2 &= [729a_{30}^6 + 81a_{11}^2a_{30}^5 - 243a_{21}^2a_{30}^4 + 36a_{21}^2a_{11}^2a_{30}^3 + 4a_{21}^4a_{11}^2a_{30}\\
&\quad + 4a_{21}^6,\ I_2a_{20} + 2a_{21}a_{11}(81a_{30}^4 + 27a_{11}^2a_{30}^3 - 9a_{21}^2a_{30}^2\\
&\quad - 2a_{21}^2a_{11}^2a_{30} - 6a_{21}^4)a_{30},\ T_3,\ P_1,\ P_2],\\
\mathbb{T}_3 &= [a_{21},\ a_{11},\ a_{03}],\\
\mathbb{T}_4 &= [a_{21},\ a_{30},\ a_{20},\ a_{11}a_{02} + 3a_{03},\ 9a_{12} + 2a_{11}^2],\\
\mathbb{T}_5 &= [a_{21},\ a_{30},\ 9a_{20}^2 + 2a_{11}^2,\ a_{11}a_{02} + 3a_{03} + a_{11}a_{20},\ -9a_{11}a_{12} + 9a_{11}a_{20}a_{02}\\
&\quad + 54a_{20}a_{03} - 2a_{11}^3],\\
\mathbb{T}_6 &= [a_{11},\ 9a_{30}^2 + a_{21}^2,\ a_{20},\ 3a_{03} + a_{21},\ a_{02},\ a_{12} + 3a_{30}],\\
\mathbb{T}_7 &= [a_{11},\ 9a_{30}^2 - 2a_{21}^2,\ a_{20}^2 + 3a_{30},\ 3a_{03} + a_{21},\ a_{02} + 2a_{20},\ a_{12} + 2a_{20}^2 + 6a_{30}],
\end{aligned}$$

$$\begin{aligned}
\mathbb{T}_8 = {} & [32a_{11}^8 + 981a_{21}^2a_{11}^4 - 324a_{21}^4, T, 729a_{21}^3a_{20} - 64a_{11}^7 - 2034a_{21}^2a_{11}^3, T_3, \\
& P_1, P_2], \\
\mathbb{T}_9 = {} & [4a_{11}^8 + 36a_{21}^2a_{11}^4 - 81a_{21}^4, T, 1114656730a_{11}^5a_{20} - 2077680789a_{21}^2a_{11}a_{20} \\
& + 1576363572a_{21}a_{11}^4 - 2938274496a_{21}^3, T_3, P_1, P_2],
\end{aligned}$$

and

$$\begin{aligned}
T &= -(128a_{11}^{12} - 2430a_{21}^2a_{11}^8 + 6885a_{21}^4a_{11}^4 - 8748a_{21}^6)a_{11}^2a_{30} \\
&\quad + 3a_{21}^2(972a_{21}^6 - 675a_{11}^4a_{21}^4 + 570a_{11}^8a_{21}^2 - 80a_{11}^{12}), \\
T_3 &= I_3a_{03} + 9a_{11}^3a_{20}^3 + 27a_{11}^3a_{30}a_{20} + 2a_{11}^5a_{20} + 4a_{21}a_{11}^4 + 9a_{21}^3; \\
I_2 &= 81a_{11}^2a_{30}^5 - 54a_{21}^2a_{11}^2a_{30}^3 - 18a_{21}^4a_{30}^2 + 4a_{21}^6, \\
I_3 &= 27(a_{21}a_{11}a_{20} - a_{11}^2a_{30} + a_{21}^2).
\end{aligned}$$

For $i = 6, \ldots, 9$, the triangular set $\mathbb{T}_i$ contains more than five polynomials and thus need not be considered for the variety decomposition by Lemma 6.2.9. Let $\mathbb{V}_i$ be the prime basis of $\mathbb{T}_i$ under the ordering ω_1 for $i = 3, 4, 5$. Obviously $\mathbb{T}_3$ already defines an irreducible variety, so $\mathbb{V}_3 = \mathbb{T}_3$. It remains to determine the prime bases from $\mathbb{T}_1$, $\mathbb{T}_2$, $\mathbb{T}_4$, and $\mathbb{T}_5$ according to Lemma 6.2.3. One may find that $\mathbb{V}_4 = \mathbb{T}_4$ and $\mathbb{V}_5$ is the same as the set obtained by replacing the last polynomial in $\mathbb{T}_5$ with $9a_{12} + 9a_{20}a_{02} - 2a_{11}^2$. A prime basis of $\mathbb{T}_1$ under ω_1 contains 20 polynomials. To reduce the number of elements, we convert this prime basis into a Gröbner basis with respect to another variable ordering ω_2: $a_{20} \prec a_{11} \prec a_{02} \prec a_{30} \prec a_{21} \prec a_{12} \prec a_{03}$. The new basis $\mathbb{V}_1$ consists of 10 polynomials as follows

$$\mathbb{V}_1 = \begin{bmatrix}
81a_{30}^3 + 72a_{11}^2a_{30}^2 + 16a_{11}^4a_{30} + 90a_{20}^2a_{11}^2a_{30} + 4a_{20}^2a_{11}^4 + 18a_{20}^4a_{11}^2, \\
6a_{20}a_{11}^2a_{21} + 9a_{20}^3a_{21} - 9a_{11}a_{30}^2 \\
\quad - 4a_{11}^3a_{30} + 9a_{20}^2a_{11}a_{30} + 2a_{20}^2a_{11}^3 + 9a_{20}^4a_{11}, \\
9a_{30}a_{21} + 4a_{11}^2a_{21} + 9a_{20}^2a_{21} + 18a_{20}a_{11}a_{30} + 2a_{20}a_{11}^3 + 9a_{20}^3a_{11}, \\
a_{21}^2 + a_{20}a_{11}a_{21} - a_{11}^2a_{30}, \\
9a_{20}^3a_{12} - 6a_{20}a_{11}a_{02}a_{21} - 12a_{20}^2a_{11}a_{21} + 9a_{02}a_{30}^2 + 18a_{20}a_{30}^2 \\
\quad + 4a_{11}^2a_{02}a_{30} - 9a_{20}^2a_{02}a_{30} + 8a_{20}a_{11}^2a_{30} - 2a_{20}^2a_{11}^2a_{02} - 2a_{20}^3a_{11}^2, \\
9a_{11}a_{12} + 9a_{02}a_{21} + 18a_{20}a_{21} + 18a_{11}a_{30} + 9a_{20}a_{11}a_{02} + 2a_{11}^3 \\
\quad + 18a_{20}^2a_{11}, \\
9a_{30}a_{12} + 9a_{20}^2a_{12} - 4a_{11}a_{02}a_{21} - 8a_{20}a_{11}a_{21} + 18a_{30}^2 \\
\quad - 9a_{20}a_{02}a_{30} + 2a_{11}^2a_{30} - 2a_{20}a_{11}^2a_{02} - 2a_{20}^2a_{11}^2, \\
9a_{21}a_{12} - 6a_{11}^2a_{21} - 18a_{20}^2a_{21} + 9a_{11}a_{02}a_{30} - 18a_{20}a_{11}a_{30} \\
\quad - 4a_{20}a_{11}^3 - 18a_{20}^3a_{11}, \\
81a_{12}^2 + 81a_{20}a_{02}a_{12} - 162a_{20}^2a_{12} + 108a_{11}a_{02}a_{21} + 216a_{20}a_{11}a_{21} \\
\quad - 324a_{30}^2 - 81a_{02}^2a_{30} + 162a_{20}a_{02}a_{30} - 72a_{11}^2a_{30} \\
\quad + 54a_{20}a_{11}^2a_{02} - 4a_{11}^4 + 36a_{20}^2a_{11}^2, \\
P_1
\end{bmatrix}.$$

As for $\mathbb{T}_2$, the difficult case, let T_i denote the ith polynomial of $\mathbb{T}_2$ and I_i the initial of T_i for $1 \le i \le 5$. The nonconstant initials are

$$I_2,\ I_3,\ \text{and}\ I_4 = I_5 = a_{11}.$$

Thus, it is necessary to determine a prime basis from $\mathbb{T}_2$ by computing a Gröbner basis of the enlarged polynomial set, for instance, $\mathbb{T}_2 \cup \{z_1 I_4 - 1, z_2 I_3 - 1, z_3 I_2 - 1\}$ or $\mathbb{T}_2 \cup \{z I_2 I_3 I_4 - 1\}$. Nevertheless, the Gröbner basis cannot be easily computed in either case. We have tried some of the most powerful Gröbner bases packages without success. For this reason, we apply Norm to normalize $\mathbb{T}_2$ to get another triangular set $\mathbb{T}_2^*$: it is obtained from $\mathbb{T}_2$ by replacing T_2 and T_3 respectively with

$$\begin{aligned}
T_2^* &= -4a_{21}^3 a_{11} a_{20} + 81 a_{30}^4 + 9a_{11}^2 a_{30}^3 - 9a_{21}^2 a_{30}^2 + 6a_{21}^2 a_{11}^2 a_{30} - 2a_{21}^4,\\
T_3^* &= 972 a_{21}^7 a_{03} + 729(2a_{11}^4 + 27a_{21}^2) a_{11}^2 a_{30}^5 + 81(2a_{11}^8 + 9a_{21}^2 a_{11}^4\\
&\quad - 81 a_{21}^4) a_{30}^4 - 648 a_{21}^2 (a_{11}^4 + 9a_{21}^2) a_{11}^2 a_{30}^3 + 9a_{21}^2 (8a_{11}^8\\
&\quad + 180 a_{21}^2 a_{11}^4 + 81 a_{21}^4) a_{30}^2 - 36 a_{21}^4 (2a_{11}^4 + 27 a_{21}^2) a_{11}^2 a_{30}\\
&\quad + 2a_{21}^4 (4a_{11}^8 + 90 a_{21}^2 a_{11}^4 + 243 a_{21}^4).
\end{aligned}$$

$\mathbb{T}_2^*$ and $\mathbb{T}_2$ have the same set of generic zeros, so the prime bases constructed from them define the same irreducible algebraic variety. $\mathbb{T}_2^*$ possesses the property that the initials of its polynomials only involve the parameters a_{21} and a_{11}.

A prime basis of $\mathbb{T}_2^*$ can be easily determined by computing the corresponding Gröbner basis with respect to the variable ordering ω_1 or ω_2 according to Lemma 6.2.3. The basis under ω_2 contains nine elements and is as follows

$$\mathbb{V}_2 = \begin{bmatrix}
81 a_{20}^3 a_{02}^2 + 16 a_{11}^4 a_{02} + 108 a_{20}^2 a_{11}^2 a_{02} + 324 a_{20}^4 a_{02}\\
\quad + 20 a_{20} a_{11}^4 + 144 a_{20}^3 a_{11}^2 + 324 a_{20}^5,\\
144 a_{11}^2 a_{30} + 729 a_{20}^2 a_{30} + 81 a_{20}^3 a_{02} + 16 a_{11}^4\\
\quad + 144 a_{20}^2 a_{11}^2 + 405 a_{20}^4,\\
4a_{02} a_{30} + 5a_{20} a_{30} + a_{20}^2 a_{02} + a_{20}^3,\\
4a_{11} a_{21} + 27 a_{20} a_{30} + 2a_{20} a_{11}^2 + 9a_{20}^3,\\
18 a_{02} a_{21} + 36 a_{20} a_{21} - 18 a_{11} a_{30} + 9a_{20} a_{11} a_{02} - 2a_{11}^3,\\
972 a_{20} a_{30} a_{21} + 324 a_{20}^3 a_{21} - 1296 a_{11} a_{30}^2 - 405 a_{20}^2 a_{11} a_{30}\\
\quad + 81 a_{20}^3 a_{11} a_{02} + 16 a_{11}^5 + 108 a_{20}^2 a_{11}^3 + 243 a_{20}^4 a_{11},\\
144 a_{21}^2 + 1296 a_{30}^2 - 81 a_{20}^2 a_{30} - 81 a_{20}^3 a_{02} - 16 a_{11}^4\\
\quad - 144 a_{20}^2 a_{11}^2 - 405 a_{20}^4,\\
6a_{12} + 18 a_{30} + 3a_{20} a_{02} + 2a_{11}^2 + 12 a_{20}^2,\\
P_1
\end{bmatrix}.$$

It is easy to verify that both $\operatorname{Zero}(\mathbb{V}_4)$ and $\operatorname{Zero}(\mathbb{V}_5)$ are subvarieties of $\operatorname{Zero}(\mathbb{V}_1)$. Therefore, the variety defined by $\mathbb{P}$ is decomposed into three irreducible subvarieties defined by $\mathbb{V}_1$, $\mathbb{V}_2$ and $\mathbb{V}_3$. Symbolically,

$$\operatorname{Zero}(\mathbb{P}) = \operatorname{Zero}(\mathbb{V}_1) \cup \operatorname{Zero}(\mathbb{V}_2) \cup \operatorname{Zero}(\mathbb{V}_3), \tag{6.2.11}$$

where $\operatorname{Zero}(\mathbb{V}_i)$ is irreducible for $i = 1, 2, 3$.

The above example comes from the qualitative study of plane differential systems. We shall discuss the background and use the obtained decomposition in Sect. 7.6.

Division of varieties

We now show how to remove a subvariety from a given algebraic variety by division. This is a generalization of the division of one polynomial by another. Such a division is particularly useful for polynomial factorization in which a factor can readily be removed from the polynomial being factorized when the factor is found. However, the removal of subvarieties appears much more difficult computationally. The removing technique can be incorporated into the decomposition algorithms according to the following theorem.

Theorem 6.2.18. Let $\mathbb{P}$ and $\mathbb{Q} = \{F_1, \ldots, F_t\}$ be two polynomial sets in $\boldsymbol{K}[\boldsymbol{x}]$ with $\mathrm{Zero}(\mathbb{Q}) \subset \mathrm{Zero}(\mathbb{P})$ and $\mathfrak{I}$ be the ideal generated by

$$\mathbb{P} \cup \{zF_1 + \cdots + z^t F_t - 1\} \quad \text{in } \boldsymbol{K}[\boldsymbol{x}, z] \tag{6.2.12}$$

or by

$$\mathbb{P} \cup \{z_1 F_1 + \cdots + z_t F_t - 1\} \quad \text{in } \boldsymbol{K}[\boldsymbol{x}, z_1, \ldots, z_t], \tag{6.2.13}$$

where $z, z_1, \ldots, z_t$ are new variables. Then

$$\mathrm{Zero}(\mathbb{P}) = \mathrm{Zero}(\mathbb{Q}) \cup \mathrm{Zero}(\mathfrak{I} \cap \boldsymbol{K}[\boldsymbol{x}]).$$

Proof. Consider the case in which $\mathfrak{I} = \mathrm{Ideal}(\mathbb{P} \cup \{zF_1 + \cdots + z^t F_t - 1\})$. Let $\bar{\boldsymbol{x}} \in \mathrm{Zero}(\mathbb{P})$. For any $P \in \mathfrak{I} \cap \boldsymbol{K}[\boldsymbol{x}]$, there exists a polynomial $Q \in \boldsymbol{K}[\boldsymbol{x}, z]$ such that

$$P - Q(zF_1 + \cdots + z^t F_t - 1) \in \mathrm{Ideal}(\mathbb{P}) \subset \boldsymbol{K}[\boldsymbol{x}, z].$$

Hence

$$P(\bar{\boldsymbol{x}}) = Q(\bar{\boldsymbol{x}}, z)[zF_1(\bar{\boldsymbol{x}}) + \cdots + z^t F_t(\bar{\boldsymbol{x}}) - 1] \tag{6.2.14}$$

for arbitrary z. Suppose that $\bar{\boldsymbol{x}} \notin \mathrm{Zero}(\mathbb{Q})$. Then there exists some j such that $F_j(\bar{\boldsymbol{x}}) \neq 0$. So there is a $\bar{z} \in \tilde{\boldsymbol{K}}$ such that $\bar{z}F_1(\bar{\boldsymbol{x}}) + \cdots + \bar{z}^t F_t(\bar{\boldsymbol{x}}) - 1 = 0$. Plunging $\bar{z}$ into (6.2.14), we get $P(\bar{\boldsymbol{x}}) = 0$. Therefore, $\mathrm{Zero}(\mathbb{P}) \subset \mathrm{Zero}(\mathbb{Q}) \cup \mathrm{Zero}(\mathfrak{I} \cap \boldsymbol{K}[\boldsymbol{x}])$.

It is obvious that $\mathbb{P} \subset \mathfrak{I}$ in $\boldsymbol{K}[\boldsymbol{x}, z]$. Since none of the polynomials in $\mathbb{P}$ does involve z, we have $\mathbb{P} \subset \mathfrak{I} \cap \boldsymbol{K}[\boldsymbol{x}, z]$. Hence $\mathrm{Zero}(\mathfrak{I} \cap \boldsymbol{K}[\boldsymbol{x}]) \subset \mathrm{Zero}(\mathbb{P})$.

The case in which $\mathfrak{I} = \mathrm{Ideal}(\mathbb{P} \cup \{z_1 F_1 + \cdots + z_t F_t - 1\})$ is proved analogously, observing that if $F_1(\bar{\boldsymbol{x}}), \ldots, F_t(\bar{\boldsymbol{x}})$ are not all 0, then there exist $\bar{z}_1, \ldots, \bar{z}_t$ such that $\bar{z}_1 F_1(\bar{\boldsymbol{x}}) + \cdots + \bar{z}_t F_t(\bar{\boldsymbol{x}}) - 1 = 0$. □

This theorem suggests a way to remove any subvariety $\mathrm{Zero}(\mathbb{Q})$ from the given variety $\mathrm{Zero}(\mathbb{P})$ by determining a finite basis $\mathbb{H}$ for the ideal $\mathfrak{I} \cap \boldsymbol{K}[\boldsymbol{x}]$. The latter can be done, for instance, by computing a Gröbner basis of (6.2.12) or of (6.2.13) with respect to the purely lexicographical ordering determined by

$x_j \prec z$ or $x_j \prec z_l$ together with its elimination property (Theorem 5.3.5). Thus, decomposing Zero($\mathbb{P}$) is reduced to decomposing Zero($\mathbb{Q}$) and Zero($\mathbb{H}$). We have tested this technique. Nevertheless, the Gröbner bases computation in this case is too inefficient and we had no gain from the experiments. One can make use of the technique only when a more effective procedure for determining the finite bases is available.

In fact, the removal of Zero($\mathbb{Q}$) from Zero($\mathbb{P}$) corresponds to computing the quotient Ideal($\mathbb{P}$) : Ideal($\mathbb{Q}$) (see Definition 6.4.2). The latter can be done by a possibly more efficient algorithm described in Cox et al. (1992, pp. 193–195).

6.3 Ideal and radical ideal membership

A fundamental problem in polynomial-ideal theory is the membership test, that is, to determine whether a given polynomial belongs to an ideal with given generators (see van der Waerden 1950, p. 58). One of the most remarkable applications of Gröbner bases is an algorithmic solution to this problem. In concrete terms, we state the following theorem.

Theorem 6.3.1. Let $\mathbb{P} \subset \boldsymbol{K}[\boldsymbol{x}]$ be a polynomial set and $\mathbb{G}$ a Gröbner basis of $\mathbb{P}$. Then for any polynomial $P \in \boldsymbol{K}[\boldsymbol{x}]$,

$$P \in \mathrm{Ideal}(\mathbb{P}) \iff \mathrm{rem}(P, \mathbb{G}) = 0.$$

The theorem follows from the definition of a Gröbner basis of $\mathbb{P}$ and Theorem 5.3.2 b.

Corollary 6.3.2. Let $\mathbb{P}, \mathbb{Q} \subset \boldsymbol{K}[\boldsymbol{x}]$ be two polynomial sets and $\mathbb{G}$ a Gröbner basis of $\mathbb{P}$. Then $\mathrm{Ideal}(\mathbb{Q}) \subset \mathrm{Ideal}(\mathbb{P}) \iff \mathrm{rem}(\mathbb{Q}, \mathbb{G}) = \{0\}$.

Example 6.3.1. Consider the following two polynomials

$$\begin{aligned} G_1 &= x_1x_4^2 + x_2x_3 - 3x_1x_2^2 + 3x_1x_2 - x_1, \\ G_2 &= 2x_2x_4 + x_3 - 2x_1x_2^2 - 2x_2 - 1, \end{aligned}$$

and let $\mathbb{P}$ be as in Example 2.2.3. A Gröbner basis $\mathbb{G}$ of $\mathbb{P}$ has been computed in Example 5.3.1. One can verify that $\mathrm{rem}(G_1, \mathbb{G}) = 0$ and $\mathrm{rem}(G_2, \mathbb{G}) \neq 0$. Hence, $G_1 \in \mathrm{Ideal}(\mathbb{P})$, $G_2 \notin \mathrm{Ideal}(\mathbb{P})$, and $\mathrm{Ideal}(\{G_1, G_2\}) \not\subset \mathrm{Ideal}(\mathbb{P})$.

In contrast to the membership test of polynomial ideals, there are a number of methods for solving the membership problem of radical ideals. We summarize the various methods introduced previously in this book in the form of the following theorem. Let SS($\mathfrak{P}$) and RS($\mathfrak{P}$) stand for any *simple series* and *regular series* of a polynomial set or system $\mathfrak{P}$ in $\boldsymbol{K}[\boldsymbol{x}]$, respectively.

Theorem 6.3.3. Let P be any polynomial and $\mathbb{P}$ a polynomial set in $\boldsymbol{K}[\boldsymbol{x}]$, and $\mathbb{P}^* = \mathbb{P} \cup \{zP - 1\}$, where z is a new variable. Then the following are equivalent:

a. $P \in \sqrt{\mathrm{Ideal}(\mathbb{P})}$;
b. $\mathrm{Zero}(\mathbb{P}) \subset \mathrm{Zero}(P)$;
c. $\mathrm{GB}(\mathbb{P}^*) = [1]$;
d. $\mathrm{ITS}([\mathbb{P}, \{P\}]) = \mathrm{ITS}(\mathbb{P}^*) = \emptyset$;

e. $\mathrm{SS}([\mathbb{P}, \{P\}]) = \mathrm{SS}(\mathbb{P}^*) = \emptyset$;
f. $\mathrm{RS}([\mathbb{P}, \{P\}]) = \mathrm{RS}(\mathbb{P}^*) = \emptyset$;
g. $\mathrm{TriSerP}(\mathbb{P}, \{P\}, 0) = \mathrm{TriSerP}(\mathbb{P}^*, \emptyset, 0) = \emptyset$;
h. $\mathrm{prem}(P, \mathbb{T}) = 0$ for all $\mathbb{T} \in \mathrm{ITS}(\mathbb{P})$;
i. $\mathrm{prem}(P, \mathbb{T}) = 0$ for all $[\mathbb{T}, \tilde{\mathbb{T}}] \in \mathrm{SS}(\mathbb{P})$;
j. $\mathrm{op}(2, \mathrm{Split}(\mathbb{T}, P, n)) = \emptyset$ for all $\mathbb{T} \in \mathrm{RS}(\mathbb{P})$.

Proof. Note that $\mathrm{Zero}(\mathbb{P}) \subset \mathrm{Zero}(P)$ if and only if $\mathrm{Zero}(\mathbb{P}/P) = \emptyset$ if and only if $\mathrm{Zero}(\mathbb{P}^*) = \emptyset$.
a $\Longleftrightarrow$ b: Theorem 1.6.3 and the definition of $\sqrt{\mathrm{Ideal}(\mathbb{P})}$.
b $\Longleftrightarrow$ c: Corollary 5.3.4.
b $\Longleftrightarrow$ d: Corollary 4.3.6.
b $\Longleftrightarrow$ e: Theorem 3.4.3 a.
b $\Longleftrightarrow$ f: Corollary 5.1.15.
b $\Longleftrightarrow$ g: Algorithm TriSerP conditions a and c.
b $\Longleftrightarrow$ h: Definition 2.2.7 and Corollary 4.3.9.
b $\Longleftrightarrow$ i: Definition 3.3.3 and Theorem 3.4.4.
b $\Longleftrightarrow$ j: Corollary 5.1.15. □

Direct consequences of the above theorem are various methods for examining containment relationship between algebraic varieties.

Example 6.3.2. Recall the polynomial set $\mathbb{P}$ in Example 2.2.3 and the polynomials G_1 and G_2 in Example 6.3.1. As the characteristic set of $\mathbb{P} \cup \{zG_1 - 1\}$ with respect to the ordering $x_1 \prec \cdots \prec x_4 \prec z$ is contradictory, $G_1 \in \sqrt{\mathrm{Ideal}(\mathbb{P})}$ (in this case further decomposition is not required). To determine that

$$G_2 \notin \sqrt{\mathrm{Ideal}(\mathbb{P})} \qquad (6.3.1)$$

according to Theorem 6.3.3 d, an irreducible decomposition is however needed.

The same conclusion can be reached by using other algorithms. When (6.3.1) is determined by using Theorem 6.3.3 h, one also knows that the membership relation does not hold for the components $\mathbb{C}_1'$, $\mathbb{C}_2''$ and $\mathbb{C}_4$ (which are given in Example 4.2.1).

Example 6.3.3. Let the ideal $\mathfrak{I}$ be generated by three polynomials

$$\begin{aligned} P_1 &= def - abc, \\ P_2 &= 4e^2 f + 3a^2 c, \\ P_3 &= 175bd^2 ef + 192ad^3 f - 108b^3 ce. \end{aligned}$$

With respect to the total degree ordering determined by $b \prec d \prec a \prec e \prec f \prec c$,

$$\begin{aligned} \mathbb{G} = [&4b^3e^2c + 3b^2daec, 4baec + 3da^2c, \\ &-108b^3ec + 175b^2dac + 192d^3af, P_2, P_1] \end{aligned}$$

is a Gröbner basis for $\mathfrak{I}$. Let $G = 8b^2ac - 20bdef - 9d^2af$. One may verify that $\mathrm{rem}(G, \mathbb{G}) \neq 0$ and $\mathrm{rem}(G^2, \mathbb{G}) = 0$. Hence, $G \notin \mathfrak{I}$ and $G \in \sqrt{\mathfrak{I}}$. The conclusion $G \in \sqrt{\mathfrak{I}}$ can be drawn in different ways by using other methods according to Theorem 6.3.3.

An important application of the radical ideal membership test is to automated theorem proving in geometry. This will be discussed in detail in Sect. 7.2.

6.4 Primary decomposition of ideals

Decomposing polynomial ideals into primary components is very classical in commutative algebra. In this section, we explain how to construct a primary decomposition of any polynomial ideal from an irreducible decomposition of the corresponding algebraic variety. The techniques of localization and extraction we use are suggested by Shimoyama and Yokoyama (1996).

Definition 6.4.1. The *intersection* of two ideals $\mathfrak{I}$ and $\mathfrak{J}$ in $\boldsymbol{K}[\boldsymbol{x}]$, denoted as $\mathfrak{I} \cap \mathfrak{J}$, is the set of polynomials which belong to both $\mathfrak{I}$ and $\mathfrak{J}$.

Definition 6.4.2. Let $\mathfrak{I}$ and $\mathfrak{J}$ be two ideals in $\boldsymbol{K}[\boldsymbol{x}]$. The infinite set of polynomials $\mathfrak{I} : \mathfrak{J} \triangleq \{F \in \boldsymbol{K}[\boldsymbol{x}] :\ FG \in \mathfrak{I} \text{ for all } G \in \mathfrak{J}\}$ is called the *ideal quotient* of $\mathfrak{I}$ by $\mathfrak{J}$.

It is easy to show that in $\boldsymbol{K}[\boldsymbol{x}]$ the intersection of two ideals is an ideal, and so is their quotient (see, e.g., Cox et al. 1992, pp. 185 and 193). Clearly, $\mathfrak{I} : \mathfrak{J}$ contains $\mathfrak{I}$. For any polynomial F, we write $\mathfrak{I} : F$ instead of $\mathfrak{I} : \mathrm{Ideal}(\{F\})$.

Lemma 6.4.1. Let $\mathfrak{I}$ be an ideal and F a polynomial in $\boldsymbol{K}[\boldsymbol{x}]$, and let k be an integer ≥ 1. Then

$$\mathfrak{I} : F^{\infty} = \mathfrak{I} : F^{k} \iff \mathfrak{I} : F^{k} = \mathfrak{I} : F^{k+1}.$$

As a consequence, the minimal k can be determined by computing $\mathfrak{I} : F^{i}$ with i increasing from 1.

Proof. Exercise in Cox et al. (1992, p. 196). □

Definition 6.4.3. An ideal $\mathfrak{I} \subset \boldsymbol{K}[\boldsymbol{x}]$ is said to be *pseudo-primary* if $\sqrt{\mathfrak{I}}$ is prime.

$\mathfrak{I}$ is said to be *primary* if $FG \in \mathfrak{I}$ and $F \notin \mathfrak{I}$ imply that there exists an integer $q > 0$ such that $G^{q} \in \mathfrak{I}$.

Definition 6.4.4. Let $\mathfrak{I}$ be an ideal in $\boldsymbol{K}[\boldsymbol{x}]$ and $\{\boldsymbol{u}\}$ a subset of $\{\boldsymbol{x}\}$. $\{\boldsymbol{u}\}$ is called a *maximally independent set* modulo $\mathfrak{I}$ if

$$\mathfrak{I} \cap \boldsymbol{K}[\boldsymbol{u}] = \{0\} \text{ and } \mathfrak{I} \cap \boldsymbol{K}[\boldsymbol{u}, x] \neq \{0\}, \ \ \forall x \in \{\boldsymbol{x}\} \setminus \{\boldsymbol{u}\}.$$

Lemma 6.4.2. Let $\mathfrak{I}$ be a prime ideal in $\boldsymbol{K}[\boldsymbol{x}]$ and $\mathbb{G}$ a Gröbner basis for $\mathfrak{I}$ with respect to any admissible ordering. Then $\{\boldsymbol{u}\}$ is a maximally independent set modulo $\mathfrak{I}$ if and only if

$$\mathrm{lt}(\mathbb{G}) \cap \mathrm{ter}(\boldsymbol{u}) = \emptyset \text{ and } \mathrm{lt}(\mathbb{G}) \cap \mathrm{ter}(\boldsymbol{u}, x) \neq \emptyset, \ \ \forall x \in \{\boldsymbol{x}\} \setminus \{\boldsymbol{u}\},$$

where $\mathrm{lt}(\mathbb{G}) \triangleq \{\mathrm{lt}(G) :\ G \in \mathbb{G}\}$ and $\mathrm{ter}(\boldsymbol{u})$ denotes the set of all the terms in $\boldsymbol{u}$, and similarly for $\mathrm{ter}(\boldsymbol{u}, x)$.

Proof. Definition A.9 and lemma A.12 in Shimoyama and Yokoyama (1996). □

From the irreducible variety decomposition (6.2.10) or (6.2.9), one immediately gets the following decomposition of the radical ideal generated by $\mathbb{P}$

$$\sqrt{\mathfrak{J}} = \bigcap_{i=1}^{e} \mathfrak{J}_i,$$

where $\mathfrak{J} = \text{Ideal}(\mathbb{P})$ and $\mathfrak{J}_i = \text{Ideal}(\mathbb{P}_i)$ for each i. From the algorithmic construction, one also knows that each $\mathbb{P}_i$ is given as a Gröbner basis and $\mathfrak{J}_i$ is prime. In what follows, we shall construct a pseudo-primary ideal $\mathfrak{Z}_i$ such that $\mathfrak{J}_i$ is the prime ideal associated with $\mathfrak{Z}_i$ for $1 \leq i \leq e$. An additional ideal $\mathfrak{J}^*$ will also be constructed, so that we have the following decomposition

$$\mathfrak{J} = \bigcap_{i=1}^{e} \mathfrak{Z}_i \cap \mathfrak{J}^*. \tag{6.4.1}$$

If $e = 1$, then $\mathfrak{J}$ is already pseudo-primary. Now assume that $e > 1$, take a polynomial $S_{ij} \in \mathbb{P}_j \setminus \mathfrak{J}_i$ for each pair $i \neq j$, and let

$$S_i = \prod_{\substack{1 \leq j \leq e \\ j \neq i}} S_{ij}$$

for each i. Then $\mathfrak{Z}_i = \mathfrak{J} : S_i^\infty$ is the pseudo-primary ideal we wanted to determine. To obtain the additional ideal $\mathfrak{J}^*$, let k_i be an integer such that $\mathfrak{J} : S_i^{k_i} = \mathfrak{Z}_i$ for each i. Then

$$\mathfrak{J}^* = \text{Ideal}(\mathbb{P} \cup \{S_1^{k_1}, \ldots, S_e^{k_e}\}).$$

From each pseudo-primary ideal $\mathfrak{Z}$ generated by a Gröbner basis $\mathbb{G}$, one can determine a primary ideal by extraction as follows.

Let $\{\boldsymbol{u}\}$ be a maximally independent set modulo $\sqrt{\mathfrak{Z}}$ which can be computed according to Lemma 6.4.2 and $\{\boldsymbol{y}\} = \{\boldsymbol{x}\} \setminus \{\boldsymbol{u}\}$. Compute a Gröbner basis $\bar{\mathbb{G}}$ of $\mathbb{G}$ with respect to the purely lexicographical ordering ω determined with $u_j \prec y_l$ for any $u_j \in \{\boldsymbol{u}\}$, $y_l \in \{\boldsymbol{y}\}$ and the extractor

$$F = \text{lcm}(\{\text{lc}(G) : G \in \bar{\mathbb{G}}\}),$$

where $\text{lc}(G)$ is the leading coefficient of G considered as a polynomial in $\boldsymbol{K}(\boldsymbol{u})[\boldsymbol{y}]$ with respect to the ordering ω.

Let $\bar{\mathfrak{Z}} = \text{Ideal}(\mathbb{G}) : F^\infty$. According to Lemma 6.4.1, one can compute an integer k such that

$$\text{Ideal}(\mathbb{G}) : F^k = \bar{\mathfrak{Z}}.$$

Thus

$$\mathfrak{Z} = \bar{\mathfrak{Z}} \cap \text{Ideal}(\mathbb{G} \cup \{F^k\}),$$

and $\bar{\mathfrak{Z}}$ is a primary ideal.

Applying the above process to the ideal $\mathfrak{J}^*$ and Ideal($\mathbb{G} \cup \{F^k\}$) recursively, we shall get further decompositions of the form (6.4.1). This procedure will terminate, resulting in an ideal decomposition of the form

$$\mathfrak{J} = \bigcap_{i=1}^{h} \mathfrak{J}_i,$$

where each $\mathfrak{J}_i$ is primary.

The above decomposition procedure is presented in the form of the following algorithm.

Algorithm PriIdeDec: $\Psi \leftarrow$ PriIdeDec($\mathbb{P}$). Given a nonempty polynomial set $\mathbb{P} \subset \boldsymbol{K}[\boldsymbol{x}]$, this algorithm computes a finite set Ψ of polynomial sets $\mathbb{P}_1, \ldots, \mathbb{P}_h$ such that Ideal($\mathbb{P}$) $= \bigcap_{i=1}^{h}$ Ideal($\mathbb{P}_i$) and Ideal($\mathbb{P}_i$) is primary for each i.

P1. Set $\Phi \leftarrow \{\mathbb{P}\}$, $\Psi \leftarrow \emptyset$.

P2. While $\Phi \neq \emptyset$, do:

P2.1. Let $\mathbb{F}$ be an element of Φ and set $\Phi \leftarrow \Phi \setminus \{\mathbb{F}\}$.

P2.2. Compute a set of defining sets $\mathbb{F}_1, \ldots, \mathbb{F}_e$ (given as Gröbner bases) from $\mathbb{F}$ by algorithm IrrVarDec. If $e = 0$, then go to P2.

P2.3. For $i = 1, \ldots, e$ do:

P2.3.1. Set $\mathbb{S} \leftarrow \emptyset$. If $e = 1$, then set $S \leftarrow 1, \mathbb{G} \leftarrow \mathbb{F}_1$ and go to P2.3.3. Otherwise, select $S_j \in \mathbb{F}_j \setminus$ Ideal($\mathbb{F}_i$) for $1 \leq j \leq e$ and $j \neq i$ and set

$$S \leftarrow \prod_{\substack{1 \leq j \leq e \\ j \neq i}} S_j.$$

P2.3.2. Compute a finite basis for Ideal($\mathbb{F}$) : S^∞ according to Lemma 6.2.3 and let it be given as a Gröbner basis $\mathbb{G}$.

P2.3.3. Compute a maximally independent set $\{\boldsymbol{u}\}$ modulo Ideal($\mathbb{F}_i$) according to Lemma 6.4.2 and let $\{\boldsymbol{y}\} \leftarrow \{\boldsymbol{x}\} \setminus \{\boldsymbol{u}\}$.

P2.3.4. Compute a Gröbner basis $\bar{\mathbb{G}}$ of $\mathbb{G}$ with respect to the purely lexicographical ordering ω determined with $u_k \prec y_l$ for any $u_k \in \{\boldsymbol{u}\}$, $y_l \in \{\boldsymbol{y}\}$ and the extractor

$$F \leftarrow \mathrm{lcm}(\{\mathrm{lc}(G)\colon\ G \in \bar{\mathbb{G}} \subset \boldsymbol{K}(\boldsymbol{u})[\boldsymbol{y}]\})$$

with respect to the ordering ω.

P2.3.5. Compute a finite basis for Ideal($\mathbb{G}$) : F^∞ according to Lemma 6.2.3, let it be given as a Gröbner basis $\mathbb{G}^*$, and set

$$\Psi \leftarrow \Psi \cup \{\mathbb{G}^*\}.$$

P2.3.6. Compute two integers k and l according to Lemma 6.4.1 such that

$$\mathrm{Ideal}(\mathbb{G}) : F^k = \mathrm{Ideal}(\mathbb{G}^*), \quad \mathrm{Ideal}(\mathbb{F}) : S^l = \mathrm{Ideal}(\mathbb{G})$$

and set

$$\Phi \leftarrow \Phi \cup \{\mathbb{G} \cup \{F^k\}\}, \ \ \mathbb{S} \leftarrow \mathbb{S} \cup \{S^l\}.$$

P2.4. Set $\Phi \leftarrow \Phi \cup \{\mathbb{F} \cup \mathbb{S}\}$.

Table 1. Generating sets for $\mathfrak{I}_i$ and their associated prime ideals[a]

$\mathfrak{I}_i$	Generating set for $\mathfrak{I}_i$	Generating set for prime associated with $\mathfrak{I}_i$
$\mathfrak{I}_1$	$[a, e]$	$[a, e]$
$\mathfrak{I}_2$	$[f, c]$	$[f, c]$
$\mathfrak{I}_3$	$[a^2, F_1, ae, e^2, P_1, F_2^2]$	$[a, e, F_2]$
$\mathfrak{I}_4$	$[a^2, 27be - 64da, ae, e^2, 27b^2c - 64d^2f, P_1]$	$[a, e, 27b^2c - 64d^2f]$
$\mathfrak{I}_5$	$[F_1, F_2, P_1, F_3]$	$[F_1, F_2, P_1, F_3]$
$\mathfrak{I}_6$	$[F_1^3, F_1f, f^2, F_2, P_1, F_3, F_1c, fc, c^2]$	$[F_1, f, c]$
$\mathfrak{I}_7$	$[d^2, F_1e, de^2, e^3, dc, P_1, F_3, ec, c^2]$	$[d, e, c]$
$\mathfrak{I}_8$	$\begin{bmatrix} b^8, b^7a, b^6a^2, b^5a^3, b^4a^4, b^3a^5, b^2a^6, ba^7, a^8, \\ b^2F_1, aF_1, b^6f, b^5af, b^4a^2f, b^3a^3f, b^2a^4f, \\ ba^5f, a^6f, F_1f, b^4f^2, b^3af^2, b^2a^2f^2, ba^3f^2, \\ a^4f^2, b^2f^3, baf^3, a^2f^3, f^4, bF_2, P_1, F_3, F_2f \end{bmatrix}$	$[b, a, f]$

[a] $F_1 = 4be + 3da$, $F_2 = 4b^2c + 3d^2f$, $F_3 = 3a^2c + 4e^2f$; P_1 is given in Example 6.3.3.

The interested reader may refer to Shimoyama and Yokoyama (1996) for a formal proof of PriIdeDec and various techniques and strategies to improve the algorithm.

Example 6.4.1. The ideals generated by $\mathbb{P}$ in Examples 6.2.1, 6.2.2, and 6.3.1 are all radical and each of them contains two primary components.

Example 6.4.2. The ideal $\mathfrak{I}$ given in Example 6.3.3 may be decomposed into 8 primary ideals $\mathfrak{I}_1, \ldots, \mathfrak{I}_8$ (with respect to the variable ordering $b \prec d \prec a \prec e \prec f \prec c$). The generating sets for $\mathfrak{I}_i$ and their associated prime ideals are shown in Table 1.

Remark 6.4.1. Finally, we point out that the various decomposition algorithms developed in this book enjoy evident parallel features and can be easily parallelized. Most of the algorithms compute decomposition trees, for which different branches can be treated individually by parallel processors. Discussions on the aspects of parallel computation are beyond the scope of this book, but it is almost sure that the power of these algorithms will be multiplied when they are brought to suitably parallelized versions and implemented on parallel machines. Some preliminary experiments on parallelizing some of the characteristic-set-based algorithms on workstation networks were reported in Wang (1991b).

7 Applications

Elimination methods have diverse applications in many areas of science, engineering, and industry. A full account of such applications could be the contents of another book. The applications discussed in this chapter are limited to a few selected problems, some of which are geometry related.

7.1 Solving polynomial systems

The various zero decompositions presented in the previous chapters apply naturally to solving systems of polynomial equations and inequations. We give a few theorems – which are consequences of already proved results – as principles for polynomial-system solving, and apply the general methods to some nontrivial examples.

Principles

All the polynomials in what follows are assumed to be in $\boldsymbol{x} = (x_1, \ldots, x_n)$ with coefficients in $\boldsymbol{K} = \mathbf{Q}(\boldsymbol{u}) = \mathbf{Q}(u_1, \ldots, u_d)$ unless specified otherwise. We are now concerned with systems of simultaneous polynomial equations and inequations of the form

$$P_1 = 0, \ldots, P_s = 0, \quad Q_1 \neq 0, \ldots, Q_t \neq 0. \tag{7.1.1}$$

Let $\mathbb{P} = \{P_1, \ldots, P_s\}$, $\mathbb{Q} = \{Q_1, \ldots, Q_t\}$ and $\mathfrak{P} = [\mathbb{P}, \mathbb{Q}]$. We often write (7.1.1) simply as

$$\mathbb{P} = 0, \quad \mathbb{Q} \neq 0. \tag{7.1.2}$$

The system (7.1.1) or (7.1.2) is said to be *solvable* in some field $\tilde{\boldsymbol{K}} \supset \boldsymbol{K}$ if it has solutions in $\tilde{\boldsymbol{K}}$.

Lemma 7.1.1. Let $[\mathbb{T}, \mathbb{U}]$ be a triangular system in $\boldsymbol{K}[\boldsymbol{x}]$ with $|\mathbb{T}| = n$. Then

$$\mathbb{T} = 0, \quad \mathbb{U} \neq 0 \tag{7.1.3}$$

has at most finitely many solutions in any extension field of $\boldsymbol{K}$. All the solutions of (7.1.3) in $\boldsymbol{K}$ can be exactly computed.

If, in particular, $d = 0$, then all the solutions of (7.1.3) in $\mathbf{R}$ (the field of real numbers) and in $\mathbf{C}$ can be approximately computed.

Proof. As $|\mathbb{T}| = n$, the ith polynomial T_i in $\mathbb{T}$ can be written in the form

$$T_i = T_i(x_1, \ldots, x_i)$$

with $\mathrm{lv}(T_i) = x_i$. Hence $x_1 = \bar{x}_1$ is a solution of $T_1 = 0$ for x_1 in $\boldsymbol{K}$ if and only if $x_1 - \bar{x}_1$ is a divisor of T_1 over $\boldsymbol{K}$. Therefore, all the solutions of $T_1 = 0$ for x_1 in $\boldsymbol{K}$ can be found by computing all the linear factors of T_1 over $\boldsymbol{K}$.

If for every solution $x_1 = \bar{x}_1$ of $T_1 = 0$ there is a $U \in \mathbb{U}$ such that $U(\bar{x}_1, x_2, \ldots, x_n) = 0$, then (7.1.3) has no solution in $\boldsymbol{K}$. Otherwise, consider those solutions $x_1 = \bar{x}_1$ of $T_1 = 0$ for which $U(\bar{x}_1, x_2, \ldots, x_n) \neq 0$ for any $U \in \mathbb{U}$. The polynomial $T_2(\bar{x}_1, x_2)$ is clearly in $\boldsymbol{K}[x_2]$, so all the solutions of $T_2(\bar{x}_1, x_2) = 0$ for x_2 in $\boldsymbol{K}$ can be found in the same way by computing all the linear factors of $T_2(\bar{x}_1, x_2)$ over $\boldsymbol{K}$.

If for every solution $x_1 = \bar{x}_1, x_2 = \bar{x}_2$ of $T_1 = 0, T_2 = 0$ and $I_2 \neq 0$ there exists a $U \in \mathbb{U}$ such that $U(\bar{x}_1, \bar{x}_2, x_3, \ldots, x_n) = 0$, then (7.1.3) has no solution in $\boldsymbol{K}$. Otherwise, we take those solutions for which $U(\bar{x}_1, \bar{x}_2, x_3, \ldots, x_n) \neq 0$ for any $U \in \mathbb{U}$. Then the polynomial $T_2(\bar{x}_1, \bar{x}_2, x_3)$ is in $\boldsymbol{K}[x_3]$ and all the solutions of $T_2(\bar{x}_1, \bar{x}_2, x_3) = 0$ for x_3 in $\boldsymbol{K}$ can be found by computing all the linear factors of $T_3(\bar{x}_1, \bar{x}_2, x_3)$ over $\boldsymbol{K}$.

In this way, we shall either end up with the conclusion that (7.1.3) has no solution, or find all the solutions of (7.1.3) in $\boldsymbol{K}$.

When $d = 0$, $\boldsymbol{K}$ becomes the rational-number field $\mathbf{Q}$. In this case, the polynomials T_i all have rational coefficients. Thus, one can solve $T_1 = 0$ for x_1 in $\mathbf{R}$ or $\mathbf{C}$ approximately by any numerical method.

If for every solution $x_1 = \bar{x}_1$ of $T_1 = 0$ there is a $U \in \mathbb{U}$ such that $U(\bar{x}_1, x_2, \ldots, x_n) = 0$ approximately, then (7.1.3) has no solution in $\mathbf{R}$ or $\mathbf{C}$ approximately. Otherwise, we consider such solutions $x_1 = \bar{x}_1$ of $T_1 = 0$ for which $U(\bar{x}_1, x_2, \ldots, x_n) \neq 0$ for any $U \in \mathbb{U}$ and solve $T_2(\bar{x}_1, x_2) = 0$ for x_2 in $\mathbf{R}$ or $\mathbf{C}$ approximately. In other words, the problem of solving polynomial systems is reduced to that of solving univariate polynomial equations or inequations. The latter can be done in $\mathbf{R}$ or $\mathbf{C}$ approximately by known methods of numerical analysis. □

Lemma 7.1.2. Let $[\mathbb{T}, \mathbb{U}]$ be a regular system or a simple system or an irreducible triangular system or a triangular system possessing the projection property in $\boldsymbol{K}[\boldsymbol{x}]$. Then the system (7.1.3) must have solutions in some extension field of $\boldsymbol{K}$. If the number of solutions is finite in an algebraic closure of $\boldsymbol{K}$, then $|\mathbb{T}| = n$.

Proof. The first claim follows from Theorems 3.4.1, 4.3.3, and 5.1.12 and Definition 3.1.3.

If $|\mathbb{T}| < n$, then infinitely many sets of values can be chosen from $\boldsymbol{K}$ for the parameters of $\mathbb{T}$ so that, after the parameters are substituted by any such set of values, $[\mathbb{T}, \mathbb{U}]$ remains perfect (see, e.g., the proofs of Theorems 4.3.3 and 5.1.12). So, in this case (7.1.3) has an infinite number of solutions in an algebraic closure of $\boldsymbol{K}$. □

For any triangular set $\mathbb{T}$, $[\mathbb{T}, \mathrm{ini}(\mathbb{T})]$ is a (special) triangular system. Thus, the above two lemmas lead to the consequent results for triangular sets. Moreover, if $\mathbb{T} = [T_1, \ldots, T_n]$ and any solution of $\mathbb{T}^{\{i\}} = 0$ does not make the vanishing of all the coefficients of T_{i+1} in x_{i+1} for every i, then $\mathbb{T} = 0$ also has at most a finite number of solutions in any extension field of $\boldsymbol{K}$.

Theorem 7.1.3. Let Ψ be a regular series or a simple series or an irreducible triangular series of any polynomial system $[\mathbb{P}, \mathbb{Q}]$ in $\boldsymbol{K}[\boldsymbol{x}]$ or a triangular series of $[\mathbb{P}, \mathbb{Q}]$ computed by algorithm TriSerP with $k = 0$. Then:

a. (7.1.2) has no solution in any extension field of $\boldsymbol{K}$ if and only if $\Psi = \emptyset$;

b. (7.1.2) has at most finitely many solutions if and only if $|\mathbb{T}| = n$ for every $[\mathbb{T}, \mathbb{U}] \in \Psi$. In this case, the solutions of (7.1.2) may be found by means of computing the solutions of $\mathbb{T} = 0$, $\mathbb{U} \neq 0$ for all $[\mathbb{T}, \mathbb{U}] \in \Psi$.

Proof. (a) Theorem 3.4.3 a, Corollaries 4.3.6 and 5.1.14, and TriSerP conditions a and c; (b) Lemmas 7.1.1 and 7.1.2 (see also Theorem 3.4.3 b). □

The process of solving arbitrary systems of polynomial equations and inequations by reducing them to triangular systems generalizes the Chinese matrix method (Boyer 1968, pp. 218 f) and the well-known Gaussian elimination for sets of linear equations. A Gröbner basis is not necessarily a triangular set, but the elimination property of Gröbner bases (Theorem 5.3.5) ensures the separation of variables. So the solutions to a set of polynomial equations can be found from its Gröbner basis (under the lexicographical ordering), possibly with some additional GCD computations. For details, see the reference given below.

Theorem 7.1.4. Let $\mathbb{P}$ be a polynomial set in $\boldsymbol{K}[\boldsymbol{x}]$ and $\mathbb{G} = \mathrm{GB}(\mathbb{P})$. Then:

a. $\mathbb{P} = 0$ has no solution in any extension field of $\boldsymbol{K}$ if and only if $\mathbb{G} = [1]$;

b. $\mathbb{P} = 0$ has at most finitely many solutions if and only if for all i $(1 \leq i \leq n)$ there exist an integer m_i and a polynomial $G_i \in \mathbb{G}$ such that $\mathrm{lt}(G_i) = x_i^{m_i}$;

c. if $\mathbb{P} = 0$ has only finitely many solutions and $\mathbb{G}$ is computed with respect to the purely lexicographical term ordering, then all the solutions in $\boldsymbol{K}$ can be exactly computed from $\mathbb{G}$. If moreover $d = 0$, then can all the solutions in $\mathbf{R}$ and $\mathbf{C}$ be computed approximately from $\mathbb{G}$ as well.

Proof. (a) Corollary 5.3.4, (b) Method 6.9 in Buchberger (1985), (c) Method 6.10 in Buchberger (1985) and Lemma 7.1.1. □

Theorem 7.1.5. Let Ψ be a simple series of $\mathfrak{P}$ in $\mathbf{Q}[\boldsymbol{u}, \boldsymbol{x}]$, or a triangular series of $\mathfrak{P}$ computed by algorithm TriSerP with projection for $x_n, \ldots, x_1$ (i.e., $k = d$) and assume that $\Psi \neq \emptyset$. Then

a. for any $[\mathbb{T}, \mathbb{U}] \in \Psi$ and $\bar{\boldsymbol{u}} \in \tilde{\mathbf{Q}}^d$ (where $\tilde{\mathbf{Q}} \supset \mathbf{Q}$), the system

$$(\mathbb{T} \setminus \mathbf{Q}[\boldsymbol{u}])|_{\boldsymbol{u}=\bar{\boldsymbol{u}}} = 0, \quad (\mathbb{U} \setminus \mathbf{Q}[\boldsymbol{u}])|_{\boldsymbol{u}=\bar{\boldsymbol{u}}} \neq 0$$

has solutions for $\boldsymbol{x}$ in $\mathbf{C}$ if and only if $\boldsymbol{u} = \bar{\boldsymbol{u}}$ is a solution of

$$\mathbb{T} \cap \mathbf{Q}[\boldsymbol{u}] = 0, \quad \mathbb{U} \cap \mathbf{Q}[\boldsymbol{u}] \neq 0;$$

b. $\mathrm{Proj}_{\boldsymbol{u}}\mathrm{Zero}(\mathfrak{P}) = \bigcup_{\mathfrak{T} \in \Psi} \mathrm{Proj}_{\boldsymbol{u}}\mathrm{Zero}(\mathfrak{T}) = \bigcup_{[\mathbb{T},\mathbb{U}] \in \Psi} \mathrm{Zero}(\mathbb{T} \cap \mathbf{Q}[\boldsymbol{u}]/\mathbb{U} \cap \mathbf{Q}[\boldsymbol{u}])$.

Proof. (a) The condition follows from b; (b) Corollary 3.4.2, Definition 3.3.3, and TriSerP conditions b and c. □

This theorem permits us to solve parametric polynomial systems: by computing simple systems or triangular systems with projection, one knows for what values of the parameters $\boldsymbol{u}$ the system $\mathbb{P} = 0$, $\mathbb{Q} \neq 0$ has solutions for the unknowns $\boldsymbol{x}$ (cf. Gao and Chou 1992). For any given parametric values $\bar{\boldsymbol{u}}$, the solutions may be computed from or represented by the simple or triangular systems

$$[(\mathbb{T} \setminus \mathbf{Q}[\boldsymbol{u}])|_{\boldsymbol{u}=\bar{\boldsymbol{u}}}, (\mathbb{U} \setminus \mathbf{Q}[\boldsymbol{u}])|_{\boldsymbol{u}=\bar{\boldsymbol{u}}}], \quad [\mathbb{T}, \mathbb{U}] \in \Psi,$$

where $\mathfrak{P} = [\mathbb{P}, \mathbb{Q}]$ and Ψ are as in Theorem 7.1.5.

Examples

Example 7.1.1. We start with a small set of polynomial equations

$$\left.\begin{array}{l} x_1x_2 - 1 = 0, \\ x_3^2 + bx_1x_2 = 0, \\ bx_1x_3 + x_2^2 - x_1 = 0, \\ bx_2x_3 - x_2 + x_1^2 = 0. \end{array}\right\} \tag{7.1.4}$$

Let $\mathbb{P}$ be the set of the four polynomials on the left-hand side of (7.1.4) and the variables be ordered as $b \prec x_1 \prec x_2 \prec x_3$. From $\mathbb{P}$:

- a characteristic series computed by CharSer consists of two ascending sets

$$\mathbb{C}_1 = [b^3 + 4, x_1^3 + 1, x_1x_2 - 1, 2x_3 + b^2],$$
$$\mathbb{C}_2 = [b, x_1^3 - 1, x_1x_2 - 1, x_3];$$

- a triangular series computed by TriSerS consists of two triangular systems $[\mathbb{C}_1, \{b, x_1\}]$ and $[\mathbb{C}_2, \{x\}]$; when computed by TriSer, the series consists of $[\mathbb{T}_1, \{b, x_1\}]$ and $[\mathbb{T}_2, \{x\}]$ with

$$\mathbb{T}_1 = [b^3 + 4, x_1^3 + 1, x_1x_2 - 1, bx_3 - 2],$$
$$\mathbb{T}_2 = [b, x_1^3 - 1, x_2 - x_1^2, x_3],$$

  where $\mathbb{T}_1$ differs from $\mathbb{C}_1$ only in their fourth elements, and so does $\mathbb{T}_2$ from $\mathbb{C}_2$ in their third elements;
- a regular series computed by RegSer and a simple series computed by SimSer are the same, consisting of $[\mathbb{T}_1, \emptyset]$ and $[\mathbb{C}_2, \emptyset]$;
- a Gröbner basis of $\mathbb{P}$ is

$$\mathbb{G} = [b^5 + 4b^2, 2x_1^3 - b^3 - 2, 2x_2 - b^3x_1^2 - 2x_1^2, 2bx_3 + b^3, x_3^2 + b].$$

In any of the above cases, one can find all the 12 solutions of (7.1.4) for b, x_1, x_2, x_3 successively from the triangularized polynomial sets. These solutions $[b, x_1, x_2, x_3]$ are listed below

$$\begin{array}{lll} [0, 1, 1, 0], & [0, -\alpha, -\beta, 0], & [0, -\beta, -\alpha, 0], \\ [-\gamma, -1, -1, -\gamma^2/2], & [-\gamma, \alpha, \beta, -\gamma^2/2], & [-\gamma, \beta, \alpha, -\gamma^2/2], \\ [\alpha\gamma, -1, -1, \beta\gamma^2/2], & [\alpha\gamma, \alpha, \beta, \beta\gamma^2/2], & [\alpha\gamma, \beta, \alpha, \beta\gamma^2/2], \\ [\beta\gamma, -1, -1, \alpha\gamma^2/2], & [\beta\gamma, \alpha, \beta, \alpha\gamma^2/2], & [\beta\gamma, \beta, \alpha, \alpha\gamma^2/2], \end{array}$$

where

$$\alpha = \frac{1-\sqrt{-3}}{2}, \quad \beta = \frac{1+\sqrt{-3}}{2}, \quad \gamma = \sqrt[3]{4}.$$

Example 7.1.2. Consider the following set of 8 polynomial equations

$$\begin{aligned}
P_1 &= u_3^2 g_{00} + u_3^2 h_{00} + u_3^3 + u_2^2 u_3 - u_1^2 u_3 = 0,\\
P_2 &= u_3^2 h_{11} + u_3^2 g_{11} = 0,\\
P_3 &= u_3^2 h_{10} + u_3^2 g_{10} = 0,\\
P_4 &= u_3^2 h_{01} + u_3^2 g_{01} = 0,\\
P_5 &= u_3^2 g_{00} h_{10} + u_3^2 g_{10} h_{00} + u_1^2 u_3^2 g_{01} h_{11} + u_1^2 u_3^2 g_{11} h_{01} - 2u_1^4 u_3 g_{11} h_{11}\\
&\quad - 2u_1^2 u_3 g_{10} h_{10} - 2u_1 u_2 u_3 g_{10} h_{10} - 2u_1^3 u_2 u_3 g_{11} h_{11} = 0,\\
P_6 &= 2u_1 u_2 u_3 g_{01} h_{11} - 2u_1^2 u_3 g_{11} h_{01} - 2u_1^2 u_3 g_{01} h_{11} + 2u_1 u_2 u_3 g_{11} h_{01}\\
&\quad + u_3^2 g_{01} h_{10} + u_3^2 g_{00} h_{11} + u_3^2 g_{11} h_{00} + u_3^2 g_{10} h_{01} - 2u_1^2 u_3 g_{11} h_{10}\\
&\quad - 2u_1^2 u_3 g_{10} h_{11} - 2u_1 u_2 u_3 g_{10} h_{11} - 4u_1^2 u_2^2 g_{11} h_{11}\\
&\quad - 2u_1 u_2 u_3 g_{11} h_{10} + 4u_1^4 g_{11} h_{11} = 0,\\
P_7 &= u_1^2 u_3^2 g_{01} h_{01} + u_1^2 u_3^2 g_{10} h_{10} + u_1^4 u_3^2 g_{11} h_{11} + u_3^2 g_{00} h_{00} + u_1^2 u_3^2 = 0,\\
P_8 &= u_3^2 g_{01} h_{00} + 2u_1 u_2 u_3 g_{01} h_{01} - 2u_1^2 u_3 g_{01} h_{01} + u_3^2 g_{00} h_{01}\\
&\quad + 2u_1^3 u_2 u_3 g_{11} h_{11} + u_1^2 u_3^2 g_{10} h_{11} - 2u_1^4 u_3 g_{11} h_{11}\\
&\quad + u_1^2 u_3^2 g_{11} h_{10} = 0.
\end{aligned} \tag{7.1.5}$$

We want to find one solution of (7.1.5) for h_{ij} and g_{ij} in $\mathbf{Q}(u_1, u_2, u_3)$. To achieve this, let us compute a modified weak-characteristic set $\mathbb{C}$ of $\{P_1, \ldots, P_8\}$ with respect to the variable ordering

$$h_{01} \prec h_{11} \prec h_{10} \prec h_{00} \prec g_{01} \prec g_{00} \prec g_{11} \prec g_{10}.$$

It is found that

$$\mathbb{C} = \begin{bmatrix}
4u_1^2 h_{01}^2 - u_2^2 - 2u_1 u_2 - u_1^2,\\
u_1 u_2 h_{11} + u_1^2 h_{11} - u_3 h_{01},\\
u_2 h_{10} + u_1 h_{10} + u_2 h_{01} - u_1 h_{01},\\
2u_3 h_{01} h_{00} + 2u_1{}^2 u_3 h_{11} h_{10} + 2u_1^3 u_2 h_{11}^2 - 2u_1^4 h_{11}^2\\
\quad + 2u_1 u_2 h_{01}^2 - 2u_1^2 h_{01}^2 + u_3^2 h_{01} + u_2^2 h_{01} - u_1^2 h_{01},\\
g_{01} + h_{01},\\
u_3 g_{00} + u_3 h_{00} + u_3^2 + u_2^2 - u_1^2,\\
g_{11} + h_{11},\\
g_{10} + h_{10}
\end{bmatrix},$$

which is quasilinear. The first polynomial in $\mathbb{C}$ factors over $\mathbf{Q}$ into

$$(2u_1 h_{01} - u_2 - u_1)(2u_1 h_{01} + u_2 + u_1).$$

The only initial not in $\mathbf{Q}(u_1, u_2, u_3)$ is h_{01}. Thus, two solutions are found easily from the triangular set by solving univariate linear equations. We list one of the solutions as follows for later use:

$$\begin{aligned} g_{11} &= \frac{u_3}{2u_1^2}, & h_{11} &= -\frac{u_3}{2u_1^2}, \\ g_{01} &= \frac{u_1+u_2}{2u_1}, & h_{01} &= -\frac{u_1+u_2}{2u_1}, \\ g_{10} &= \frac{u_1-u_2}{2u_1}, & h_{10} &= -\frac{u_1-u_2}{2u_1}, \\ g_{00} &= \frac{2u_1^2-2u_2^2-u_3^2}{2u_3}, & h_{00} &= -\frac{u_3}{2}. \end{aligned} \tag{7.1.6}$$

By computing a triangular, characteristic or Gröbner series of $\mathbb{P}$, one may see that (7.1.5) has no other solution for h_{ij} and g_{ij} in $\mathbf{Q}(u_1, u_2, u_3)$.

Example 7.1.3. Refer to the polynomial set $\mathbb{P}$ and its decomposition into simple systems in Example 3.3.5. It is not difficult to verify that

$$\begin{aligned} \bigcup_{j=1}^{13} \mathrm{Zero}(\mathbb{T}_j^{(1)}/\bar{\mathbb{T}}_j^{(1)}) = \bigcup_{j=1}^{5} \mathrm{Zero}(\emptyset/\bar{\mathbb{T}}_j^{(1)}) \cup \mathrm{Zero}(H_1) \cup \mathrm{Zero}(H_2) \\ \cup \mathrm{Zero}(c) \cup \mathrm{Zero}(2c^3-27) = \tilde{\mathbf{Q}}. \end{aligned}$$

Hence, the set of polynomial equations $\mathbb{P} = 0$ has solutions for any value of c, considered as a parameter. When a concrete value of c is given, the solutions for z, y, x may be determined from the corresponding simple systems.

Example 7.1.4. Let $\mathbb{P} = \{P_1, \ldots, P_4\}$, where

$$\begin{aligned} P_1 &= x_2(x_3 - x_4) - x_1 + c, \\ P_2 &= x_3(x_4 - x_1) - x_2 + c, \\ P_3 &= x_4(x_1 - x_2) - x_3 + c, \\ P_4 &= x_1(x_2 - x_3) - x_4 + c. \end{aligned}$$

With the ordering $c \prec x_1 \prec \cdots \prec x_4$, $\mathbb{P}$ can be decomposed by TriSerS into 21 quasi-irreducible triangular systems $\mathfrak{T}_i$ such that (2.1.8) holds with $\mathfrak{P} = [\mathbb{P}, \emptyset]$ and $e = 21$. An irreducible triangular series of $\mathbb{P}$ computed by IrrTriSer consists of 13 irreducible triangular sets.

The polynomial set $\mathbb{P}$ in the above example arises from the dynamical system of a chaotic attractor considered by E. Lorenz. It was investigated by Z. Liu and used in Gao and Chou (1992). $\mathbb{P}$ in the following example, communicated to S. R. Czapor and K. O. Geddes by G. Fee, may be found in Wang (1993).

For any polynomial $P \in \boldsymbol{K}[\boldsymbol{x}]$ we use an *index triple* $[t\ \mathrm{lv}(P)\ \mathrm{ldeg}(P)]$ to characterize P, where t is the number of terms of P.

Example 7.1.5. Let $\mathbb{P} = \{P_1, \ldots, P_4\}$, where

$$\begin{aligned}
P_1 &= 2(b-1)^2 + 2(q - pq + p^2) + c^2(q-1)^2 - 2bq \\
&\quad + 2cd(1-q)(q-p) + 2bpqd(d-c) + b^2d^2(1-2p) \\
&\quad + 2bd^2(p-q) + 2bdc(p-1) + 2bpq(c+1) \\
&\quad + (b^2 - 2b)p^2d^2 + 2b^2p^2 + 4b(1-b)p + d^2(p-q)^2, \\
P_2 &= d(2p+1)(q-p) + c(p+2)(1-q) + b(b-2)d \\
&\quad + b(1-2b)pd + bc(q+p-pq-1) + b(b+1)p^2d, \\
P_3 &= -b^2(p-1)^2 + 2p(p-q) - 2(q-1), \\
P_4 &= b^2 + 4(p-q^2) + 3c^2(q-1)^2 - 3d^2(p-q)^2 \\
&\quad + 3b^2d^2(p-1)^2 + b^2p(p-2) + 6bdc(p+q+pq-1).
\end{aligned}$$

Consider b as a parameter and order the other variables as $p \prec d \prec c \prec q$. An irreducible triangular series of $\mathbb{P}$, which may be easily computed by IrTriSer, consists of two irreducible triangular sets. One of them is very simple: $[p-1, d, bc+2, q-1]$; the other consists of four polynomials, of which the first three have the following index triples: $[625\ p\ 23]$, $[373\ d\ 1]$, and $[17\ c\ 1]$, and the last is P_3.

7.2 Automated geometry theorem proving

Since the pioneering work of Wu (1978), automated theorem proving in geometry has been an active area of research for two decades. There is a rich literature on the subject. We recommend the comprehensive exposition by Wu (1994) for a thorough understanding of his method and the subject and the popular book by Chou (1988) for an easy presentation and many examples. The reader may also look at the survey by Wang (1996b) and references therein for the state of the art.

Elementary approach

Most of the successful methods for proving geometric theorems developed by Wu and his followers are algebraic in character. They can be considered as one major application of the various elimination techniques presented in the preceding chapters. The first step of proving geometric theorems by algebraic methods is to algebraize the geometric problems in question. For this purpose, one chooses a coordinate system and denotes the coordinates of points as well as other involved geometric entities like areas of triangles and squares of distances by the indeterminates $x_1, \ldots, x_n$. Then the hypotheses and the conclusions of most geometric theorems can be expressed by means of polynomial equations ($=$), inequations ($\neq$), and inequalities ($\leq$, $<$) in $x_1, \ldots, x_n$. This is illustrated by the following example.

Example 7.2.1 (Simson's theorem). From a point D draw three perpendiculars to the three sides of an arbitrary triangle ABC (Fig. 5). Then the three perpendicular feet P, Q, and R are collinear if and only if D lies on the circumscribed circle of $\triangle ABC$.

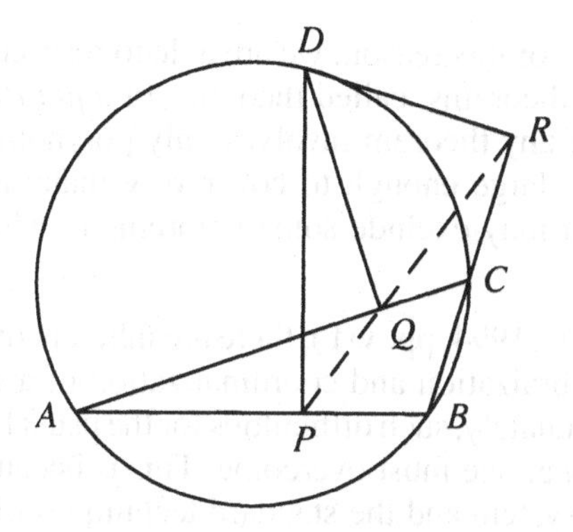

Fig. 5. Example 7.2.1

Consider the "if" part of the theorem. Without loss of generality, we take a Cartesian coordinate system with AB as its first axis and the perpendicular bisector of AB as its second axis. Let the points be assigned coordinates as follows

$$A(-x_1, 0),\quad B(x_1, 0),\quad C(x_2, x_3),\quad D(x_4, x_5),$$
$$P(x_4, 0),\quad Q(x_6, x_7),\quad R(x_8, x_9).$$

Then the hypothesis HYP of the theorem consists of the following relations:

– D lies on the circumscribed circle of ΔABC

$$\Longleftrightarrow H_1 = x_1x_3x_5^2 - x_1x_3^2x_5 - x_1x_2^2x_5 + x_1^3x_5 + x_1x_3x_4^2 - x_1^3x_3 = 0;$$

– Q is the foot of the perpendicular drawn from point D to line AC

$$\Longleftrightarrow \begin{cases} H_2 = (x_2 + x_1)(x_6 - x_4) + x_3(x_7 - x_5) = 0, \\ H_3 = (x_2 + x_1)x_7 - x_3(x_6 + x_1) = 0; \end{cases}$$

– R is the foot of the perpendicular drawn from point D to line BC

$$\Longleftrightarrow \begin{cases} H_4 = (x_2 - x_1)(x_8 - x_4) + x_3(x_9 - x_5) = 0, \\ H_5 = (x_2 - x_1)x_9 - x_3(x_8 - x_1) = 0. \end{cases}$$

That P is the foot of the perpendicular drawn from D to AB is ensured by the special choice of the coordinates for point P.

Someone careful might observe that the theorem may become meaningless if the triangle ABC is flat. This degenerate case can be ruled out: The three points A, B, C are not collinear if and only if $D_1 = x_1x_3 \neq 0$. The exclusion of this degenerate case is not substantial. We will see that *nondegeneracy conditions* may be found automatically by Wu's method.

The conclusion CON of the theorem to be proved is: The three points P, Q, R are collinear, i.e., $G = (x_6 - x_4)x_9 - x_7(x_8 - x_4) = 0$.

The algebraic expressions of most ordinary geometric relations like collinearity, perpendicularity, and congruence involve only polynomial equations – an observation made by Wu that is of special significance for the theory and methods of geometry theorem proving. Also for this reason, we are able to restrict our consideration to an important class of theorems, called theorems of *equality type*, in which the algebraic formulation of any theorem involves only polynomial equations and inequations. The class is large enough to cover very many nontrivial and interesting theorems, though it may exclude some theorems in which order relations are involved.

Remark 7.2.1. As pointed out by Wu (1994, pp. vi f), there are inherent difficulties along the path to arrive at the algebraization and coordinatization of a geometry starting from its axiom system. Fortunately, such difficulties for the usual Euclidean geometry do not appear seriously that one must overcome. This is because of our knowledge about the real-number system and the standard techniques of analytic geometry. It is for this reason that one may be supposed to know how to transform ordinary geometric relations into algebraic expressions by introducing coordinate systems as in analytic geometry, without going through the correctness proof of the algebraization.

The algebraic formulation of Simson's theorem in Example 7.2.1 is of equality type. However, with this formulation one may fail in proving the logical implication (HYP $\Rightarrow$ CON). For in the statement of a geometric theorem the considered figures are usually implicitly assumed to be in a *generic* position. For example, while speaking about a triangle, we mean a real triangle which does not degenerate into a line or a point. In the above formulation, this degenerate case has been excluded a priori, but other degenerate cases may still be included that might make the implication (HYP $\Rightarrow$ CON) logically false. Therefore, one has to determine some subsidiary (nondegeneracy) conditions so that the theorem becomes true under these conditions. We do not give a precise definition of *degenerate cases* and *nondegeneracy conditions* here. Actually, it is rather difficult to give such a definition because of the unclosedness of stating geometric theorems and the different understandings of the word "degenerate." For the moment the reader is only assumed to have a rough impression of the concept of degeneracy. More explanations will be given later.

Let $\wedge$, $\vee$, and $\Rightarrow$ denote the logical "and," "or," and "imply" respectively. We propose the following algebraic formulation for the decision problem of geometry theorem proving.

Formulation α. Suppose that we are given a geometry $\mathfrak{G}$, a geometry-associated field $\boldsymbol{K}$ of characteristic 0 and an appropriate coordinate system $\mathfrak{D}$ under which a correspondence between statements in $\mathfrak{G}$ and algebraic expressions over $\boldsymbol{K}$ may be established. Let the hypothesis of a theorem $\mathbb{T}$ in $\mathfrak{G}$ be expressed under $\mathfrak{D}$ as a finite set of polynomial equations and inequations

$$\text{HYP}: \begin{cases} H_1(\boldsymbol{x}) = 0, \dots, H_s(\boldsymbol{x}) = 0, \\ D_1(\boldsymbol{x}) \neq 0, \dots, D_t(\boldsymbol{x}) \neq 0 \end{cases} \tag{7.2.1}$$

(where each $D_i = 0$ corresponds usually to a degenerate case determined a priori from some analysis or observation of the theorem), and the conclusion be expressed as a single polynomial equation

$$\text{CON:}\quad G(\boldsymbol{x}) = 0. \tag{7.2.2}$$

All the polynomials are in the indeterminates $\boldsymbol{x} = (x_1, \ldots, x_n)$ – which are coordinates of points and other geometric entities involved in the theorem – with coefficients in $\boldsymbol{K}$. Decide

a. whether the formula

$$(\forall \boldsymbol{x})[H_1(\boldsymbol{x}) = 0 \wedge \cdots \wedge H_s(\boldsymbol{x}) = 0 \wedge D_1(\boldsymbol{x}) \neq 0 \wedge \cdots \wedge D_t(\boldsymbol{x}) \neq 0 \Longrightarrow G(\boldsymbol{x}) = 0] \tag{7.2.3}$$

is valid; and if not,

b. find "appropriate" subsidiary conditions $D_1^*(\boldsymbol{x}) \neq 0, \ldots, D_{t^*}^*(\boldsymbol{x}) \neq 0$ so that the formula

$$\begin{aligned}(\forall \boldsymbol{x})[H_1(\boldsymbol{x}) = 0 \wedge \cdots \wedge H_s(\boldsymbol{x}) = 0 \wedge \\ D_1(\boldsymbol{x}) \neq 0 \wedge \cdots \wedge D_t(\boldsymbol{x}) \neq 0 \wedge \\ D_1^*(\boldsymbol{x}) \neq 0 \wedge \cdots \wedge D_{t^*}^*(\boldsymbol{x}) \neq 0 \Longrightarrow G(\boldsymbol{x}) = 0]\end{aligned}$$

becomes valid over $\boldsymbol{K}$ or some extension field of $\boldsymbol{K}$.

The additional inequations $D_j^*(\boldsymbol{x}) \neq 0$ are determined to ensure the configuration of the geometric hypotheses to be in a generic position. In the proof algorithms presented below,

$$\mathbb{P} = \{H_1, \ldots, H_s\}, \quad \mathbb{Q} = \{D_1, \ldots, D_t\}.$$

For any geometric statement or theorem $\mathbb{T}$, we write

HC($\mathbb{T}$) for "the hypothesis of $\mathbb{T}$ is self-contradictory";
NC($\mathbb{T}$) for "$\mathbb{T}$ is not confirmed";
True($\mathbb{T}$)/SC for "$\mathbb{T}$ is true under the subsidiary conditions SC."

It is possible that the subsidiary conditions are not explicitly provided; in this case SC is not set to any value. If SC $= \emptyset$, then the theorem $\mathbb{T}$ is *universally* true; otherwise, $\mathbb{T}$ is *conditionally* true.

The following elementary method is very efficient for confirming geometric theorems, in particular when N-characteristic sets and principal triangular systems are used.

Algorithm ProverA: HC, True/SC, or NC $\leftarrow$ ProverA($\mathbb{P}$, $\mathbb{Q}$, G). Given the algebraic form $\mathbb{T}$: $\mathbb{P} = 0 \wedge \mathbb{Q} \neq 0 \Rightarrow G = 0$ of a geometric theorem of equality type, this algorithm either proves True($\mathbb{T}$)/SC, or reports HC($\mathbb{T}$) or NC($\mathbb{T}$).

P1. Compute a (quasi-, weak-) medial set $\mathbb{T}$ of $\mathbb{P}$ over $\boldsymbol{K}$ by CharSetN or PriTriSys. If $\mathbb{T}$ is contradictory or $0 \in \mathrm{prem}(\mathbb{Q}, \mathbb{T})$, then report HC($\mathbb{T}$) and the algorithm terminates.

P2. Compute $R \leftarrow \mathrm{prem}(G, \mathbb{T})$. If $R \equiv 0$, then let $I_1, \ldots, I_r$ be all the distinct irreducible factors of the polynomials in ini($\mathbb{T}$) which do not divide any D_i, set

$$\text{SC} \leftarrow I_1 \neq 0 \wedge \cdots \wedge I_r \neq 0$$

and return True($\mathbb{T}$)/SC; else report NC($\mathbb{T}$).

Steps P1 and P2 may be replaced alternatively by the following three steps, in which Gröbner bases are used.

P1′. Compute a Gröbner basis $\mathbb{G}$ of $\mathbb{P} \cup \{D_1 z_1 - 1, \ldots, D_t z_t - 1\}$ over $\boldsymbol{K}$ with respect to the purely lexicographical term ordering determined by $x_1 \prec \cdots \prec x_n \prec z_1 \prec \cdots \prec z_t$, where $z_1, \ldots, z_t$ are new indeterminates. If $1 \in \mathbb{G}$, then report $\mathtt{HC}(\mathbb{T})$ and the algorithm terminates.

P2′. Compute $R \leftarrow \mathrm{rem}(G, \mathbb{G})$. If $R \equiv 0$, then return $\mathtt{True}(\mathbb{T})/\emptyset$ and the algorithm terminates.

P3′. Take a *quasi*-basic set of $\mathbb{G}$: $\mathbb{B} \leftarrow \mathrm{BasSet}(\mathbb{G})$, and compute $R \leftarrow \mathrm{prem}(R, \mathbb{B})$. If $R \equiv 0$, then let $I_1, \ldots, I_r$ be all the distinct irreducible factors of the polynomials in $\mathrm{ini}(\mathbb{B})$ which do not divide any D_i, set

$$\mathrm{SC} \leftarrow I_1 \neq 0 \wedge \cdots \wedge I_r \neq 0$$

and return $\mathtt{True}(\mathbb{T})/\mathrm{SC}$; else report $\mathtt{NC}(\mathbb{T})$.

The termination of this and other algorithms in Sects. 7.2–7.4 is obvious, so the proofs are given only for their correctness.

Proof. As the medial set $\mathbb{T}$ of $\mathbb{P}$ computed by CharSetN or PriTriSys is contained in Ideal($\mathbb{P}$), $\mathbb{P} = 0$ implies that $\mathbb{T} = 0$. Let $\bar{\boldsymbol{x}} \in \mathrm{Zero}(\mathbb{P}/\mathbb{Q})$; then there exists a $\bar{z}_i$ in some extension field of $\boldsymbol{K}$ such that $D_i(\bar{\boldsymbol{x}})\bar{z}_i - 1 = 0$ for $1 \leq i \leq t$. It follows that $G(\bar{\boldsymbol{x}}) = 0$ for any

$$G \in \mathbb{G} \cap \boldsymbol{K}[\boldsymbol{x}] \subset \mathrm{Ideal}(\mathbb{P} \cup \{D_1 z_1 - 1, \ldots, D_t z_t - 1\}),$$

wherefore $\mathbb{P} = 0$ and $\mathbb{Q} \neq 0$ imply that $\mathbb{G} \cap \boldsymbol{K}[\boldsymbol{x}] = 0$. Thus, the theorem $\mathbb{T}$ is universally true when

$$\mathrm{rem}(G, \mathbb{G}) = \mathrm{rem}(G, \mathbb{G} \cap \boldsymbol{K}[\boldsymbol{x}]) \equiv 0.$$

By the pseudo-remainder formula, if $R \equiv 0$, then

$$\mathbb{T} = 0 \wedge \mathrm{ini}(\mathbb{T}) \neq 0 \Longrightarrow G = 0;$$

this is also true when $\mathbb{T}$ is replaced by $\mathbb{B}$. Note that $\mathbb{B} \subset \mathbb{G}$. Hence $\mathbb{T}$ is conditionally true under the subsidiary conditions SC when $R \equiv 0$. □

The medial set $\mathbb{T}$ in algorithm ProverA may also be $\mathbb{F}$-modified, while the cases in which $F = 0$ for $F \in \mathbb{F}$ have to be handled separately. The following two steps, which are necessary for implementing a geometry theorem prover, are not included in the algorithms presented in this section.

P0. This is a preprocess that translates the geometric statement of a theorem into the algebraic form. It can be done automatically by implementing a translator for some commonly used geometric relations.

P∞. This is a postprocess that interprets the algebraic subsidiary conditions geometrically and determines which conditions are nondegeneracy ones. In most cases, the interpretation can be done easily and automatically (see, e.g., Chou 1988, Wang 1996a). Whether a subsidiary condition is a nondegeneracy condition may be seen from its geometric meaning, dimension analysis, etc.

It is a key insight of Wu that most geometric theorems are true only under subsidiary conditions. Without predetermining all such conditions the two steps P1′ and P2′ can prove only a limited number of theorems. In order to deal with this issue and to speak about genericness we may separate the variables $\boldsymbol{x}$ into *parameters* and *geometric dependents*. The former are free variables which can take arbitrary values, while the latter are constrained by the geometric conditions. The separation can be done rather easily when the geometric theorem is stated constructively step by step. Assume that all the parameters $\boldsymbol{u}$ are correctly identified from $\boldsymbol{x}$. Then any inequation in $\boldsymbol{u}$ can be considered as a nondegeneracy condition. So in this case the medial sets, principal triangular systems, or Gröbner bases may all be computed over $\boldsymbol{K}(\boldsymbol{u})$, i.e., only with respect to the geometrically dependent variables. Thus the theorem is proved to be true under some nondegeneracy conditions which are not necessarily provided, and step P3′ may be skipped when Gröbner bases are used (see Kutzler and Stifter 1986).

Whether or not the theorem is true in a degenerate case can be determined by the same method, regarding the degeneracy condition as an additional hypothesis of the theorem.

Unless explicitly stated, the Gröbner bases mentioned in the examples of this chapter are always with respect to the purely lexicographical term ordering determined by the indicated variable ordering.

Example 7.2.2. Refer to Example 7.2.1 and let $\mathbb{P} = \{H_1, \ldots, H_5\}$. With respect to the ordering $x_1 \prec \cdots \prec x_9$, a weak-N-characteristic set of $\mathbb{P}$ is

$$\mathbb{C} = \begin{bmatrix} I_1 x_5^2 - x_1(x_3^2 x_5 + x_2^2 x_5 - x_1^2 x_5 - x_3 x_4^2 + x_1^2 x_3), \\ I_2 x_6 - x_2 x_3 x_5 - x_1 x_3 x_5 - x_2^2 x_4 - 2x_1 x_2 x_4 - x_1^2 x_4 + x_1 x_3^2, \\ I_3 x_7 - x_3 x_6 - x_1 x_3, \\ I_4 x_8 - x_2 x_3 x_5 + x_1 x_3 x_5 - x_2^2 x_4 + 2x_1 x_2 x_4 - x_1^2 x_4 - x_1 x_3^2, \\ I_5 x_9 - x_3 x_8 + x_1 x_3 \end{bmatrix},$$

where

$$\begin{aligned} I_1 &= x_1 x_3, \\ I_2 &= x_3^2 + x_2^2 + 2x_1 x_2 + x_1^2, \\ I_3 &= x_2 + x_1, \\ I_4 &= x_3^2 + x_2^2 - 2x_1 x_2 + x_1^2, \\ I_5 &= x_2 - x_1 \end{aligned}$$

are the initials of the five polynomials $C_1, \ldots, C_5$ in $\mathbb{C}$ respectively. Clearly, $\operatorname{prem}(I_i, \mathbb{C})$ is nonzero for $1 \leq i \leq 5$, and so is $\operatorname{prem}(D_1, \mathbb{C})$. It is easy to verify that $\operatorname{prem}(G, \mathbb{C}) = 0$, so the theorem is proved to be true under the subsidiary conditions $I_i \neq 0$, for $2 \leq i \leq 5$. The geometric meanings of the four conditions, interpreted automatically by GEOTHER (Wang 1996a), are as follows:

$I_2 \neq 0 \iff AC$ is nonisotropic;
$I_3 \neq 0 \iff AC$ is not perpendicular to AB;

$I_4 \neq 0 \Longleftrightarrow BC$ is nonisotropic;
$I_5 \neq 0 \Longleftrightarrow AB$ is not perpendicular to BC.

One can examine whether the theorem is true in each of the degenerate cases by taking $I_i = 0$ as a new hypothesis. Consider the case $I_3 = 0$ for example. Let $\mathbb{P}^* = \{H_1, \ldots, H_5, I_3\}$. Then the hypothesis consists of $\mathbb{P}^* = 0$ and $D_1 \neq 0$. A characteristic set of $\mathbb{P}^*$ with the same variable ordering is

$$\mathbb{C}^* = \begin{bmatrix} x_2 + x_1, \\ x_5^2 - x_3x_5 + x_4^2 - x_1^2, \\ x_6 + x_1, \\ x_7 - x_5, \\ x_3^2x_8 + 4x_1^2x_8 + 2x_1x_3x_5 - 4x_1^2x_4 - x_1x_3^2, \\ x_3^2x_9 + 4x_1^2x_9 - x_3^2x_5 + 2x_1x_3x_4 - 2x_1^2x_3 \end{bmatrix}$$

with some factors x_1 and x_3 removed. Since $\mathrm{prem}(G, \mathbb{C}^*) = 0$, the theorem is also true in this case under the nondegeneracy condition $x_3^2 + 4x_1^2 \neq 0$ (i.e., the line BC is nonisotropic).

One can verify the other degenerate cases one by one in the same way. A systematic treatment as will be presented below is to compute a zero decomposition for $[\mathbb{P}, \{x_1, x_3\}]$ and to see for which components the conclusion holds. One should finally conclude that only the first and the third nondegeneracy condition are necessary.

A Gröbner basis $\mathbb{G}$ of $\mathbb{P}$ under the same variable ordering consists of 17 polynomials, and $\mathrm{rem}(G, \mathbb{G}) = G \not\equiv 0$. Now $\mathbb{G}$ has a quasi-basic set identical to $\mathbb{C}$ (up to a sign for some polynomials). According to the above verifications, the theorem is proved to be true under the nondegeneracy conditions $I_2 \cdots I_5 \neq 0$.

With respect to $x_5 \prec \cdots \prec x_9$ a Gröbner basis of $\mathbb{P}$ is

$$\mathbb{G}^* = [C_1/x_1, C_2, G_3, C_4, G_5],$$

where

$$G_3 = I_2x_7 - x_3^2x_5 - x_2x_3x_4 - x_1x_3x_4 - x_1x_2x_3 - x_1^2x_3,$$
$$G_5 = I_4x_9 - x_3^2x_5 - x_2x_3x_4 + x_1x_3x_4 + x_1x_2x_3 - x_1^2x_3,$$

and C_1, C_2, C_4, I_2, I_4 are as above. One can verify that $\mathrm{rem}(x_1x_3, \mathbb{G}^*) \neq 0$ and $\mathrm{rem}(G, \mathbb{G}^*) = 0$. It follows that the theorem is true under some nondegeneracy conditions.

The above method with variation has been implemented by several researchers (Chou 1988, Ko and Hussain 1985, Kusche et al. 1987, Wang and Gao 1987, Wu 1984). A large number of geometric theorems including Steiner's theorems (generalized/rediscovered), Morley's trisector theorem, and the recently confirmed conjecture of Thébault have been proved by using different implementations in a matter of seconds; some interesting "new" theorems were also discovered (see, e.g., Wu 1984, 1994; Chou 1988; Wang 1995c).

Complete method

We must note that Formulation α is not fine. First of all, there was no requirement on verifying the consistency of the hypothesis HYP before determining the validity of (7.2.3). If some H_i, for instance, is a nonzero constant, then $H_i = 0$ itself is contradictory. In this case, (7.2.3) is always a true formula. Second, no definition has been given for what we call "appropriate" and "subsidiary conditions." Apparently, adding $D_j^* \neq 0$ to HYP should not exclude interesting cases of the theorem. In particular, every $D_j^* = 0$ should not be a consequence of HYP, i.e., the addition of $D_j^* \neq 0$ to HYP does not destroy the consistency. However, it is not easy, theoretically and computationally, to completely examine the consistency of the hypothesis and to enforce the above-mentioned requirement to be fulfilled for the found subsidiary conditions.

The purpose of finding nondegeneracy conditions in the context of geometric theorem proving is to rule out some degenerate cases in which the theorem becomes false or meaningless. This aims at proving theorems even if their algebraic formulations are not logically complete due to the lack of such conditions. The problem of missing conditions is caused by the imprecise nature of human beings in expressing geometric problems and the rigorlessness of the axiom system of geometry. In practice, one may add conditions to get rid of some degenerate cases, but it is difficult and impossible to predetermine all such cases.

Even though nondegeneracy conditions have been taken into account, one may still have troubles in proving geometric theorems according to Formulation α. The reason is: some ambiguities corresponding to the reducibility of geometric configurations may occur when geometric statements are transformed into polynomial expressions. Let us come to the following example.

Example 7.2.3. The bisectors of the three angles of an arbitrary triangle, three-to-three, intersect at four points. Let the triangle be $\triangle ABC$, the two bisectors of $\angle A$ and $\angle B$ intersect at point D, and the bisector of $\angle C$ meet line AB at point E (Fig. 6). We need to show that D lies on CE.

To simplify calculation, and without loss of generality, we take the coordinates of the points as

$$A(x_1, 0), \quad B(x_2, 0), \quad C(0, x_3), \quad D(x_4, x_5), \quad E(x_6, 0).$$

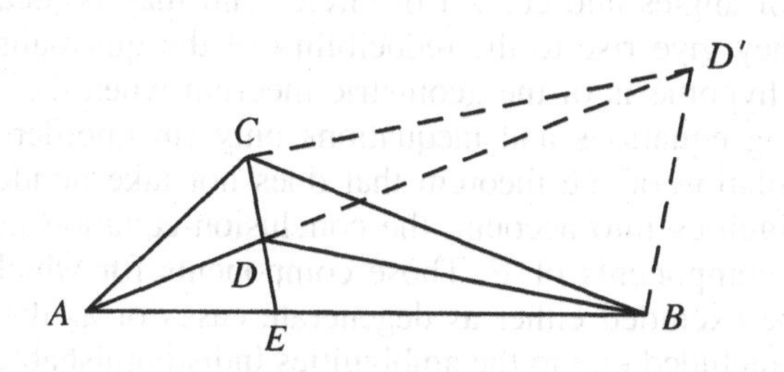

Fig. 6. Example 7.2.3

The hypothesis of the theorem consists of the following three relations

$$\text{HYP:}\quad \begin{cases} H_1 = x_3[x_5^2 - (x_4 - x_1)^2] - 2x_1x_5(x_4 - x_1) = 0, \\ \qquad\qquad \twoheadleftarrow \ DA \text{ is the bisector of } \angle CAB \\ H_2 = x_3[x_5^2 - (x_4 - x_2)^2] - 2x_2x_5(x_4 - x_2) = 0, \\ \qquad\qquad \twoheadleftarrow \ DB \text{ is the bisector of } \angle ABC \\ H_3 = x_3[(x_1 - x_6)(x_3^2 + x_2x_6) + (x_2 - x_6)(x_3^2 + x_1x_6)] = 0. \\ \qquad\qquad \twoheadleftarrow \ EC \text{ is the bisector of } \angle BCA \end{cases}$$

We add the condition

$$D_1 = x_3 \neq 0, \qquad \twoheadleftarrow \ C \text{ does not lie on } AB$$

to eliminate the trivial degenerate case. The conclusion to be proved is

$$\text{CON: } G = x_3x_4 + x_5x_6 - x_3x_6 = 0. \quad \twoheadleftarrow \ D \text{ lies on } CE$$

At first sight, one might not see any problem in the above formulation. Looking over the theorem and its formulation carefully, one may be aware of the fact that the bisectors may be internal and external; both of them are represented by the same polynomial equations. Without using inequalities, the two kinds of bisectors cannot be distinguished from each other. If the bisector of one angle of $\triangle ABC$ is external and those of the two others are internal, then the three bisectors are certainly not concurrent. So the theorem could not be proved to be generically true with the above formulation. To deal with this situation, let us slightly modify the formulation (cf. Wu 1994, pp. 197–199).

Example 7.2.4. Instead of the collinearity of D, C, and E, we may prove that

$$\text{CON}^*: \quad \begin{aligned} G^* = \ & [x_1(x_5 - x_3) + x_3x_4][x_3(x_5 - x_3) - x_2x_4] \\ & + [x_2(x_5 - x_3) + x_3x_4][x_3(x_5 - x_3) - x_1x_4] = 0. \\ & \qquad \twoheadleftarrow \ DC \text{ is the bisector of } \angle BCA \end{aligned}$$

Then point E need not be introduced, and the third relation $H_3 = 0$ in Example 7.2.3 becomes redundant. Now the four possibilities for which the three bisectors are not concurrent have been excluded.

Ambiguities of this kind also appear inherently in other geometric relations like trisection of angles and contact of circles and may be dealt with by using inequalities. They give rise to the reducibility of the quasi-algebraic variety $\mathcal{V}$ defined by the hypothesis of the geometric theorem when the hypothesis is expressed by using equations and inequations only (in unordered geometry). In a natural formulation of the theorem that does not take nondegeneracy conditions and ambiguities into account, the conclusion-equation holds true usually only for some components of $\mathcal{V}$. Those components for which the theorem is false have to be excluded either as degenerate cases or as the unwanted cases that have been included due to the ambiguities indistinguishable in the algebraic formulation.

Although there are special techniques dealing with reducibility (see, e.g., Wu 1986c, Wang and Gao 1987), a complete and systematic treatment of the problem is to decompose $\mathcal{V}$ into irreducible components.

Formulation β. Let $\mathfrak{G}$, $\boldsymbol{K}$ and $\mathfrak{D}$ be as in Formulation α, and let the hypothesis of a theorem $\mathbb{T}$ in $\mathfrak{G}$ be expressed under $\mathfrak{D}$ as a finite set of polynomial equations and inequations (7.2.1), and the conclusion be expressed as one polynomial equation (7.2.2). Set $\mathbb{P} = \{H_1, \ldots, H_s\}$ and $\mathbb{Q} = \{D_1, \ldots, D_t\}$. Decide

a. whether $\text{Zero}(\mathbb{P}/\mathbb{Q}) = \emptyset$; and if not,
b. on which components of $\text{Zero}(\mathbb{P}/\mathbb{Q})$ G vanishes (and thus $\mathbb{T}$ is true).

More precisely, let Ψ be a regular series of $[\mathbb{P}, \mathbb{Q}]$ and define the set of *regular zeros* of $[\mathbb{P}, \mathbb{Q}]$ to be

$$\text{RegZero}(\mathbb{P}/\mathbb{Q}) \triangleq \bigcup_{\mathfrak{T}\in\Psi} \text{RegZero}(\mathfrak{T}).$$

Then problem b consists in separating $\text{RegZero}(\mathbb{P}/\mathbb{Q})$ into

$$\mathcal{Z}^+ = \{\boldsymbol{\xi} \in \text{RegZero}(\mathbb{P}/\mathbb{Q}) :\ G(\boldsymbol{\xi}) = 0\}, \quad \text{and}$$
$$\mathcal{Z}^- = \{\boldsymbol{\xi} \in \text{RegZero}(\mathbb{P}/\mathbb{Q}) :\ G(\boldsymbol{\xi}) \neq 0\}.$$

The theorem $\mathbb{T}$ is universally true if and only if $\mathcal{Z}^- = \emptyset$ and $\mathcal{Z}^+ \neq \emptyset$. If $\mathcal{Z}^+ = \emptyset$ and $\mathcal{Z}^- \neq \emptyset$, we say that "$\mathbb{T}$ is *generically* false," which is denoted by `False`($\mathbb{T}$). Otherwise, $\mathbb{T}$ is conditionally true. The subsidiary conditions SC are provided by excluding those components of $\text{Zero}(\mathbb{P}/\mathbb{Q})$ for which $\mathbb{T}$ is generically false.

The following algorithm is directed to Formulation β.

Algorithm ProverB: `HC`, `True`/SC, or `False` $\leftarrow$ ProverB ($\mathbb{P}$, $\mathbb{Q}$, G). Given the algebraic form $\mathbb{T}:\ \mathbb{P} = 0 \wedge \mathbb{Q} \neq 0 \Rightarrow G = 0$ of a geometric theorem of equality type, this algorithm either proves `True`($\mathbb{T}$)/SC, or determines `False`($\mathbb{T}$), or reports `HC`($\mathbb{T}$).

P1. Compute a characteristic series or triangular series Ψ of $[\mathbb{P}, \mathbb{Q}]$ over $\boldsymbol{K}$ by CharSer, TriSer, or TriSerS. If $\Psi = \emptyset$, then report `HC`($\mathbb{T}$) and the algorithm terminates.

P2. Let all the triangular systems in Ψ be $[\mathbb{T}_1, \mathbb{U}_1], \ldots, [\mathbb{T}_e, \mathbb{U}_e]$. Compute

$$R_i \leftarrow \text{prem}(G, \mathbb{T}_i), \quad 1 \leq i \leq e,$$

and set

$$\Delta \leftarrow \{i:\ R_i \not\equiv 0, 1 \leq i \leq e\}, \quad \mathcal{Z} \leftarrow \bigcup_{\substack{1 \leq i \leq e \\ i \notin \Delta}} \text{Zero}(\mathbb{T}_i/\mathbb{U}_i).$$

If $\Delta = \emptyset$, then

$$\begin{cases} \text{report HC}(\mathbb{T}) & \text{when } \mathcal{Z} = \emptyset, \\ \text{return True}(\mathbb{T})/\emptyset & \text{otherwise,} \end{cases}$$

and the algorithm terminates.

P3. Compute an irreducible triangular series Ψ_i of $[\mathbb{T}_i, \mathbb{U}_i]$ over $\boldsymbol{K}$ by IrrTriSer, IrrCharSer, or IrrCharSerE for each $i \in \Delta$ and set $\Psi^* \leftarrow \bigcup_{i\in\Delta} \Psi_i$. If $\Psi^* = \emptyset$, then

$$\begin{cases} \text{report } \mathtt{HC}(\mathbb{T}) & \text{when } |\Delta| = e \text{ or } \mathcal{Z} = \emptyset, \\ \text{return } \mathtt{True}(\mathbb{T})/\emptyset & \text{otherwise,} \end{cases}$$

and the algorithm terminates.

P4. Let $[\mathbb{T}_1^*, \mathbb{U}_1^*], \ldots, [\mathbb{T}_{e^*}^*, \mathbb{U}_{e^*}^*]$ be all the irreducible triangular systems in Ψ^*. Compute

$$R_j^* \leftarrow \operatorname{prem}(G, \mathbb{T}_j^*), \quad 1 \le j \le e^*,$$

and set $\Delta^* \leftarrow \{j: \ R_j^* \not\equiv 0, 1 \le j \le e^*\}$.
If $\Delta^* = \emptyset$, then return $\mathtt{True}(\mathbb{T})/\emptyset$ and the algorithm terminates.
If $|\Delta| = e$ or $\mathcal{Z} = \emptyset$ and $|\Delta^*| = e^*$, then return $\mathtt{False}(\mathbb{T})$ and the algorithm terminates.

P5. Set

$$\mathrm{SC} \leftarrow \bigwedge_{j\in\Delta^*} \Big(\bigvee_{T\in\mathbb{T}_j^*} T \neq 0 \vee \bigvee_{I\in \operatorname{ini}(\mathbb{T}_j^*)\setminus\mathbb{Q}} I = 0 \Big)$$

and return $\mathtt{True}(\mathbb{T})/\mathrm{SC}$.

Proof. The triangular series Ψ and Ψ^* give rise to a zero decomposition

$$\operatorname{Zero}(\mathbb{P}/\mathbb{Q}) = \mathcal{Z} \cup \mathcal{Z}^+ \cup \mathcal{Z}^-$$

such that

$$\mathcal{Z} \cup \mathcal{Z}^+ \subset \operatorname{Zero}(G);$$
$$G(\boldsymbol{\xi}) \neq 0, \quad \forall \boldsymbol{\xi} \in \mathcal{Z}^- \text{ that is regular,}$$

where

$$\mathcal{Z}^+ = \bigcup_{\substack{1 \le j \le e^* \\ j \notin \Delta^*}} \operatorname{Zero}(\mathbb{T}_j^*/\mathbb{U}_j^*), \quad \mathcal{Z}^- = \bigcup_{j\in\Delta^*} \operatorname{Zero}(\mathbb{T}_j^*/\mathbb{U}_j^*).$$

Note that $\mathbb{T}_j^*$ is irreducible for $1 \le j \le e^*$. Thus,

$$\operatorname{Zero}(\mathbb{P}/\mathbb{Q}) = \emptyset \iff \mathcal{Z} = \emptyset \text{ and } \Psi^* = \emptyset.$$

Suppose that $\operatorname{Zero}(\mathbb{P}/\mathbb{Q}) \neq \emptyset$. Then the theorem is universally true, i.e., $\operatorname{Zero}(\mathbb{P}/\mathbb{Q}) \subset \operatorname{Zero}(G)$, if and only if $\Delta^* = \emptyset$. It is generically false if and only if $|\Delta| = e$ or $\mathcal{Z} = \emptyset$ and $|\Delta^*| = e^*$. Otherwise, the theorem is conditionally true under the subsidiary condition SC (cf. Theorem 4.3.11 b). □

Remark 7.2.2. For the sake of practical efficiency some redundant triangular systems, for example, those $[\mathbb{T}, \mathbb{U}]$ for which $|\mathbb{T}| > |\mathbb{P}|$, should be removed from Ψ

and Ψ_i in ProverB (see Lemma 6.2.9). The algorithm starts by computing a triangular series, not an irreducible one, mainly for bypassing unnecessary (algebraic) polynomial factorization. It may be simplified by computing directly an irreducible triangular series of $[\mathbb{P}, \mathbb{Q}]$. The computation of triangular series in the algorithm may also be performed over $\boldsymbol{K}(\boldsymbol{u})$ when the parameters $\boldsymbol{u}$ are correctly identified from the variables $\boldsymbol{x}$ and the theorem is considered only for the nondegenerate cases.

To confirm theorems, one may also use algorithm ProverC below, in which an irreducible triangular series of $[\mathbb{P}, \mathbb{Q} \cup \{G\}]$ is computed. Assume for simplicity that $x_1, \ldots, x_d$ are the parameters and $x_{d+1}, \ldots, x_n$ the geometric dependents, which are correctly specified. We use a bar over SC to indicate that the subsidiary conditions have been identified as nondegeneracy conditions. Thus, $\mathtt{True}(\mathbb{T})/\overline{\mathrm{SC}}$ means that "the theorem $\mathbb{T}$ is *generically* true under the nondegeneracy conditions $\overline{\mathrm{SC}}$." And, we can talk about "$\mathbb{T}$ is not *generically* true," which is denoted by $\mathtt{NGT}(\mathbb{T})$. It means that there exist $\bar{x}_{d+1}, \ldots, \bar{x}_n$ in some algebraic-extension field of $\boldsymbol{K}(\boldsymbol{x}^{\{d\}})$ such that $(\boldsymbol{x}^{\{d\}}, \bar{x}_{d+1}, \ldots, \bar{x}_n)$ is a zero of $[\mathbb{P}, \mathbb{Q}]$ but not a zero of G.

Algorithm ProverC: HC, $\mathtt{True}/\overline{\mathrm{SC}}$, or NGT $\leftarrow$ ProverC($\mathbb{P}, \mathbb{Q}, G$). Given the algebraic form $\mathbb{T}$: $\mathbb{P} = 0 \wedge \mathbb{Q} \neq 0 \Rightarrow G = 0$ of a geometric theorem of equality type, this algorithm either proves $\mathtt{True}(\mathbb{T})/\overline{\mathrm{SC}}$, or determines $\mathtt{NGT}(\mathbb{T})$, or reports $\mathtt{HC}(\mathbb{T})$.

P1. Determine whether $\mathrm{Zero}(\mathbb{P}/\mathbb{Q}) = \emptyset$ in $\bar{\boldsymbol{K}}$ by algorithm TriSerP, SimSer, RegSer, RegSer*, IrrCharSer, IrrCharSerE, or IrrTriSer. If so, then report $\mathtt{HC}(\mathbb{T})$ and the algorithm terminates.

P2. Compute over $\boldsymbol{K}$ a triangular series Ψ of $[\mathbb{P}, \mathbb{Q} \cup \{G\}]$ by TriSerP with projection for $x_n, \ldots, x_d$, or an irreducible triangular series Ψ of $[\mathbb{P}, \mathbb{Q} \cup \{G\}]$ by IrrCharSer, IrrCharSerE, or IrrTriSer.
If $\Psi = \emptyset$, then return $\mathtt{True}(\mathbb{T})/\emptyset$ and the algorithm terminates.
Let $[\mathbb{T}_1, \mathbb{U}_1], \ldots, [\mathbb{T}_e, \mathbb{U}_e]$ be all the triangular systems in Ψ. If $\mathbb{T}_i^{(d)} \neq \emptyset$ for all $1 \leq i \leq e$, then let D_i^* be any polynomial in $\mathbb{T}_i^{(d)}$, set

$$\overline{\mathrm{SC}} \leftarrow \bigwedge_{i=1}^{e} D_i^* \neq 0,$$

and return $\mathtt{True}(\mathbb{T})/\overline{\mathrm{SC}}$; else return $\mathtt{NGT}(\mathbb{T})$.

Proof. If $\Psi = \emptyset$, then $\mathrm{Zero}(\mathbb{P}/\mathbb{Q} \cup \{G\}) = \emptyset$. It follows that

$$\mathrm{Zero}(\mathbb{P}/\mathbb{Q}) \subset \mathrm{Zero}(G),$$

so the theorem is universally true. If $\mathbb{T}_i^{(d)} \neq \emptyset$ for all $1 \leq i \leq e$, then according to the selection of D_i^* we have

$$\mathrm{Zero}(\mathbb{P}/\mathbb{Q} \cup \{D_1^*, \ldots, D_e^*, G\}) = \emptyset.$$

This implies that

$$\mathrm{Zero}(\mathbb{P}/\mathbb{Q} \cup \{D_1^*, \ldots, D_e^*\}) \subset \mathrm{Zero}(G).$$

Hence, the theorem is conditionally true under the subsidiary conditions SC. Otherwise, there exists an i, $1 \le i \le e$, such that $\mathbb{T}_i^{(d)} = \emptyset$. Note that $[\mathbb{T}_i, \mathbb{U}_i]$ is perfect and thus has a regular/generic zero $\boldsymbol{\xi}$. Now

$$\boldsymbol{\xi} \in \mathrm{Zero}(\mathbb{P}/\mathbb{Q} \cup \{G\}),$$

so $\boldsymbol{\xi}$ is a zero of $[\mathbb{P}, \mathbb{Q}]$ but not a zero of G. Therefore, the theorem is not generically true. □

As an alternative, one may determine the vacancy of $\mathrm{Zero}(\mathbb{P}/\mathbb{Q})$ and the subsidiary conditions under which $[\mathbb{P}, \mathbb{Q} \cup \{G\}]$ has no zero by computing Gröbner bases according to Theorem 6.3.3 c (see Kapur 1988, for details).

Examples

In this section we use the formulations in Examples 7.2.3–7.2.4 and Steiner's theorem to illustrate different aspects of proving geometric theorems by the algorithms described above.

Example 7.2.5. See Examples 7.2.4 and 7.2.4. Determine when the following algebraic form of the theorem is true:

$$(\forall x_1, \dots, x_5)[H_1 = 0 \wedge H_2 = 0 \wedge D_1 \neq 0 \Longrightarrow G^* = 0].$$

Application of ProverA

Compute a characteristic set $\mathbb{C}$ of $\mathbb{P} = \{H_1, H_2\}$ with respect to the ordering $x_1 \prec \cdots \prec x_5$: $\mathbb{C} = [(x_2 - x_1)x_3C_1, (x_2 - x_1)C_2]$ with

$$\begin{aligned} C_1 = {} & 4x_4^4 - 8x_2x_4^3 - 8x_1x_4^3 - 4x_3^2x_4^2 + 4x_2^2x_4^2 + 12x_1x_2x_4^2 + 4x_1^2x_4^2 \\ & + 4x_2x_3^2x_4 + 4x_1x_3^2x_4 - 4x_1x_2^2x_4 - 4x_1^2x_2x_4 - x_2^2x_3^2 - 2x_1x_2x_3^2 - x_1^2x_3^2, \\ C_2 = {} & 2x_4x_5 - 2x_2x_5 - 2x_1x_5 - 2x_3x_4 + x_2x_3 + x_1x_3. \end{aligned}$$

The initials of the two polynomials in $\mathbb{C}$ are

$$I_1 = 4(x_2 - x_1)x_3, \quad I_2 = 2(x_2 - x_1)(x_4 - x_2 - x_1)$$

respectively. Simple computation shows that $\mathrm{prem}(G^*, \mathbb{C}) = 0$. Hence, the theorem is proved to be true under the subsidiary conditions

$$D_1^* = x_2 - x_1 \neq 0, \quad D_2^* = x_4 - x_2 - x_1 \neq 0.$$

The first condition has evident geometric meaning: A and B do not coincide, so it can be considered as a nondegeneracy condition.

To see whether the theorem is true when $D_2^* = 0$, we form an enlarged set $\mathbb{P}^* = \mathbb{P} \cup \{D_2^*\}$ of hypothesis-polynomials. Proceeding in the same way, one should prove that the theorem is also true in this case under the nondegeneracy condition $D_1^* \neq 0$.

In the above proof, the consistency of the hypothesis is not examined. For the examination, one has to see whether or not

$$\mathrm{Zero}(\mathbb{P}/x_3D_1^*D_2^*) = \mathrm{Zero}(\mathbb{C}/x_3D_1^*D_2^*) = \emptyset.$$

Application of ProverB
Instead of verifying the degenerate cases one by one, we compute a characteristic series of $[\mathbb{P}, \{x_3\}]$ in order to determine when the theorem is true. With the same variable ordering, the series consists of three ascending sets

$$\mathbb{C}_1 = [C_1, C_2],$$
$$\mathbb{C}_2 = [x_2 - x_1, C_2'],$$
$$\mathbb{C}_3 = [x_2^2 - x_1^2, x_4 - x_2 - x_1, x_3x_5^2 - 2x_1x_2x_5 - x_1^2x_3],$$

where C_1, C_2 are given above and

$$C_2' = x_3x_5^2 - 2x_1x_4x_5 + 2x_1^2x_5 - x_3x_4^2 + 2x_1x_3x_4 - x_1^2x_3.$$

As $\operatorname{prem}(G^*, \mathbb{C}_1) = 0$, the theorem is true for $\mathbb{C}_1$. However, $\operatorname{prem}(G^*, \mathbb{C}_i) \neq 0$ for $i = 2, 3$. It is easy to verify that $\mathbb{C}_2$ is irreducible and $\mathbb{C}_3$ is reducible. Therefore, the theorem is not true for $\mathbb{C}_2$, and one does not know whether it is true for $\mathbb{C}_3$ without going further.

It is trivial to see the consistency of the hypothesis, i.e., $\operatorname{Zero}(\mathbb{P}/x_3) \neq \emptyset$, because $\operatorname{Zero}(\mathbb{C}_2/\operatorname{ini}(\mathbb{C}_2) \cup \{x_3\}) \neq \emptyset$, for instance.

If $x_2^2 - x_1^2 \in \mathbb{C}_3$ is factorized as to compute an irreducible zero decomposition, one can get three irreducible ascending sets of which one is

$$\mathbb{C}_{3'} = [x_2 + x_1, x_4, x_3x_5^2 + 2x_1^2x_5 - x_1^2x_3],$$

and the two others are identical to $\mathbb{C}_1$ and $\mathbb{C}_2$. For computing the decomposition factorization does not need to be over algebraic-extension fields. It is again easy to verify that $\operatorname{prem}(G^*, \mathbb{C}_{3'}) = 0$.

Therefore, we can conclude that the hypothesis of the theorem is consistent, the theorem is true under the nondegeneracy condition

$$x_2 - x_1 \neq 0 \vee C_2' \neq 0,$$

and in the degenerate case $x_2 - x_1 = C_2' = 0$ the theorem is not true.

Here the disjunction of inequations is used to represent the nondegeneracy condition. This is to keep the excluded part of $\operatorname{Zero}(\mathbb{P}/x_3)$ (for which the theorem is false) minimal. One may take $D_1^* = x_2 - x_1 \neq 0$ as the nondegeneracy condition for simplicity, but this condition also excludes, for example, the degenerate case $x_1 = x_2 = x_4 \neq 0, x_5 = 0$ in which the theorem is true.

By Theorem 6.2.8, we have

$$\operatorname{Zero}(\mathbb{P}/x_3) = \operatorname{Zero}(\mathrm{PB}(\mathbb{C}_1)/x_3) \cup \operatorname{Zero}(\mathrm{PB}(\mathbb{C}_2)/x_3).$$

Therefore, the geometric configuration – quasi-algebraic variety – defined by the hypothesis is decomposed into two irreducible components. The conclusion-polynomial G vanishes on one of them but not on the other. Hence, the theorem is true only for one component – the case in which $\triangle ABC$ is located in a generic position. The other component for which the theorem is false corresponds to the case when $\triangle ABC$ degenerates.

Application of ProverC

Instead of $\mathrm{Zero}(\mathbb{P}/x_3)$, let us compute an (irreducible) decomposition for $\mathrm{Zero}(\mathbb{P}/x_3G^*)$ under the same variable ordering: we get the ascending set $\mathbb{C}_2$ given above,

$$\mathbb{C}_{3''} = [x_2 - x_1, x_4 - 2x_1, x_3x_5^2 - 2x_1^2x_5 - x_1^2x_3],$$

and two polynomials

$$\begin{aligned} G_2 &= x_3(x_3^2x_4^2x_5 + x_1^2x_4^2x_5 - 4x_1x_3^2x_4x_5 - 4x_1^3x_4x_5 + 4x_1^2x_3^2x_5 + 4x_1^4x_5 \\ &\quad - x_3^3x_4^2 - x_1^2x_3x_4^2 + 3x_1x_3^3x_4 + 3x_1^3x_3x_4 - 2x_1^2x_3^3 - 2x_1^4x_3), \\ G_{3''} &= x_1x_3(x_3^2 + x_1^2), \end{aligned}$$

such that

$$\mathrm{Zero}(\mathbb{P}/x_3G^*) = \bigcup_{i=2,3''} \mathrm{Zero}(\mathbb{C}_i/G_i).$$

One sees that $x_2 - x_1$ is contained in both of the ascending sets. If we assume $x_2 \neq x_1$ and consider it as a nondegeneracy condition of the theorem, then $\mathrm{Zero}(\mathbb{P}/x_3G^*)$ becomes empty; i.e., $\mathrm{Zero}(\mathbb{P}/(x_2 - x_1)x_3G^*) = \emptyset$. Hence, the theorem is proved to be true under the given nondegeneracy condition $x_3 \neq 0$ and the found nondegeneracy condition $x_2 - x_1 \neq 0$.

Example 7.2.6. Refer to Example 7.2.3. We want to show that

$$(\forall x_1, \ldots, x_6)[H_1 = 0 \wedge H_2 = 0 \wedge H_3 = 0 \wedge x_3 \neq 0 \Longrightarrow G = 0].$$

For this purpose, let $\mathbb{P} = \{H_1, H_2, H_3\}$.

Application of ProverB

With respect to $x_1 \prec \cdots \prec x_6$, a characteristic set of $\mathbb{P}$ (with two factors x_3 and $x_2 - x_1$ removed during the computation) is $\mathbb{C} = [C_1, C_2, C_3]$, where C_1, C_2 are as in Example 7.2.5 and

$$C_3 = x_2x_6^2 + x_1x_6^2 + 2x_3^2x_6 - 2x_1x_2x_6 - x_2x_3^2 - x_1x_3^2.$$

Now $\mathrm{prem}(G, \mathbb{C}) \neq 0$, so one cannot tell if the theorem is true or not. It is then necessary to determine whether $\mathbb{C}$ is irreducible or not. By the methods explained in Sect. 7.5, one may find that over the extension field $\mathbf{Q}(x_1, \ldots, x_4)$ – where x_1, x_2, x_3 are adjoined to $\mathbf{Q}$ as transcendental elements and x_4 an algebraic element with C_1 as minimal polynomial – C_3 is reducible and factors as

$$\begin{aligned} C_3 &\doteq \frac{C_3'C_3''}{x_2 + x_1}, \quad \text{where} \\ C_3' &= x_2x_6 + x_1x_6 + 2x_4^2 - 2x_2x_4 - 2x_1x_4, \\ C_3'' &= x_2x_6 + x_1x_6 - 2x_4^2 + 2x_2x_4 + 2x_1x_4 + 2x_3^2 - 2x_1x_2. \end{aligned}$$

In fact, decomposing $[\mathbb{P}, \{x_3\}]$ results in seven irreducible triangular sets $\mathbb{T}_1, \ldots, \mathbb{T}_7$ as given in Example 4.2.4. One may verify that $\text{prem}(G, \mathbb{T}_i) = 0$ for $i = 1, 3, 5$, but not for the others.

Moreover, from the obtained triangular sets one can compute an irreducible decomposition of the quasi-algebraic variety defined by $[\mathbb{P}, \{x_3\}]$ into four irreducible components. This decomposition actually corresponds to (4.2.8) with $\mathbb{T}_3, \mathbb{T}_4, \mathbb{T}_5$ removed. It follows that the theorem is true only for the component that corresponds to $\mathbb{T}_1$. The component corresponding to $\mathbb{T}_2$ represents the cases such as two bisectors are internal, whereas the third is external, which are not degenerate cases at all. The remaining two components for which the theorem is false can be interpreted as corresponding to some degenerate cases.

If we specify x_1, x_2, x_3 as parameters (as to ensure $\triangle ABC$ to be generic) and x_4, x_5, x_6 as geometric dependents and consider any inequations in x_1, x_2, x_3 as nondegeneracy conditions of the theorem, then an irreducible decomposition may be computed over the functional field $\mathbf{Q}(x_1, x_2, x_3)$. The inequations can be collected as to give the exact nondegeneracy conditions during the computation if desirable. In this case, the irreducible characteristic series contains only the two triangular sets $\mathbb{T}_1$ and $\mathbb{T}_2$; now $\text{prem}(G, \mathbb{T}_1) = 0$ and $\text{prem}(G, \mathbb{T}_2) \neq 0$. Hence, the theorem is generically true for one component and false for the other and thus is conditionally true.

Application of ProverC

Now compute an irreducible characteristic series of $[\mathbb{P}, \{x_3, G\}]$, yielding one irreducible ascending set, that is $\mathbb{T}_2$ in Example 4.2.4, with a polynomial

$$G_2 = (x_2 + x_1)x_3(x_4 - x_2 - x_1)G$$

such that

$$\text{Zero}(\mathbb{P}/x_3G) = \text{Zero}(\mathbb{T}_2/G_2) \neq \emptyset.$$

Without further consideration and analysis, it is hardly possible to figure out from this ascending set whether the theorem is true or false. Similarly, if one computes an irreducible triangular series of $\mathbb{P} \cup \{x_3Gz - 1\}$ (with respect to $x_4 \prec x_5 \prec x_6 \prec z$), then the series contains only one triangular set that is $\mathbb{T}_2 \cup [T_4]$ with

$$T_4 = x_3(2x_4^2 - 2x_2x_4 - 2x_1x_4 - x_3^2 + x_1x_2)z - x_4 + x_2 + x_1$$

such that

$$\text{Zero}(\mathbb{P} \cup \{Gz - 1\}) = \text{Zero}(\mathbb{T}_2 \cup [T_4]/\text{ini}(\mathbb{T}_2 \cup [T_4])).$$

From this decomposition one cannot conclude the conditional truth of the theorem either. This is why ProverC is considered incomplete. There is some possibility for determining the conditional truth of the theorem via a detailed analysis of the computed ascending set, for example, by interpreting its polynomials geometrically. In general this type of analysis is difficult.

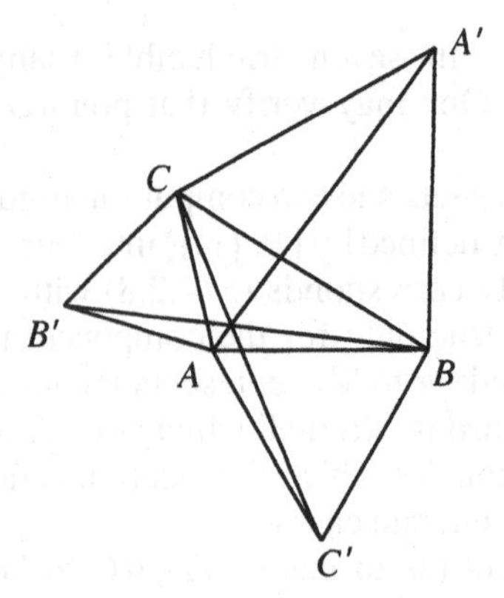

Fig. 7. Example 7.2.7

Algebraic factorization may be avoided for Example 7.2.6 when reflection of points is used instead of bisection of angles to formulate the theorem. See Wu (1994, pp. 199–201), for details.

The above examples should have illustrated the following point: For a given geometric theorem there are numerous ways to state it and to formulate it algebraically. The difference between two formulations of the theorem may be negligible in principle yet remarkable in practice. A simple modification on one formulation may considerably reduce the practical computational complexity, may yield a proof of the theorem that appears to be beyond the applicability of a method, or may bypass some time-consuming steps in the algebraic algorithm.

Example 7.2.7 (Steiner's theorem). Let ABC', BCA' and CAB' be three equilateral triangles drawn all inward or all outward on the three sides of an arbitrary triangle ABC (Fig. 7). Then the three lines AA', BB', and CC' are concurrent.

Without loss of generality, let the points be located as

$$A(0,0),\ \ B(1,0),\ \ C(u_1,u_2),\ \ C'(y_1,y_2),\ \ A'(y_3,y_4),\ \ B'(y_5,y_6).$$

Then the theorem can be transformed into the following algebraic form

$$\text{HYP:}\begin{cases} H_1 = 2y_1 - 1 = 0, & \leftarrow |AC'| = |BC'| \\ H_2 = y_1^2 + y_2^2 - 1 = 0, & \leftarrow |AC'| = |AB| \\ H_3 = y_3^2 + y_4^2 - u_1^2 - u_2^2 = 0, & \leftarrow |AB'| = |AC| \\ H_4 = y_3^2 + y_4^2 - (y_3 - u_1)^2 \\ \qquad - (y_4 - u_2)^2 = 0, & \leftarrow |AB'| = |CB'| \\ H_5 = (y_5 - 1)^2 + y_6^2 \\ \qquad - (u_1 - 1)^2 - u_2^2 = 0, & \leftarrow |BA'| = |BC| \\ H_6 = (y_5 - 1)^2 + y_6^2 - (y_5 - u_1)^2 \\ \qquad - (y_6 - u_2)^2 = 0, & \leftarrow |BA'| = |CA'| \\ D_1 = u_2 \neq 0, & \leftarrow C \text{ is not on } AB \end{cases}$$

$$\text{CON:}\ \begin{cases} G = y_1y_4y_6 - u_1y_4y_6 - u_1y_2y_3y_6 + u_2y_1y_3y_6 + u_1y_2y_6 \\ \quad - u_2y_1y_6 + u_1y_2y_4y_5 - y_2y_4y_5 - u_2y_1y_4y_5 \\ \quad + u_2y_4y_5 = 0. \\ \qquad\qquad \twoheadleftarrow\ AA',\ BB' \text{ and } CC' \text{ are concurrent} \end{cases}$$

Here the square of distance is used instead of distance to avoid radicals and the case in which $\triangle ABC$ degenerates into a line is eliminated by $D_1 \neq 0$. The variables u_1, u_2 are regarded as *parameters* which are arbitrary, and $y_1, \ldots, y_6$ are *geometric dependents* constrained by the algebraic conditions $H_i = 0$ for $1 \leq i \leq 6$.

Set

$$\mathbb{P} = \{H_1, \ldots, H_6\}, \quad \mathbb{Q} = \{u_2\}, \quad \mathbb{Q}^* = \{u_2, G\}$$

and order the variables as $u_1 \prec u_2 \prec y_1 \prec \cdots \prec y_6$. Using ProverB, we compute an irreducible decomposition for Zero($\mathbb{P}/\mathbb{Q}$) over $\mathbf{Q}$. The output Ψ of IrrTriSer consists of nine triangular systems $[\mathbb{T}_i, \mathbb{U}_i]$, so we have

$$\mathrm{Zero}(\mathbb{P}/u_2) = \bigcup_{i=1}^{9} \mathrm{Zero}(\mathbb{T}_i/u_2),$$

where

$$\begin{aligned}
\mathbb{T}_1 &= [T_1, T_2, T_3, T_4, T_5, T_6],\\
\mathbb{T}_2 &= [T_1, T_2, T_3', T_4, T_5', T_6],\\
\mathbb{T}_3 &= [T_1, T_2, T_3', T_4, T_5, T_6],\\
\mathbb{T}_4 &= [T_1, T_2, T_3, T_4, T_5', T_6],\\
\mathbb{T}_5 &= [u_2^2 + u_1^2, T_1, T_2, T_4, T_5, T_6],\\
\mathbb{T}_6 &= [u_2^2 + u_1^2, T_1, T_2, T_4, T_5', T_6],\\
\mathbb{T}_7 &= [u_2^2 + u_1^2 - 2u_1 + 1, T_1, T_2, T_3, T_4, T_6],\\
\mathbb{T}_8 &= [u_2^2 + u_1^2 - 2u_1 + 1, T_1, T_2, T_3', T_4, T_6],\\
\mathbb{T}_9 &= [2u_1 - 1, 4u_2^2 + 1, T_1, T_2, T_4, T_6];
\end{aligned}$$

$$\begin{aligned}
T_1 &= 2y_1 - 1,\\
T_2 &= 4y_2^2 - 3,\\
T_3 &= 2y_3 - 2u_2y_2 - u_1,\\
T_3' &= 2y_3 + 2u_2y_2 - u_1,\\
T_4 &= 2u_2y_4 + 2u_1y_3 - u_2^2 - u_1^2,\\
T_5 &= 2y_5 + 2u_2y_2 - u_1 - 1,\\
T_5' &= 2y_5 - 2u_2y_2 - u_1 - 1,\\
T_6 &= 2u_2y_6 + 2u_1y_5 - 2y_5 - u_2^2 - u_1^2 + 1.
\end{aligned}$$

Hence the hypotheses of the theorem are consistent. To see for which components the theorem is true, we compute prem$(G, \mathbb{T}_i)$ for $1 \leq i \leq 9$. From this, one may find that the theorem is true only for $\mathbb{T}_1$ and false for all the other components. Therefore, the theorem is conditionally true with the subsidiary condition given as

$$\bigwedge_{i=2}^{9}\Big(\bigvee_{T\in\mathbb{T}_i} T\neq 0\Big).$$

When the theorem is considered for $\mathbb{T}_1$, we have $T_1 = \cdots = T_6 = 0$ and $u_2 \neq 0$. Hence, the above subsidiary condition can be simplified to

$$T_3' \neq 0 \wedge T_5' \neq 0 \wedge u_2^2 + u_1^2 \neq 0 \wedge u_2^2 + (u_1 - 1)^2 \neq 0.$$

If the variables u_1 and u_2 are specified as parameters, then

$$u_2^2 + u_1^2 \neq 0 \wedge u_2^2 + (u_1 - 1)^2 \neq 0$$

is clearly a (minimal) nondegeneracy condition for the theorem, as it is composed of polynomial inequations in u_1 and u_2 only. Under this nondegeneracy condition the components $\mathbb{T}_5, \ldots, \mathbb{T}_9$ are all excluded. Therefore, the decomposition, if computed over $\mathbf{Q}(u_1, u_2)$, should become

$$\mathrm{Zero}(\mathbb{P}/u_2) = \mathrm{Zero}(\mathbb{P}) = \bigcup_{i=1}^{4} \mathrm{Zero}(\mathbb{T}_i).$$

This can be confirmed by computing the decomposition directly. From either of the two decompositions together with the pseudo-remainder verification, we can conclude that the theorem is not generically true.

The geometric meanings of the two inequations for the nondegeneracy condition are easy to explain: AC and BC are both nonisotropic. However, neither $T_3' = 0$ nor $T_5' = 0$ corresponds to a degenerate case of the theorem, so the subsidiary condition $T_3' \neq 0 \wedge T_5' \neq 0$ cannot be considered as a nondegeneracy condition. It turns out to be nontrivial to explain the geometric meaning of this condition merely from the two polynomials.

Note that T_3', T_5' are taken from the (nondegenerate) triangular sets as to exclude three components in the irreducible decomposition. Since for any given values of u_1 and u_2, the values of $y_1, \ldots, y_6$ for each component can be determined from the corresponding triangular set, the geometric meaning of each component can be observed by some geometric means such as drawing a figure. This would help us understand the ambiguity of drawing triangles on a segment. It is not difficult to figure out that $T_3' = 0$ if and only if one of $\triangle ABC'$ and $\triangle CAB'$ is drawn inward and the other outward, and $T_5' = 0$ if and only if one of $\triangle ABC'$ and $\triangle BCA'$ is drawn inward and the other outward. The theorem is true if and only if $\triangle ABC'$, $\triangle CAB'$ and $\triangle BCA'$ are drawn all inward or all outward.

Using ProverC, we compute an irreducible decomposition for $\mathrm{Zero}(\mathbb{P}/\mathbb{Q}^*)$ over $\mathbf{Q}$ and obtain eight triangular sets, which are $\mathbb{T}_2, \ldots, \mathbb{T}_9$ as given above. If the decomposition is computed over $\mathbf{Q}(u_1, u_2)$, one gets the three triangular sets $\mathbb{T}_2, \mathbb{T}_3, \mathbb{T}_4$. From either of the two decompositions, one can reach the same conclusion that the theorem is not generically true.

We have found a generalization of Steiner's theorem: the lines AA', BB' and CC' are still concurrent when ABC', BCA' and CAB' are isosceles triangles (not necessarily equilateral) whose altitudes are proportional to the corresponding bases $|AB|$, $|BC|$, and $|CA|$ by the same factor.

7.3 Automatic derivation of unknown relations

In the case of theorem proving, there is a known conclusion whose truth one wishes to confirm. Now consider another situation where we want to derive some possible conclusion or relation we do not know. We discuss two example applications of elimination methods to deal with the situation.

Geometric formulas

We want to derive automatically unknown algebraic relations among some geometric entities, where an adequate description of the geometric hypotheses among the geometric entities is given. The idea is first to algebraize the geometric hypotheses as a system of polynomial equations and inequations, then to compute a triangular set, triangular series or Gröbner basis of the corresponding polynomial set or system using an appropriate variable ordering, and finally to get the desired relations from the triangularized sets. A typical example is the automated derivation of the Qin–Heron formula (representing the area of a triangle in terms of its three sides).

The problem of deriving unknown algebraic relations and its solution may be formulated in the form of the following algorithm.

Algorithm DeriveA: HC, NO, or $R \leftarrow \text{DeriveA}(\mathbb{P}, \mathbb{Q})$. Given a set HYP of geometric hypotheses expressed as a system of polynomial equations and inequations

$$\mathbb{P} = \{P_1(\boldsymbol{u}, \boldsymbol{x}), \ldots, P_s(\boldsymbol{u}, \boldsymbol{x})\} = 0,$$
$$\mathbb{Q} = \{Q_1(\boldsymbol{u}, \boldsymbol{x}), \ldots, Q_t(\boldsymbol{u}, \boldsymbol{x})\} \neq 0$$

in two sets of geometric entities $\boldsymbol{u} = (u_1, \ldots, u_d)$ and $\boldsymbol{x} = (x_1, \ldots, x_n)$ with coefficients in $\boldsymbol{K}$ and given a fixed integer k, without loss of generality, say $k = 1$, this algorithm either reports HC(HYP) or determines whether there exists a polynomial relation $R(\boldsymbol{u}, x_1) = 0$ between $\boldsymbol{u}$ and x_1 such that $\text{Zero}(\mathbb{P}/\mathbb{Q}) \subset \text{Zero}(R)$, and if so, finds such a $R(\boldsymbol{u}, x_1)$; otherwise, the algorithm reports NO.

D1. Compute over $\boldsymbol{K}$ a (quasi-, weak-) medial set $\mathbb{T}$ of $\mathbb{P}$ by CharSetN or PriTriSys, or a Gröbner basis $\mathbb{T}$ of $\mathbb{P} \cup \{Q_1 z_1 - 1, \ldots, Q_t z_t - 1\}$ with respect to the purely lexicographical ordering under $u_1 \prec \cdots \prec u_d \prec x_1 \prec \cdots \prec x_n \prec z_1 \prec \cdots \prec z_t$, where $z_1, \ldots, z_t$ are new indeterminates. If $\mathbb{T} \cap \boldsymbol{K} \neq \emptyset$ or $0 \in \text{prem}(\mathbb{Q}, \mathbb{T})$, then return HC(HYP) and the algorithm terminates.

D2. Set $\mathbb{T}^{\langle 1 \rangle} \leftarrow \mathbb{T} \cap (\boldsymbol{K}[\boldsymbol{u}, x_1] \setminus \boldsymbol{K}[\boldsymbol{u}])$. If $\mathbb{T}$ is a Gröbner basis computed in D1, then go to D4. If there exists a polynomial $R(\boldsymbol{u}, x_1) \in \mathbb{T}^{\langle 1 \rangle}$ and $\mathbb{T} \cap \boldsymbol{K}[\boldsymbol{u}]$ is empty or irreducible as a triangular set, then return $R(\boldsymbol{u}, x_1)$ and the algorithm terminates.

D3. Compute an irreducible triangular series $\Psi = \{\mathbb{T}_1, \ldots, \mathbb{T}_e\}$ of $[\mathbb{P}, \mathbb{Q}]$ over $\boldsymbol{K}$. If $\Psi = \emptyset$, then return HC(HYP) and the algorithm terminates. Set

$$\mathbb{T}_i^{\langle 1 \rangle} \leftarrow \mathbb{T}_i \cap (\boldsymbol{K}[\boldsymbol{u}, x_1] \setminus \boldsymbol{K}[\boldsymbol{u}]), \quad 1 \leq i \leq e.$$

If for every $1 \leq i \leq e$ there exists a polynomial $R_i(\boldsymbol{u}, x_1) \in \mathbb{T}_i^{\langle 1 \rangle}$, then return

$$R(\boldsymbol{u}, x_1) \leftarrow \prod_{i=1}^{e} R_i(\boldsymbol{u}, x_1);$$

else return NO. The algorithm terminates.

D4. If $\mathbb{T}^{\langle 1 \rangle} \neq \emptyset$, then return the polynomial $R(\boldsymbol{u}, x_1) \in \mathbb{T}^{\langle 1 \rangle}$ that has minimal degree in x_1 else return NO.

Proof. The equality $R(\boldsymbol{u}, x_1) = 0$, if computed, is clearly a polynomial relation between $\boldsymbol{u}$ and x_1. Since $\mathbb{T}$ is a medial set computed by CharSetN or PriTriSys from $\mathbb{P}$ or a Gröbner basis of $\mathbb{P}^* = \mathbb{P} \cup \{Q_1 z_1 - 1, \ldots, Q_t z_t - 1\}$, $\mathbb{T} \subset \mathrm{Ideal}(\mathbb{P}^*)$. It follows that $\mathrm{Zero}(\mathbb{P}/\mathbb{Q}) \subset \mathrm{Zero}(R)$.

If there exists an i, $1 \leq i \leq e$, such that $\mathbb{T}_i^{\langle 1 \rangle} = \emptyset$, then x_1 is a parameter of $\mathbb{T}_i$. In this case, the scope of x_1 in $\mathrm{Zero}(\mathbb{P}/\mathbb{Q})$ for a fixed $\boldsymbol{u} = \bar{\boldsymbol{u}}$ covers any extension field of $\boldsymbol{K}$. Hence, there is no algebraic relation between $\boldsymbol{u}$ and x_1 in general. This is also true when $\mathbb{T}$ is a Gröbner basis of $\mathbb{P}^*$ and $\mathbb{T}^{\langle 1 \rangle} = \emptyset$. □

The following postprocess may be incorporated into the algorithm.

D∞. When NO is returned, analyze the computed irreducible triangular series or Gröbner basis and try to get possible relations by providing appropriate subsidiary conditions of the form $D_i \neq 0$ and adding the D_i to $\mathbb{Q}$ to exclude some components.

The triangular sets/series and the Gröbner bases may also be computed over $\mathbf{Q}(\boldsymbol{u})$ when the variables $\boldsymbol{u}$ are specified to be independent parameters. Then, any case in which $\boldsymbol{u}$ are constrained by a polynomial equation is considered as a degenerate case. The algorithm either detects the algebraic dependency of $\boldsymbol{u}$ or derives a relation that holds generically; it does not necessarily hold in the degenerate cases.

Example 7.3.1 (Qin–Heron formula; Wu 1986b, Chou and Gao 1990a, Wang 1995b). Determine the area Δ of an arbitrary triangle ABC in terms of its three sides a, b, c (Fig. 8).

Let the vertices of the triangle be located as $A(x_1, 0)$, $B(0, 0)$, $C(x_2, x_3)$. Then the geometric hypotheses may be expressed as the following polynomial

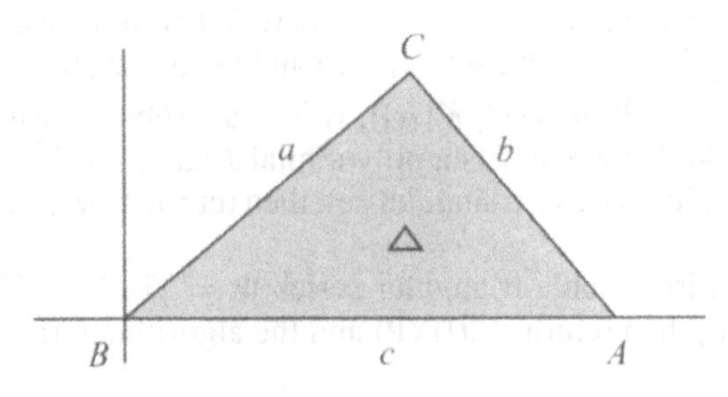

Fig. 8. Example 7.3.1

equations

$$\text{HYP:}\quad \begin{cases} H_1 = x_1^2 - c^2 = 0, & \leftarrow\!\!\!\leftarrow\ c = |AB| \\ H_2 = x_2^2 + x_3^2 - a^2 = 0, & \leftarrow\!\!\!\leftarrow\ a = |BC| \\ H_3 = (x_2 - x_1)^2 + x_3^2 - b^2 = 0, & \leftarrow\!\!\!\leftarrow\ b = |AC| \\ H_4 = x_3^2 x_1^2 - 4\Delta^2 = 0. & \leftarrow\!\!\!\leftarrow\ \Delta = \frac{1}{2}|AB| \cdot |x_3| \end{cases}$$

Let $\mathbb{P} = \{H_1, \ldots, H_4\}$ and the variables be ordered as $a \prec b \prec c \prec \Delta \prec x_1 \prec x_2 \prec x_3$. It is easy to compute a principal triangular system $[\mathbb{T}, \mathbb{U}]$ of $\mathbb{P}$:

$$\mathbb{T} = [R, H_1, T, H_2], \quad \mathbb{U} = \{x_1\},$$

where

$$R = 16\Delta^2 + c^4 - 2b^2c^2 - 2a^2c^2 + b^4 - 2a^2b^2 + a^4,$$
$$T = 2x_1x_2 - c^2 + b^2 - a^2.$$

Actually, $\mathbb{T}$ is a weak-characteristic set of $\mathbb{P}$. A Gröbner basis of $\mathbb{P}$ is

$$\mathbb{G} = [R, H_1, 2c^2x_2 - (c^2 - b^2 + a^2)x_1, T, H_2].$$

In either case, $R = 0$ gives the algebraic relation we wanted to derive. Let $p = (a + b + c)/2$; we have

$$\Delta^2 = p(p - a)(p - b)(p - c).$$

This is the well-known Qin–Heron formula (Wu 1986b).

Example 7.3.2 (Brahmagupta formula; Chou and Gao 1990a, Wang 1995b). Let $ABCD$ be a cyclic quadrilateral (Fig. 9). Determine the hatched area of the oriented quadrilateral $ABCD$ in terms of its four sides.

Let the coordinates of the points be chosen as

$$A(0, 0), \quad B(a, 0), \quad C(x_1, x_2), \quad D(x_3, x_4),$$

and

$$b = |BC|, \quad c = |CD|, \quad d = |DA|.$$

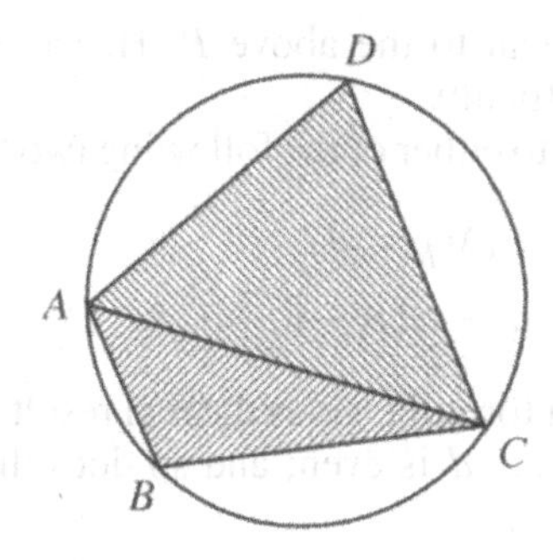

Fig. 9. Example 7.3.2

Denote the sum of the signed areas of $\triangle ABC$ and $\triangle ACD$ by Θ. Then the conditions relating these geometric entities can be expressed as

$$\begin{aligned}
H_1 &= x_2^2 + x_1^2 - 2ax_1 - b^2 + a^2 = 0,\\
H_2 &= x_4^2 - 2x_2x_4 + x_3^2 - 2x_1x_3 + x_2^2 + x_1^2 - c^2 = 0,\\
H_3 &= x_4^2 + x_3^2 - d^2 = 0,\\
H_4 &= ax_2x_4^2 - a(x_2^2 + x_1^2 - ax_1)x_4 + ax_2x_3^2 - a^2x_2x_3 = 0,\\
H_5 &= x_1x_4 - x_2x_3 + ax_2 - 2\Theta = 0.
\end{aligned}$$

We wish to find a relation among $a, \ldots, d$ and Θ. To this end, set $\mathbb{P} = \{H_1, \ldots, H_5\}$ and compute a quasi-N-characteristic set $\mathbb{C}$ of $\mathbb{P}$ with respect to the ordering $a \prec \cdots \prec d \prec \Theta \prec x_1 \prec \cdots \prec x_4$: $\mathbb{C}$ may be found to contain five polynomials with the following index triples

$$[46\ \Theta\ 4],\quad [35\ x_1\ 1],\quad [6\ x_2\ 1],\quad [10\ x_3\ 1],\quad [4\ x_4\ 1],$$

with three factors a, x_1 and $F = d^2+c^2-b^2-a^2$ removed during the computation. Thus, we have the following zero relation

$$\mathrm{Zero}(\mathbb{P}/ax_1F) \subset \mathrm{Zero}(\mathbb{C}).$$

It may be verified with ease that $ax_1F = 0$ corresponds to some degenerate cases of the geometric problem. The first polynomial R in $\mathbb{C}$ may be factorized as

$$R = (R_0 + 8abcd)(R_0 - 8abcd),$$

where

$$R_0 = 16\Theta^2 + d^4 - 2(c^2 + b^2 + a^2)d^2 + c^4 - 2(b^2 + a^2)c^2 + (b^2 - a^2)^2.$$

Therefore, we get the algebraic relation $R = 0$ under some nondegeneracy conditions. In fact, by computing a characteristic series we have verified that $R = 0$ holds in all the degenerate cases; namely, the relation follows from the geometric hypotheses universally.

A Gröbner basis of $\mathbb{P}$ under $\Theta \prec x_1 \prec \cdots \prec x_4$ may be found to consist of five polynomials with index triples

$$[46\ \Theta\ 4],\quad [26\ x_1\ 1],\quad [13\ x_2\ 1],\quad [26\ x_3\ 1],\quad [13\ x_4\ 1].$$

The first polynomial in the basis is identical to the above R. Hence, the same relation $R = 0$ is derived without much difficulty.

Set $p = (a+b+c+d)/2$; $R = 0$ leads to either of the following two equalities

$$\begin{aligned}
\Theta^2 &= (p-a)(p-b)(p-c)(p-d),\\
\Theta^2 &= p(p-a-b)(p-a-c)(p-a-d).
\end{aligned}$$

The first, which is the known Brahmagupta formula, gives the real result when the number t of positive variables among $a, \ldots, d$ is even; and so does the second when t is odd (see Chou and Gao 1990a).

Locus equations

The method of formula derivation may be generalized to derive the locus equations of a motion whose geometric description is given. The difference is that now one needs to determine one or several sets of algebraic relations between n variables $\boldsymbol{x} = (x_1, \ldots, x_n)$ and $\boldsymbol{u}$, and projection is required.

By locus equations we mean a system or the disjunction of several systems of polynomial equations and inequations in $\boldsymbol{x}$ with $\boldsymbol{u}$ as parameters such that not only the system is a formal consequence of the geometric hypotheses, but also for any point on the locus there is at least one configuration which satisfies the geometric hypotheses.

Before stating the problem and its solution in the form of an algorithm, let us make the following convention. For any set union $S = \bigcup_{A \in \Delta} S_A$, by *removing redundant sets* from S we mean determining a subset Δ' of Δ such that $\bigcup_{A \in \Delta \setminus \Delta'} S_A = S$. By *simplifying* S we mean finding another set Ω such that $\bigcup_{A \in \Omega} S_A = S$ and $\bigcup_{A \in \Omega} S_A$ as a representation of S is *simpler* than $\bigcup_{A \in \Delta} S_A$. We have indicated in Sect. 6.2 some possibilities of removing redundant zero sets. Other techniques have been given in some implementation-related articles, for example, Chou and Gao (1990b) and Wang (1995a). A satisfactory discussion on how to simplify the union of zero sets is much beyond the scope of this section. See Examples 7.4.1 and 7.4.2 for two concrete instances of such simplification.

Algorithm DeriveB: $\Psi \leftarrow \mathrm{DeriveB}(\mathbb{P}, \mathbb{Q})$. Given a set HYP of geometric constraints expressed as a system of polynomial equations and inequations

$$\mathbb{P} = \{P_1(\boldsymbol{u}, \boldsymbol{x}, \boldsymbol{y}), \ldots, P_s(\boldsymbol{u}, \boldsymbol{x}, \boldsymbol{y})\} = 0,$$
$$\mathbb{Q} = \{Q_1(\boldsymbol{u}, \boldsymbol{x}, \boldsymbol{y}), \ldots, Q_t(\boldsymbol{u}, \boldsymbol{x}, \boldsymbol{y})\} \neq 0$$

in $\boldsymbol{u}, \boldsymbol{x}$ and $\boldsymbol{y}$ for a point $\boldsymbol{x} = (x_1, \ldots, x_n)$ to move in an n-dimensional affine space $\mathbf{A}^n_{\boldsymbol{K}}$, where $\boldsymbol{u} = (u_1, \ldots, u_d)$ is a set of (geometric) parameters and $\boldsymbol{y} = (y_1, \ldots, y_m)$ a set of other geometric entities, this algorithm computes a finite set Ψ of polynomial systems $[\mathbb{P}_1, \mathbb{Q}_1], \ldots, [\mathbb{P}_e, \mathbb{Q}_e]$ in $\boldsymbol{K}(\boldsymbol{u})[\boldsymbol{x}]$ such that

a. for any $(\bar{\boldsymbol{x}}, \bar{\boldsymbol{y}}) \in \mathrm{Zero}(\mathbb{P}/\mathbb{Q})$, there exists an i, $1 \leq i \leq e$, such that $\bar{\boldsymbol{x}} \in \mathrm{Zero}(\mathbb{P}_i/\mathbb{Q}_i)$;
b. for any $1 \leq i \leq e$ and any $\bar{\boldsymbol{x}} \in \mathrm{Zero}(\mathbb{P}_i/\mathbb{Q}_i)$, there exists a $\bar{\boldsymbol{y}} \in \tilde{\boldsymbol{K}}^m$ such that $(\bar{\boldsymbol{x}}, \bar{\boldsymbol{y}}) \in \mathrm{Zero}(\mathbb{P}/\mathbb{Q})$.

The disjunction $\bigvee_{i=1}^{e} (\mathbb{P}_i = 0 \wedge \mathbb{Q}_i \neq 0)$ is called the *locus equations* of point $\boldsymbol{x}$ (in terms of $\boldsymbol{u}$).

D1. Compute a characteristic, triangular, or Gröbner series Ψ of $[\mathbb{P}, \mathbb{Q}]$ with projection for $\boldsymbol{y}$ with respect to the variable ordering

$$x_1 \prec \cdots \prec x_n \prec y_1 \prec \cdots \prec y_m.$$

If $\Psi = \emptyset$, i.e., $\mathrm{Zero}(\mathbb{P}/\mathbb{Q}) = \emptyset$, then either the geometric conditions are self-contradictory or the motion is free (i.e., for any $\bar{\boldsymbol{x}}$ there is a $\bar{\boldsymbol{y}}$ such that $(\bar{\boldsymbol{x}}, \bar{\boldsymbol{y}}) \in \mathrm{Zero}(\mathbb{P}/\mathbb{Q})$, so the locus fills up the whole space); thus the procedure terminates.

D2. Remove redundant sets from

$$\bigcup_{[\mathbb{T}, \mathbb{U}] \in \Psi} \mathrm{Zero}(\mathbb{T} \cap \boldsymbol{K}(\boldsymbol{u})[\boldsymbol{x}] / \mathbb{U} \cap \boldsymbol{K}(\boldsymbol{u})[\boldsymbol{x}]),$$

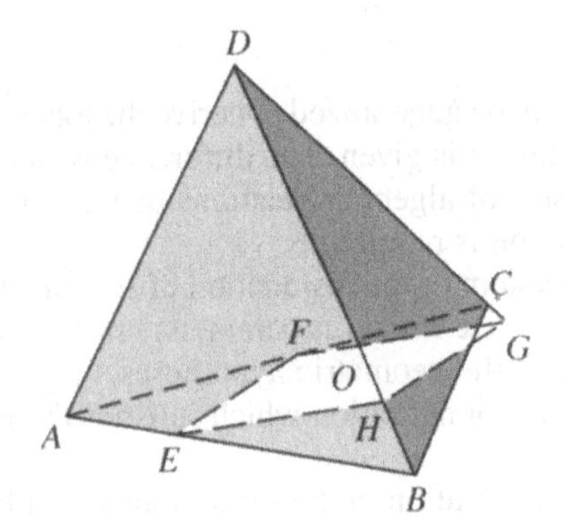

Fig. 10. Example 7.3.3

simplify it and let the obtained zero set be $\bigcup_{i=1}^{e} \mathrm{Zero}(\mathbb{P}_i/\mathbb{Q}_i)$. Return

$$\Psi \leftarrow \{[\mathbb{P}_1, \mathbb{Q}_1], \ldots, [\mathbb{P}_e, \mathbb{Q}_e]\}.$$

Proof. It follows from the definition of characteristic, triangular or Gröbner series and the projection property of $[\mathbb{T}, \mathbb{U}] \in \Psi$. □

In D1 the series Ψ may also be computed in $\boldsymbol{K}[\boldsymbol{u}, \boldsymbol{x}, \boldsymbol{y}]$. Actually, one needs to perform the elimination only for $\boldsymbol{y}$ because it is sufficient when one has already obtained the equations and inequations in $\boldsymbol{u}$ and $\boldsymbol{x}$ – they do not have to be in triangular form. We do not enter into the technical details of projection for Gröbner bases (see Example 7.4.1).

Example 7.3.3. Let a plane intersect the four edges AB, AC, DC, and DB of a tetrahedron $ABCD$ at points E, F, G, and H, respectively, such that $EFGH$ is a parallelogram. Determine the locus equations of the center O of $\square EFGH$.

Let the points be located as

$$A(0,0,0),\quad B(u_1,0,0),\quad C(u_2,u_3,0),\quad D(u_4,u_5,u_6),\quad E(y_1,0,0),$$
$$F(y_2,y_3,0),\quad G(y_4,y_5,y_6),\quad H(y_7,y_8,y_9),\quad O(X,Y,Z).$$

We have the following relations

$$H_1 = u_2y_3 - u_3y_2 = 0, \qquad \twoheadleftarrow F \text{ lies on } AC$$

$$\left.\begin{aligned} H_2 &= u_4y_6 - u_2y_6 - u_6y_4 + u_2u_6 = 0,\\ H_3 &= u_4y_5 - u_2y_5 - u_5y_4 + u_3y_4 + u_2u_5\\ &\quad - u_3u_4 = 0, \end{aligned}\right\} \twoheadleftarrow G \text{ lies on } CD$$

$$\left.\begin{aligned} H_4 &= u_4y_8 - u_1y_8 - u_5y_7 + u_1u_5 = 0,\\ H_5 &= u_4y_9 - u_1y_9 - u_6y_7 + u_1u_6 = 0, \end{aligned}\right\} \twoheadleftarrow H \text{ lies on } BD$$

$$\left.\begin{aligned} H_6 &= y_7 - y_4 + y_2 - y_1 = 0,\\ H_7 &= y_8 - y_5 + y_3 = 0,\\ H_8 &= y_9 - y_6 = 0, \end{aligned}\right\} \twoheadleftarrow \overrightarrow{FE} = \overrightarrow{GH}$$

$$\left.\begin{aligned} H_9 &= 2X - y_4 - y_1 = 0,\\ H_{10} &= 2Y - y_5 = 0,\\ H_{11} &= 2Z - y_6 = 0. \end{aligned}\right\} \twoheadleftarrow O \text{ is the center of } \square EFGH$$

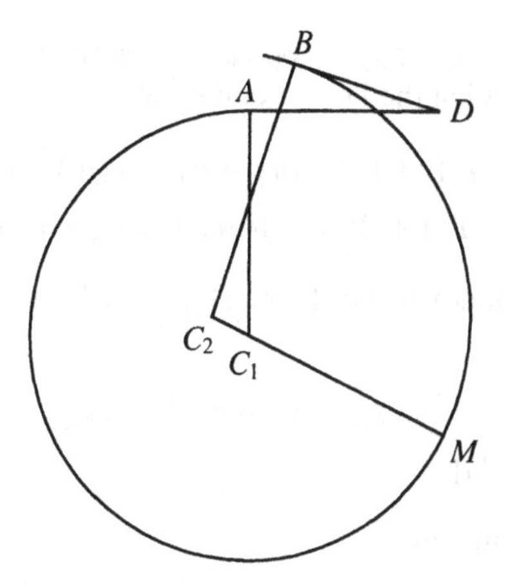

Fig. 11. Example 7.3.4

Let $\mathbb{P} = \{H_1, \ldots, H_{11}\}$ and the variables be ordered as

$$X \prec Y \prec Z \prec y_1 \prec \cdots \prec y_9.$$

Either of the characteristic, triangular, and Gröbner series of $\mathbb{P}$ contains only one element (triangular system, ascending set, or Gröbner basis). Projection onto X, Y, Z yields the first (and the same) two polynomials of the corresponding set:

$$\begin{aligned}
P_1 &= 2(u_3 - u_5)X - 2(u_1 + u_2 - u_4)Y + (u_1 + u_2)u_5 - u_3u_4,\\
P_2 &= 2u_6X + 2(u_1 + u_2 - u_4)Z - (u_1 + u_2)u_6.
\end{aligned}$$

This is because all the initials are in the parameters u_i. Hence the locus equations are $P_1 = 0 \wedge P_2 = 0$, which represents the intersection line of the two planes defined by $P_1 = 0$ and $P_2 = 0$ respectively.

Example 7.3.4 (Biarcs; Wang 1995b). Given two points A and B of two different circular arcs which have given tangent directions at A and B, determine the locus of an intermediate point M at which the two circular arcs join together with a common tangent.

This example originates from a book by A. W. Nutbourne and R. R. Martin (see Wang 1995b), in which one branch of the locus is proved to be a circle by technical derivations. Here, we show how to derive the locus automatically by elimination methods. Let us choose the point coordinates as

$$A(0,0),\quad D(u_1,0),\quad B(u_2,u_3),\quad M(X,Y),\quad C_1(0,x_1),\quad C_2(x_2,x_3).$$

From the geometric conditions we get the following relations

$$\text{HYP:}\quad \begin{cases}
H_1 = (u_2 - u_1)(x_2 - u_2) & \\
\qquad + u_3(x_3 - u_3) = 0, & \twoheadleftarrow\ BC_2 \perp BD\\
H_2 = X^2 + (x_1 - Y)^2 - x_1^2 = 0, & \twoheadleftarrow\ |C_1A| = |C_1M|\\
H_3 = (x_2 - u_2)^2 + (x_3 - u_3)^2 & \\
\qquad - (x_2 - X)^2 - (x_3 - Y)^2 = 0, & \twoheadleftarrow\ |C_2B| = |C_2M|\\
H_4 = X(x_3 - x_1) + x_2(x_1 - Y) = 0. & \twoheadleftarrow\ M \text{ lies on } C_1C_2
\end{cases}$$

Let $\mathbb{P} = \{H_1, \ldots, H_4\}$ and $X \prec Y \prec x_1 \prec x_2 \prec x_3$. A characteristic series of $\mathbb{P}$ consists of three ascending sets, of which the largest comprises

$$\begin{aligned} R = {} & u_3(X^2+Y^2)^2 - 2u_1u_3X(X^2+Y^2) + 2(u_1u_2 - u_2^2 - u_3^2)(X^2+Y^2)Y \\ & + (2u_1u_2 - u_2^2 - u_3^2)u_3(X^2 - Y^2) + 2(u_2^3 - u_1u_2^2 + u_2u_3^2 + u_1u_3^2)XY, \end{aligned}$$

and three other polynomials having index triples $[3\ x_1\ 1]$, $[12\ x_2\ 1]$, and $[6\ x_3\ 1]$. The two simpler ascending sets are

$$\begin{aligned} & [X - u_2, Y - u_3, 2u_3x_1 - u_3^2 - u_2^2, -x_2 + u_2, x_3 - u_3], \\ & [X, Y, x_1, [4\ x_2\ 1], [5\ x_3\ 1]]. \end{aligned}$$

Projection of the three onto X, Y results in

$$\{R\}, \ \{X - u_2, Y - u_3\}, \ \{X, Y\}.$$

The last two polynomial sets correspond respectively to the points B and A which are actually on the curve $R = 0$, so they are redundant. Therefore, $R = 0$ is the locus equation of point M that we wanted to derive.

A triangular series of $\mathbb{P}$ computed with projection for x_3, x_2, x_1 is similar to the characteristic series above. A Gröbner basis of $\mathbb{P}$ consists of R and six other polynomials with index triples

$$[20\ x_1\ 1], \ [3\ x_1\ 1], \ [39\ x_2\ 1], \ [12\ x_2\ 1], \ [22\ x_2\ 1], \ [6\ x_3\ 1].$$

By computing further Gröbner bases and projection, the same locus equation $R = 0$ can be derived as well.

Using an extension of algorithm FactorA (Sect. 7.5 and Wang 1987), one can factorize R into the following two polynomials

$$\begin{aligned} R_1 &= \left(X - \frac{u_1 - \alpha}{2}\right)^2 + \left(Y - \frac{\beta + u_2\alpha}{2u_3}\right)^2 - \frac{\alpha(u_1u_2 + \beta)}{2(u_1 - u_2 + \alpha)}, \\ R_2 &= \left(X - \frac{u_1 + \alpha}{2}\right)^2 + \left(Y - \frac{\beta - u_2\alpha}{2u_3}\right)^2 - \frac{\alpha(u_1u_2 + \beta)}{2(u_2 - u_1 + \alpha)}, \end{aligned}$$

where

$$\begin{aligned} \alpha &= \sqrt{u_3^2 + (u_1 - u_2)^2} = |BD|, \\ \beta &= u_1u_2 - u_1^2 + \alpha^2. \end{aligned}$$

Hence the locus of M has two components for any fixed u_1, u_2, u_3. $R_1 = 0$ and $R_2 = 0$ represent two circles $\odot I_1$ and $\odot I_2$ passing through A and B, whose centers I_1, I_2 and radii are readily determined. We thought that one of the circles corresponds to the biarc of convex shape and the other to the biarc of S-shape; this is not true. The situation seems to be more complicated. We have observed how the two circles $\odot C_1$ and $\odot C_2$ centered at C_1 and C_2 contact at M along the locus circles $\odot I_1$ and $\odot I_2$ with numerical simulation for a particular case

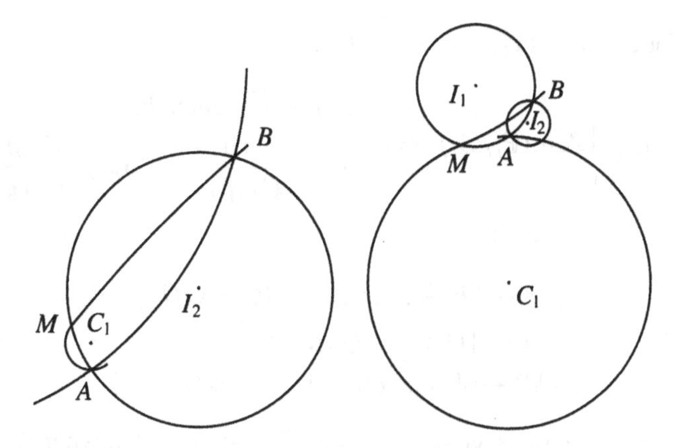

Fig. 12. Example 7.3.4: numerical simulation

$u_1 = -40, u_2 = 55, u_3 = 80$ (see Fig. 12). The circle $\odot I_1$ is divided by the two lines AD and BD into four arcs, and so is $\odot I_2$. $\odot C_1$ and $\odot C_2$ are tangent *externally* when M moves along two opposite arcs on $\odot I_1$ or $\odot I_2$, and *internally* otherwise. In the latter case, $\odot C_1$ is inside $\odot C_2$ when M moves along one of the two arcs, and so is $\odot C_2$ inside $\odot C_1$ when M moves along the other. It remains to be an interesting geometric question to show whether this is always true.

The method works also for establishing locus equations for the space biarcs. We omit the details.

7.4 Other geometric applications

Implicitization of parametric objects

Geometric objects like curves and surfaces may be represented algebraically by implicit equations or parametric equations. The advantage of each representation depends upon the type of problems to be solved. In geometric modeling, one often needs to convert one representation into the other. The rational parametrization of a geometric object in an n-dimensional affine space may be represented as

$$x_1 = \frac{P_1(\boldsymbol{y})}{Q_1(\boldsymbol{y})}, \ldots, x_n = \frac{P_n(\boldsymbol{y})}{Q_n(\boldsymbol{y})},$$

where $\boldsymbol{y} = (y_1, \ldots, y_m)$ are parametric variables. The problem of implicitization amounts to find the implicit equations in $\boldsymbol{x}$ which define the same geometric object as the parametrized representation does. This can be done by using the following algorithm. The incorporation of projection into implicitization algorithms was suggested first by Li (1989b).

Algorithm Impli: $\Psi \leftarrow \text{Impli}(\mathbb{P}, \mathbb{Q})$. Given two sets of polynomials $P_1, \ldots, P_n$ and $Q_1, \ldots, Q_n$ in $\boldsymbol{K}[\boldsymbol{y}]$, where $Q_1 \cdots Q_n \neq 0$ and $m \leq n$, this algorithm computes a finite set Ψ of polynomial systems $[\mathbb{P}_1, \mathbb{Q}_1], \ldots, [\mathbb{P}_e, \mathbb{Q}_e]$ in $\boldsymbol{K}[\boldsymbol{x}]$

such that for any $\bar{\boldsymbol{x}} = (\bar{x}_1, \ldots, \bar{x}_n) \in \tilde{\boldsymbol{K}}^n$,

$$\bar{\boldsymbol{x}} \in \bigcup_{i=1}^{e} \mathrm{Zero}(\mathbb{P}_i/\mathbb{Q}_i) \iff \begin{array}{l} \exists\, \bar{\boldsymbol{y}} \in \tilde{\boldsymbol{K}}^m \text{ such that} \\ \bar{x}_1 = \dfrac{P_1(\bar{\boldsymbol{y}})}{Q_1(\bar{\boldsymbol{y}})}, \ldots, \bar{x}_n = \dfrac{P_n(\bar{\boldsymbol{y}})}{Q_n(\bar{\boldsymbol{y}})}. \end{array}$$

I1. Let

$$\begin{aligned} \mathbb{P} &\leftarrow \{P_1 - x_1 Q_1, \ldots, P_n - x_n Q_n\}, \\ \mathbb{Q} &\leftarrow \{Q_1, \ldots, Q_n\}, \\ \mathbb{P}^* &\leftarrow \mathbb{P} \cup \{z_1 Q_1 - 1, \ldots, z_n Q_n - 1\} \end{aligned}$$

and $x_1 \prec \cdots \prec x_n \prec y_1 \prec \cdots \prec y_m \prec z_1 \prec \cdots \prec z_n$, where the z_j are new indeterminates. Compute a triangular series Ψ of $[\mathbb{P}, \mathbb{Q}]$, or a Gröbner series Ψ of $\mathbb{P}^*$ under the purely lexicographical term ordering, with projection for $z_1, \ldots, z_n$ and $\boldsymbol{y}$.

I2. Remove redundant sets from $\bigcup_{[\mathbb{T},\mathbb{U}]\in\Psi} \mathrm{Zero}(\mathbb{T} \cap \boldsymbol{K}[\boldsymbol{x}]/\mathbb{U} \cap \boldsymbol{K}[\boldsymbol{x}])$, simplify it, and let the obtained zero set be $\bigcup_{i=1}^{e} \mathrm{Zero}(\mathbb{P}_i/\mathbb{Q}_i)$. Then return

$$\Psi \leftarrow \{[\mathbb{P}_1, \mathbb{Q}_1], \ldots, [\mathbb{P}_e, \mathbb{Q}_e]\}.$$

Proof. By the definition of triangular and Gröbner series and the projection property of $[\mathbb{T}, \mathbb{U}] \in \Psi$. □

Example 7.4.1 (Buchberger 1987, Wu 1990, Wang 1995b). Consider the parametric surface in 3-dimensional affine space defined by the following equations

$$x = rt, \quad y = rt^2, \quad z = r^2.$$

Let $\mathbb{P} = \{x - rt, y - rt^2, z - r^2\}$. A Gröbner basis $\mathbb{G}$ of $\mathbb{P}$ with respect to $z \prec y \prec x \prec t \prec r$ can be easily computed:

$$\mathbb{G} = [x^4 - zy^2, zyt - x^3, xt - y, zt^2 - x^2, yr - x^2, xr - zt, tr - x, r^2 - z].$$

The equation $x^4 - zy^2 = 0$ resulting from $\mathbb{G}$ appears to be the implicit equation of the surface, but it does not strictly meet the specification of the implicitization problem as remarked by Buchberger (1987). For the y-axis is a solution to this implicit equation, whereas it does not appear in the surface defined by the parametric representation.

To get the exact implicit equations by projection, we adjoin x – the initial of the third and the sixth polynomial in $\mathbb{G}$ which have lowest degree 1 in their leading variables – to $\mathbb{P}$, compute the Gröbner basis of the obtained polynomial set, and proceed further. Finally, one may get two additional Gröbner bases

$$\mathbb{G}_1 = [y, x, t, r^2 - z], \quad \mathbb{G}_2 = [z, y, x, r],$$

such that

$$\mathrm{Zero}(\mathbb{P}) = \mathrm{Zero}(\mathbb{G}/x) \cup \mathrm{Zero}(\mathbb{G}_1) \cup \mathrm{Zero}(\mathbb{G}_2).$$

Thus

$$\begin{aligned}
&\mathrm{Proj}_{z,y,x}\mathrm{Zero}(\mathbb{P})\\
&\quad= \mathrm{Proj}_{z,y,x}\mathrm{Zero}(\mathbb{G}/x) \cup \mathrm{Proj}_{z,y,x}\mathrm{Zero}(\mathbb{G}_1) \cup \mathrm{Proj}_{z,y,x}\mathrm{Zero}(\mathbb{G}_2)\\
&\quad= \mathrm{Zero}(y^2z - x^4/xyz) \cup \mathrm{Zero}(\{x, y\}) \cup \mathrm{Zero}(\{x, y, z\})\\
&\quad= \mathrm{Zero}(y^2z - x^4/xy) \cup \mathrm{Zero}(\{x, y\}).
\end{aligned}$$

This implies that the implicit equations are

$$(y^2z - x^4 = 0 \wedge xy \neq 0) \vee (x = 0 \wedge y = 0).$$

Now compute a characteristic series of $\mathbb{P}$ with respect to the same variable ordering: it consists of three ascending sets

$$\mathbb{C}_1 = [x^4 - zy^2, xt - y, yr - x^2], \quad \mathbb{C}_2 = \mathbb{G}_1, \quad \mathbb{C}_3 = \mathbb{G}_2.$$

Projecting the corresponding zero sets, one obtains the same implicit equations for the surface.

Example 7.4.2. Find the implicit form (in the variables x and y) of the curve given by the following set of equations

$$\begin{aligned}
&(x - u)^2 + (y - v)^2 - 1 = 0,\\
&v^2 - u^3 = 0,\\
&2v(x - u) + 3u^2(y - v) = 0,\\
&(3wu^2 - 1)(2wv - 1) = 0.
\end{aligned}$$

This is a formulation of an offset to the curve $y^2 - x^3 = 0$. It has appeared in Example 3.2.2, where a triangular series with projection for w, v, u under the variable ordering $x \prec y \prec u \prec v \prec w$ has been computed. Also listed there are the five triangular systems $[\mathbb{T}_i, \mathbb{U}_i]$ contained in the series. Thus, the implicit equations may be given as

$$\bigvee_{i=1}^{5} (\mathbb{T}_i^{(2)} = 0 \wedge \mathbb{U}_i^{(2)} \neq 0), \tag{7.4.1}$$

where $\mathbb{T}_i^{(2)} = \mathbb{T}_i \cap \mathbf{Q}[x, y]$ and $\mathbb{U}_i^{(2)} = \mathbb{U}_i \cap \mathbf{Q}[x, y]$ for each i. However, the equations (7.4.1) are rather tedious. We show how they can be simplified considerably. First of all, computing a regular series of $[\mathbb{T}_i^{(2)}, \mathbb{U}_i^{(2)}]$ one finds that all the polynomials in $\mathbb{U}_i^{(2)}$ can be eliminated for $i = 2, \ldots, 5$. In other words,

$$\mathrm{Zero}(\mathbb{T}_i^{(2)}/\mathbb{U}_i^{(2)}) = \mathrm{Zero}(\mathbb{T}_i^{(2)}), \quad 2 \leq i \leq 5.$$

A regular series of $[\mathbb{T}_1^{(2)}, \mathbb{U}_1^{(2)}]$ comprises three regular systems $[\mathbb{T}_{1j}, \mathbb{U}_{1j}]$ with

$\mathbb{T}_{11} = [T_{11}]$ and

$$\begin{aligned}
\mathbb{T}_{12} &= [T_{41}, \mathrm{coef}(T_{11}, y^6)y^4 + \mathrm{coef}(T_{11}, y^4)y^2 + \mathrm{coef}(T_{11}, y^2)],\\
\mathbb{T}_{13} &= [T_{31}, 729(18x - 1)y^2 - 39366x^4 - 26244x^3\\
&\quad - 60993x^2 - 32868x - 13381],\\
\mathbb{U}_{11} &= \{x, T_{21}, T_{31}, T_{41}\}, \quad \mathbb{U}_{12} = \mathbb{U}_{13} = \emptyset.
\end{aligned}$$

See Example 3.2.2 for the polynomials T_{11}, T_{21}, etc. It is easy to verify that

$$\begin{aligned}
\mathcal{Z}_1 &= \mathrm{Zero}(\{T_{21}, T_{11}\}/x) = \mathrm{Zero}(\mathbb{T}_2^{(2)}),\\
\mathcal{Z}_2 &= \mathrm{Zero}(\{T_{31}, T_{11}\}/xT_{21}) = \mathrm{Zero}(\mathbb{T}_3^{(2)}) \cup \mathrm{Zero}(\mathbb{T}_{13}),\\
\mathcal{Z}_3 &= \mathrm{Zero}(\{T_{41}, T_{11}\}/xT_{21}T_{31}) = \mathrm{Zero}(\mathbb{T}_4^{(2)}) \cup \mathrm{Zero}(\mathbb{T}_{12}).
\end{aligned}$$

It follows that

$$\begin{aligned}
\mathrm{Zero}(T_{11}/x) &= \mathcal{Z}_1 \cup \mathcal{Z}_2 \cup \mathcal{Z}_3 \cup \mathrm{Zero}(T_{11}/\mathbb{U}_{11})\\
&= \bigcup_{i=1}^{4} \mathrm{Zero}(\mathbb{T}_i^{(2)}/\mathbb{U}_i^{(2)}).
\end{aligned}$$

Therefore,

$$\bigcup_{i=1}^{5} \mathrm{Zero}(\mathbb{T}_i^{(2)}/\mathbb{U}_i^{(2)}) = \mathrm{Zero}(T_{11}/x) \cup \mathrm{Zero}(\mathbb{T}_5^{(2)})$$

and thus the implicit equations (7.4.1) are simplified (with $E = T_{11}$) to:

$$\begin{aligned}
E &= 729x^8 + 216x^7 + 729x^6y^2 - 2900x^6 - 1458x^5y^2 - 2376x^5\\
&\quad - 2619x^4y^2 + 3870x^4 - 1458x^3y^4 - 4892x^3y^2 + 4072x^3\\
&\quad + 729x^2y^4 - 297x^2y^2 - 1188x^2 - 4158xy^4 + 5814xy^2\\
&\quad - 1656x + 427y^2 - 1685y^4 + 729y^6 + 529 = 0,\\
x &\neq 0
\end{aligned} \tag{7.4.2}$$

or

$$x = 0, \quad 729y^4 - 956y^2 - 529 = 0. \tag{7.4.3}$$

These equations may also be derived by computing a characteristic series with projection. A characteristic set of $\mathbb{P}$ is easy to compute, but the computation of characteristic series may take much time.

One can examine that the first equation $E = 0$ in (7.4.2) becomes

$$(y^2 - 1)(729y^4 - 956y^2 - 529) = 0$$

when $x = 0$. However, $(0, 1)$ and $(0, -1)$ which are solutions of $E = 0$ do not lie on the parametric curve (i.e., there are no corresponding u, v, and w such that the parametric equations are satisfied). This is why one needs (7.4.3) instead of (7.4.2) in the case of $x = 0$. In summary, we have:

- any point (x, y) on the curve defined by the parametric equations is a point on the curve defined by the implicit equation $E = 0$;
- any point (x, y) other than $(0, 1)$ and $(0, -1)$ on the curve defined by the implicit equation $E = 0$ is a point on the curve defined by the parametric equations.

Related to the implicitization of parametric objects, there are several other problems such as the independency of parameters, the propriety of parametrization, and the inversion problem. They can also be treated by elimination methods.

Existence conditions and detection of singularities

The study of singularities is not only a classical topic in algebraic geometry but also of importance for modern geometric applications. For example, while tracing an algebraic curve, one first has to detect all the singular points at which numeric methods do not work well. While studying the kinematic behavior of a robot motion, one has to determine the singular configurations as in a singular situation the robot arm has difficulties to move. We explain how to establish the sufficient and necessary conditions for parametric algebraic hypersurfaces to have singularities of an arbitrary multiplicity and to depict the structure of the singular varieties by computing their irreducible decomposition, or all the singular points when they are finite.

An algebraic hypersurface $\mathfrak{H}$ in an n-dimensional projective space $\mathbf{P}^n$ or affine space $\mathbf{A}^n$ is an algebraic variety of dimension $n-1$ given by a single homogeneous polynomial equation $F(x_0, \boldsymbol{x}) = 0$ or "ordinary" polynomial equation $F(\boldsymbol{x}) = 0$. It is called an *algebraic curve* or an *algebraic surface* respectively for $n = 2, 3$. A point $(\bar{x}_0, \bar{\boldsymbol{x}})$ of $\mathfrak{H}$ in $\mathbf{P}^n$ is said to be of *multiplicity* p if all the partial derivatives of order $< p$ of F vanish at $(\bar{x}_0, \bar{\boldsymbol{x}})$, but some of order p do not, i.e.,

$$\frac{\partial^r F}{\partial x_0^{r_0} \partial x_1^{r_1} \ldots \partial x_n^{r_n}}(\bar{x}_0, \bar{\boldsymbol{x}}) = 0 \quad \text{for all } r_0 + r_1 + \cdots + r_n = r < p,$$

$$\frac{\partial^r F}{\partial x_0^{r_0} \partial x_1^{r_1} \ldots \partial x_n^{r_n}}(\bar{x}_0, \bar{\boldsymbol{x}}) \neq 0 \quad \text{for some } r_0 + r_1 + \cdots + r_n = r = p.$$

A point $\bar{\boldsymbol{x}}$ of $\mathfrak{H}$ in $\mathbf{A}^n$ is said to be of *multiplicity* p if

$$\frac{\partial^r F}{\partial x_1^{r_1} \ldots \partial x_n^{r_n}}(\bar{\boldsymbol{x}}) = 0 \quad \text{for all } r_1 + \cdots + r_n = r < p,$$

$$\frac{\partial^r F}{\partial x_1^{r_1} \ldots \partial x_n^{r_n}}(\bar{\boldsymbol{x}}) \neq 0 \quad \text{for some } r_1 + \cdots + r_n = r = p.$$

Any point of multiplicity $p \geq 2$ is called a *singular point* of $\mathfrak{H}$.

Algorithm SinConP: $\Psi \leftarrow \mathrm{SinConP}(F, p)$. Given the homogeneous polynomial equation $F(x_0, \boldsymbol{x}) = 0$ in $\boldsymbol{K}[\boldsymbol{t}, x_0, \boldsymbol{x}]$ of an algebraic hypersurface $\mathfrak{H}$ in $\mathbf{P}^n$

with $\boldsymbol{t} = (t_1, \ldots, t_m)$ as parameters and an integer $p \geq 1$, this algorithm computes a set Ψ of $n+1$ polynomial sets $\mathbb{P}_0, \ldots, \mathbb{P}_n \subset \boldsymbol{K}[\boldsymbol{t}]$ such that $\mathfrak{H}$ has singularities of multiplicity $\geq p+1$ for $\boldsymbol{t} = \bar{\boldsymbol{t}} \in \tilde{\boldsymbol{K}}^m$ if and only if

$$\bar{\boldsymbol{t}} \in \bigcup_{i=0}^{n} \operatorname{Zero}(\mathbb{P}_i).$$

S1. Set

$$\mathbb{D} \leftarrow \left\{ \frac{\partial^p F}{\partial x_0^{r_0} \partial x_1^{r_1} \ldots \partial x_n^{r_n}} : r_0 + r_1 + \cdots + r_n = p \right\}.$$

Compute a Gröbner basis $\mathbb{G}_i$ of $\mathbb{D}|_{x_i=1}$ with respect to the purely lexicographical ordering determined by $t_1 \prec \cdots \prec t_m \prec x_0 \prec \cdots \prec x_n$ for $0 \leq i \leq n$.

S2. Let $\mathbb{P}_i \leftarrow \mathbb{G}_i \cap \boldsymbol{K}[\boldsymbol{t}]$ for $0 \leq i \leq n$ and $\Psi \leftarrow \{\mathbb{P}_0, \ldots, \mathbb{P}_n\}$.

Proof. Suppose that $\mathfrak{H}$ has a singular point $\bar{\boldsymbol{x}}$ of multiplicity $\geq p+1$ for some $\boldsymbol{t} = \bar{\boldsymbol{t}}$; then $(\bar{\boldsymbol{t}}, \bar{\boldsymbol{x}}) \in \operatorname{Zero}(\mathbb{D})$. The trivial zero $\mathbf{0}$ is not counted, so there exists an $i, 0 \leq i \leq n$, such that $\bar{x}_i \neq 0$. It follows that

$$\left(\bar{\boldsymbol{t}}, \frac{\bar{x}_0}{\bar{x}_i}, \ldots, \frac{\bar{x}_{i-1}}{\bar{x}_i}, 1, \frac{\bar{x}_{i+1}}{\bar{x}_i}, \ldots, \frac{\bar{x}_n}{\bar{x}_i}\right) \in \operatorname{Zero}(\mathbb{D}|_{x_i=1}) = \operatorname{Zero}(\mathbb{G}_i).$$

Hence

$$\bar{\boldsymbol{t}} \in \operatorname{Zero}(\mathbb{G}_i \cap \boldsymbol{K}[\boldsymbol{t}]) = \operatorname{Zero}(\mathbb{P}_i). \tag{7.4.4}$$

On the other hand, let (7.4.4) hold for some $i, 0 \leq i \leq n$; assume without loss of generality that $i = 0$. Then

$$\bar{\boldsymbol{t}} \in \operatorname{Zero}(\operatorname{Ideal}(\mathbb{G}_0) \cap \boldsymbol{K}[\boldsymbol{t}]) = \operatorname{Zero}(\operatorname{Ideal}(\mathbb{D}|_{x_0=1}) \cap \boldsymbol{K}[\boldsymbol{t}]).$$

Let $\mathbb{R}$ be the resultant system of $\mathbb{D}$ with respect to $x_0, \boldsymbol{x}$. From Lemma 1.3.1 and the construction of $\mathbb{R}$ in Sect. 5.4, one knows that, for any $R \in \mathbb{R}$, there exists an integer k such that $R x_0^k \in \operatorname{Ideal}(\mathbb{D})$. This can also be seen from (5.4.4) and van der Waerden (1950, p. 8). Hence,

$$\operatorname{Zero}(\operatorname{Ideal}(\mathbb{D}|_{x_0=1}) \cap \boldsymbol{K}[\boldsymbol{t}]) \subset \operatorname{Zero}(R), \quad \forall R \in \mathbb{R}.$$

It follows that $R(\bar{\boldsymbol{t}}) = 0$ for all $R \in \mathbb{R}$. By Theorem 5.4.3, $\mathbb{D}|_{\boldsymbol{t}=\bar{\boldsymbol{t}}}$ has a nontrivial zero $\bar{\boldsymbol{x}}$ in some extension field of $\boldsymbol{K}(\bar{\boldsymbol{t}})$ for $\boldsymbol{x}$. In other words, $\mathfrak{H}$ has a singular point $\bar{\boldsymbol{x}}$ of multiplicity $\geq p+1$ for $\boldsymbol{t} = \bar{\boldsymbol{t}}$. The proof is complete. □

Now consider hypersurfaces in the affine space $\mathbf{A}^n$. Let F be a polynomial in $\boldsymbol{K}[\boldsymbol{x}]$ of total degree m, and F_i be the homogeneous part of total degree i of F for $0 \leq i \leq m$. We define

$$\partial F/\partial 1 \triangleq F_{m-1} + 2F_{m-2} + \cdots + m F_0$$

and accordingly the successive derivatives of higher order of F with respect to 1. It is easy to verify the following Euler relation

$$\frac{\partial F}{\partial 1} = mF - \sum_{i=1}^{n} x_i \frac{\partial F}{\partial x_i}.$$

Algorithm SinConA: $\Psi \leftarrow \text{SinConA}(F, p)$. Given the polynomial equation $F(\boldsymbol{x}) = 0$ in $\boldsymbol{K}[\boldsymbol{t}, \boldsymbol{x}]$ of an algebraic hypersurface $\mathfrak{H}$ in $\mathbf{A}^n$ with $\boldsymbol{t} = (t_1, \ldots, t_m)$ as parameters and an integer $p \geq 1$, this algorithm computes a finite set Ψ of polynomial systems $[\mathbb{P}_1, \mathbb{Q}_1], \ldots, [\mathbb{P}_e, \mathbb{Q}_e]$ in $\boldsymbol{K}[\boldsymbol{t}]$ such that $\mathfrak{H}$ has singularities of multiplicity $\geq p+1$ for $\boldsymbol{t} = \bar{\boldsymbol{t}} \in \tilde{\boldsymbol{K}}^m$ if and only if $\bar{\boldsymbol{t}} \in \bigcup_{i=1}^{e} \text{Zero}(\mathbb{P}_i/\mathbb{Q}_i)$.

S1. Set

$$\mathbb{D} \leftarrow \left\{ \frac{\partial^p F}{\partial 1^{r_0} \partial x_1^{r_1} \ldots \partial x_n^{r_n}} : r_0 + r_1 + \cdots + r_n = p \right\}.$$

Compute a triangular series Ψ of $\mathbb{D}$ with projection for $\boldsymbol{x}$ with respect to the variable ordering $t_1 \prec \cdots \prec t_m \prec x_1 \prec \cdots \prec x_n$. If $\Psi = \emptyset$, then $\mathfrak{H}$ has no singularity for any $\boldsymbol{t}$ and the procedure terminates.

S2. Remove redundant sets from $\bigcup_{[\mathbb{T},\mathbb{U}] \in \Psi} \text{Zero}(\mathbb{T} \cap \boldsymbol{K}[\boldsymbol{t}]/\mathbb{U} \cap \boldsymbol{K}[\boldsymbol{t}])$, simplify it, and let the obtained zero set be $\bigcup_{i=1}^{e} \text{Zero}(\mathbb{P}_i/\mathbb{Q}_i)$. Return

$$\Psi \leftarrow \{[\mathbb{P}_1, \mathbb{Q}_1], \ldots, [\mathbb{P}_e, \mathbb{Q}_e]\}.$$

Proof. By the definition of triangular series and the projection property of $[\mathbb{T}, \mathbb{U}] \in \Psi$. □

Remark 7.4.1. Together with projection, triangular series may also be used to determine the conditions for projective hypersurfaces, and so may Gröbner bases for affine hypersurfaces.

In case the hypersurface $\mathfrak{H}$ has singular points of multiplicity $\geq p+1$ for some specialized $\boldsymbol{t}$, the structure of the singular variety may be described by computing its irreducible decomposition, from which the dimension of each component is readily determined. When the singular points are finite, computing all of them amounts to solving systems of triangularized polynomial equations and inequations.

The necessary and sufficient conditions for $\mathfrak{H}$ to have singularities of exact multiplicity $p+1$ and the structure of the corresponding singular variety for specialized $\boldsymbol{t}$ may be easily determined when these have been done for multiplicity $\geq p+1$: one simply introduces inequations.

Example 7.4.3. Consider the projective algebraic surface in $\mathbf{P}^3$ defined by the equation

$$F = x_0^3 + x_1^3 + x_2^3 + x_3^3 + 3ax_0x_1x_2 + 3bx_1x_2x_3 = 0.$$

The set of four first partial derivatives of F with the constant 3 removed is

$$\mathbb{D} = \{ax_1x_2 + x_0^2, bx_2x_3 + ax_0x_2 + x_1^2, bx_1x_3 + ax_0x_1 + x_2^2, x_3^2 + bx_1x_2\}.$$

Computing the Gröbner bases of $\mathbb{D}|_{x_i=1}$ for $0 \leq i \leq 3$, one finds that there is one and only one polynomial

$$\delta = a^6 - 2a^3b^3 + b^6 + 2a^3 + 2b^3 + 1$$

involving variables a and b only in all the four bases. Hence the projective surface has a singular point if and only if $\delta = 0$. By the same method one may find that the surface has no singularity of multiplicity ≥ 3.

Consider in particular the case when x_0 is replaced by 1:

$$\bar{F} = F|_{x_0=1} = 1 + x_1^3 + x_2^3 + x_3^3 + 3ax_1x_2 + 3bx_1x_2x_3 = 0$$

defines an algebraic surface in 3-dimensional affine space. With the ordering $a \prec b \prec x_1 \prec x_2 \prec x_3$, a characteristic series of

$$\mathbb{D}_0 = \left\{ \frac{\partial \bar{F}}{\partial 1}, \frac{\partial \bar{F}}{\partial x_1}, \frac{\partial \bar{F}}{\partial x_2}, \frac{\partial \bar{F}}{\partial x_3} \right\}$$

consists of two ascending sets

$$\begin{aligned} \mathbb{C}_1 &= [\delta, 2a^3x_1^3 + b^3 - a^3 + 1, ax_1x_2 + 1, 2a^2bx_3 + b^3 + a^3 + 1], \\ \mathbb{C}_2 &= [a^3 + 1, b, x_1^3 - 1, ax_1x_2 + 1, x_3^2]. \end{aligned}$$

Projecting $\operatorname{Zero}(\mathbb{C}_i)$ onto a, b for $i = 1, 2$, we have

$$\begin{aligned} &\operatorname{Proj}_{a,b}\operatorname{Zero}(\mathbb{D}_0) \\ &\quad = \operatorname{Proj}_{a,b}\operatorname{Zero}(\mathbb{C}_1/abx_1) \cup \operatorname{Proj}_{a,b}\operatorname{Zero}(\mathbb{C}_2/ax_1) \\ &\quad = \operatorname{Zero}(\delta/ab(a^3 - b^3 - 1)) \cup \operatorname{Zero}(\{a^3 + 1, b\}/a) \\ &\quad = \operatorname{Zero}(\delta/a). \end{aligned}$$

Therefore, the surface $\bar{F} = 0$ has singular points if and only if $\delta = 0$ and $a \neq 0$. Using the same method, one can find that the surface has no singularity of multiplicity ≥ 3.

Take, for instance, $a = b = -1/\sqrt[3]{4}$, which satisfies the condition obtained in either case. Thus the surface must have singular points. To determine all the points, one simply substitutes the values of a, b into the characteristic series or Gröbner bases. From them all the three singular points may be easily found as follows

$$\begin{aligned} &[1, \sqrt[3]{2}, \sqrt[3]{2}, 1], \\ &\left[1, -\frac{\sqrt[3]{2}(\sqrt{3}i + 1)}{2}, \frac{\sqrt[3]{2}(\sqrt{3}i - 1)}{2}, 1\right], \\ &\left[1, \frac{\sqrt[3]{2}(\sqrt{3}i - 1)}{2}, -\frac{\sqrt[3]{2}(\sqrt{3}i + 1)}{2}, 1\right]. \end{aligned}$$

If we take $a = 1$, then there are four values of b such that $\delta = 0$. For each of them the surface has three singular points. All these points have been found in Example 7.1.1.

Example 7.4.4. For the univariate quartic equation

$$F = x^4 + x_1x^3 + x_2x^2 + x_3x + x_4 = 0 \tag{7.4.5}$$

with indeterminate coefficients x_1, x_2, x_3 and x_4, the discriminant Δ_F of F has been computed in Example 5.4.1. It is a polynomial of total degree 6. $\Delta_F = 0$ defines an algebraic hypersurface, called the *discriminant surface* of F, in 4-dimensional affine space. Let us investigate its singularities. The existence of singular points, for example, $(0, \ldots, 0)$, is obvious. For the set of four first partial derivatives of Δ_F, an irreducible characteristic series consists of three ascending sets

$$\begin{aligned}
\mathbb{C}_1 &= [8x_2 - 3x_1^2, 16x_3 - x_1^3, 256x_4 - x_1^4],\\
\mathbb{C}_2 &= [8x_3 - 4x_1x_2 + x_1^3, 64x_4 - 16x_2^2 + 8x_1^2x_2 - x_1^4],\\
\mathbb{C}_3 &= [108x_3^2 - 108x_1x_2x_3 + 27x_1^3x_3 + 32x_2^3 - 9x_1^2x_2^2, 12x_4 - 3x_1x_3 + x_2^2].
\end{aligned}$$

They are of dimensions 1, 2, and 2 respectively. Since the initials of all the polynomials in $\mathbb{C}_1, \mathbb{C}_2, \mathbb{C}_3$ are constants, each ascending set itself defines an irreducible algebraic variety. We have thus accomplished an irreducible decomposition of the singular variety of the discriminant surface as well. With some inspection, one may find that

$\mathbb{C}_1 = 0 \iff$ (7.4.5) has a quadruple root;
$\mathbb{C}_2 = 0 \iff$ (7.4.5) has two double roots;
$\mathbb{C}_3 = 0 \iff$ (7.4.5) has a triple root.

The remaining points on the discriminant surface correspond to (7.4.5) having only one double root. This can also be confirmed by elimination: for example, collecting the coefficients of $F - (x^2 - ax - b)^2$ in x yields a set $\mathbb{P}$ of four polynomials in x_i and a, b. $\mathbb{C}_2$ may obtained by computing a characteristic set or series of $\mathbb{P}$ with respect to $x_1 \prec \cdots \prec x_4 \prec a \prec b$.

Furthermore, one may check with ease that the pseudo-remainders of the second partial derivatives of Δ_F are all 0 with respect to $\mathbb{C}_1$ but not with respect to $\mathbb{C}_2$ and $\mathbb{C}_3$. Hence the zeros, and in fact only those zeros, of $\mathbb{C}_1$ are singular points of multiplicity ≥ 3 of the discriminant surface. The origin $(0, \ldots, 0)$ is the only singular point of multiplicity >3 – it is of multiplicity 6. It is also easy to verify that $\mathrm{Zero}(\mathbb{C}_2) \subset \mathrm{Zero}(\mathbb{C}_i)$ for $i = 2, 3$; actually,

$$\mathrm{Zero}(\mathbb{C}_1) = \mathrm{Zero}(\mathbb{C}_2) \cap \mathrm{Zero}(\mathbb{C}_3).$$

Hence, $\mathrm{Zero}(\mathbb{C}_1)$ is a redundant component that can be removed from the decomposition.

Note incidentally that if the quintic is considered instead of quartic, the computation becomes much more complicated. We have tried the case without success.

There are numerous theoretical and practical problems in other areas related to elementary geometry and differential geometry to which elimination methods can apply. These include robot kinematics, intersection of geometric objects, perspective viewing in computer vision, and constrained geometric construction. See Buchberger (1987), MMRC (1987–1996), and Wang (1995b), for more information.

7.5 Algebraic factorization

The first method

Let $u_1, \ldots, u_d$ be d transcendental elements (indeterminates), abbreviated $\boldsymbol{u}$, and $\boldsymbol{K}_0 = \mathbf{Q}(u_1, \ldots, u_d)$ be the extension field obtained from $\mathbf{Q}$ by adjoining $u_1, \ldots, u_d$. For every $1 \le i \le r$, $\boldsymbol{K}_i = \boldsymbol{K}_0(\eta_1, \ldots, \eta_i)$ denotes the algebraic-extension field obtained from $\boldsymbol{K}_0$ by adjoining successively the algebraic elements $\eta_1, \ldots, \eta_i$, where η_i has adjoining polynomial $A_i \in \boldsymbol{K}_{i-1}[y_i]$. As usual, let $\boldsymbol{y}^{\{i\}}$ stand for $y_1, \ldots, y_i$ with $\boldsymbol{y} = \boldsymbol{y}^{\{r\}}$. When the polynomials A_i are explicitly given, we simply write $\boldsymbol{K}_0(\boldsymbol{y}^{\{i\}})$ for $\boldsymbol{K}_i$ without introducing the η_i. Assume without loss of generality that $A_i \in \boldsymbol{K}_0[\boldsymbol{y}^{\{i\}}]$ for each i. Then $\mathbb{A} = [A_1, \ldots, A_r]$ forms an irreducible adjoining ascending set of the field $\boldsymbol{K}_r$ for $\boldsymbol{y}$ (see Sect. 1.4).

Our first algebraic-factoring method may be described as follows.

Algorithm FactorA: $F^* \leftarrow \text{FactorA}(F, \mathbb{A})$. Given an irreducible ascending set $\mathbb{A} = [A_1, \ldots, A_r] \subset \boldsymbol{K}_0[\boldsymbol{y}]$ and a polynomial $F \in \boldsymbol{K}_0[\boldsymbol{y}, y]$ of degree $m \ge 1$, irreducible over $\boldsymbol{K}_0$ and reduced with respect to $\mathbb{A}$, this algorithm factorizes F into the product F^* of irreducible factors over $\boldsymbol{K}_r = \boldsymbol{K}_0(\boldsymbol{y})$ with adjoining ascending set $\mathbb{A}$ for $\boldsymbol{y}$.

F1. If $m = 1$, then go to F3. If m is even, then set $\bar{m} \leftarrow m/2$; else set $\bar{m} \leftarrow (m-1)/2$.

F2. For $s = 1, \ldots, \bar{m}$ do:

F2.1. Let $d_i \leftarrow \text{ldeg}(A_i)$ for $1 \le i \le r$ and $t \leftarrow m - s$. Set

$$G \leftarrow y^s + g_1 y^{s-1} + \cdots + g_s, \quad H \leftarrow y^t + h_1 y^{t-1} + \cdots + h_t,$$

where

$$g_i \leftarrow \sum_{\substack{0 \le k_l \le d_l - 1 \\ 1 \le l \le r}} g_{ik_1 \ldots k_r} y_1^{k_1} \cdots y_r^{k_r}, \quad 1 \le i \le s,$$

$$h_j \leftarrow \sum_{\substack{0 \le k_l \le d_l - 1 \\ 1 \le l \le r}} h_{jk_1 \ldots k_r} y_1^{k_1} \cdots y_r^{k_r}, \quad 1 \le j \le t$$

and $g_{ik_1 \ldots k_r}, h_{jk_1 \ldots k_r}$ are new indeterminates. Let the total number of $g_{ik_1 \ldots k_r}$ and $h_{jk_1 \cdots k_r}$ be M [which is equal to $(s+t)d_1 \cdots d_r$], and rename these indeterminates $x_1, \ldots, x_M$.

F2.2. Expand $R \leftarrow F - \text{lc}(F, y) \cdot G \cdot H$, compute $R \leftarrow \text{prem}(R, \mathbb{A})$ and equate the coefficients of all the terms of R in $\boldsymbol{y}$ and y to 0. Let the obtained set of M polynomial equations in $\boldsymbol{K}_0[x_1, \ldots, x_M]$ be

$$\left.\begin{aligned} P_1(x_1, \ldots, x_M) &= 0, \\ P_2(x_1, \ldots, x_M) &= 0, \\ &\cdots\cdots \\ P_M(x_1, \ldots, x_M) &= 0. \end{aligned}\right\} \tag{7.5.1}$$

F2.3. Solve the equations (7.5.1) for $x_1, \ldots, x_M$ in $\boldsymbol{K}_0$ by any of the methods presented in Sect. 7.1. If (7.5.1) has no solution in $\boldsymbol{K}_0$, then go back to F2 for next s. Otherwise, let $x_1 = \bar{x}_1, \ldots, x_M = \bar{x}_M$ be any solution of (7.5.1), set

$$G \leftarrow G|_{x_1=\bar{x}_1,\ldots,x_M=\bar{x}_M}, \quad H \leftarrow H|_{x_1=\bar{x}_1,\ldots,x_M=\bar{x}_M}$$

and go to F4 [in this case F is factorized as $F \doteq \mathrm{lc}(F, y) \cdot G \cdot H$ over $\boldsymbol{K}_r$].

F3. Return $F^* \leftarrow F$ (which is irreducible over $\boldsymbol{K}_r$) and the algorithm terminates.

F4. Factorize G and H over $\boldsymbol{K}_r$ and return

$$F^* \leftarrow \mathrm{lc}(F, y) \cdot \mathrm{FactorA}(G, \mathbb{A}) \cdot \mathrm{FactorA}(H, \mathbb{A}).$$

Proof. It is obvious. □

In the above algorithm, algebraic factoring is reduced to solving polynomial equations. In other words, whether F can be factorized into G and H over $\boldsymbol{K}_r$ is equivalent to whether (7.5.1) has a solution for $x_1, \ldots, x_M$ in $\boldsymbol{K}_0$. Hu and Wang (1986) explained how the solvability and solutions can be determined by the method of characteristic sets with Gauss' lemma.

Example 7.5.1. Let us consider the following three polynomials

$$\begin{aligned}
H_1 &= u_3 y_1^2 + 2u_1u_2y_1 + 2u_1^2y_1 - u_1^2u_3,\\
H_2 &= u_3 y_2^2 - 2u_1u_2y_2 + 2u_1^2y_2 - u_1^2u_3,\\
H_3 &= u_3 y_3^2 - u_3^2y_3 - u_2^2y_3 + u_1^2y_3 - u_1^2u_3,
\end{aligned}$$

which come from another formulation of the geometric theorem stated in Example 7.2.3 (see Wang 1994). Let $\boldsymbol{K}_0 = \mathbf{Q}(u_1, u_2, u_3)$. We first examine the irreducibility of H_2 over $\boldsymbol{K}_1 = \boldsymbol{K}_0(y_1)$, where y_1 is an algebraic element having adjoining polynomial H_1. For this purpose, let

$$\begin{aligned}
G &= y_2 + g_1y_1 + g_0,\\
H &= y_2 + h_1y_1 + h_0.
\end{aligned}$$

Then

$$R = \mathrm{prem}(H_2 - \mathrm{lc}(H_2, y_2) \cdot G \cdot H, H_1, y_1) = R_1y_1y_2 + R_2y_2 + R_3y_1 + R_4,$$

where

$$\begin{aligned}
R_1 &= u_3g_1 + u_3h_1,\\
R_2 &= u_3g_0 + u_3h_0 + 2u_1u_2 - 2u_1^2,\\
R_3 &= -2u_1u_2g_1h_1 - 2u_1^2g_1h_1 + u_3g_1h_0 + u_3g_0h_1,\\
R_4 &= u_1^2u_3g_1h_1 + u_3g_0h_0 + u_1^2u_3.
\end{aligned}$$

Let $\mathbb{P} = \{R_1, \ldots, R_4\}$. To determine whether $\mathbb{P} = 0$ has a solution for g_1, g_0 and h_1, h_0 in $\boldsymbol{K}_0$, we compute, for instance, a characteristic series of $\mathbb{P}$ under $g_0 \prec h_0 \prec h_1 \prec g_1$: it consists of two quasilinear ascending sets

$$\mathbb{C}_1 = \begin{bmatrix} (u_3^2 + u_2^2 + 2u_1u_2 + u_1^2)u_3g_0^2 + 2u_1(u_2u_3^2 - u_1u_3^2 \\ \quad + u_2^3 + u_1u_2^2 - u_1^2u_2 - u_1^3)g_0 - 4u_1^3u_2u_3, \\ u_3h_0 + u_3g_0 + 2u_1u_2 - 2u_1^2, \\ u_1(u_2 + u_1)h_1 + u_3g_0 + u_1u_2 - u_1^2, \\ u_1(u_2 + u_1)g_1 - u_3g_0 - u_1u_2 + u_1^2 \end{bmatrix},$$

$$\mathbb{C}_2 = \begin{bmatrix} u_3g_0^2 - 2u_1^2g_0 + 2u_1u_2g_0 - u_1^2u_3, \\ u_3h_0 + u_3g_0 - 2u_1^2 + 2u_1u_2, \\ h_1, \\ g_1 \end{bmatrix}.$$

The first polynomials in $\mathbb{C}_1$ and in $\mathbb{C}_2$ are both irreducible over $\mathbf{Q}$, so neither the system $\mathbb{C}_1 = 0 \wedge \mathrm{ini}(\mathbb{C}_1) \neq 0$ nor $\mathbb{C}_2 = 0 \wedge \mathrm{ini}(\mathbb{C}_2) \neq 0$ has a solution in $\boldsymbol{K}_0$. Hence, the polynomial H_2 is irreducible over $\boldsymbol{K}_1$.

Now we want to factorize H_3 over $\boldsymbol{K}_2 = \boldsymbol{K}_1(y_2)$, with adjoining polynomial H_2 for y_2. Proceeding in a similar way, let

$$G = y_3 + g_{11}y_1y_2 + g_{01}y_2 + g_{10}y_1 + g_{00},$$
$$H = y_3 + h_{11}y_1y_2 + h_{01}y_2 + h_{10}y_1 + h_{00}.$$

The polynomial

$$R = \mathrm{prem}(H_3 - \mathrm{ini}(H_3) \cdot G \cdot H, [H_1, H_2])$$

consists of 46 terms, where $y_1 \prec y_2 \prec y_3$. Equating the coefficients of R in y_1, y_2, y_3 to 0, one obtains a set of eight polynomial equations (7.1.5) given in Example 7.1.2. A solution to (7.1.5) for h_{ij} and g_{ij} has been found as in (7.1.6). Therefore, H_3 is factorized as

$$H_3 \doteq \frac{1}{4u_1^4}[2u_1^2y_3 - u_3y_1y_2 - u_1(u_2 + u_1)y_2 + u_1(u_2 - u_1)y_1 - u_1^2u_3]$$
$$\cdot [2u_1^2u_3y_3 + u_3^2y_1y_2 + u_1(u_2 + u_1)u_3y_2$$
$$- u_1(u_2 - u_1)u_3y_1 - u_1^2(u_3^2 + 2u_2^2 - 2u_1^2)].$$

The second method

The key idea underlying this method is the reduction of polynomial factorization over algebraic-extension fields to that over $\mathbf{Q}$ via linear transformation and characteristic sets computation. Let $\mathbb{A} = [A_1, \ldots, A_r]$, $\boldsymbol{K}_i$ and F be as in FactorA. Set

$$\mathbb{A}^+ = [A_1, \ldots, A_r, F].$$

With respect to $y_1 \prec \cdots \prec y_r \prec y$, $\mathbb{A}^+$ is clearly an ascending set and F is irreducible over $\boldsymbol{K}_r$ if and only if $\mathbb{A}^+$ is irreducible. When we say that G is a factor of F over $\boldsymbol{K}_r$, we always mean that $\deg(G, y) > 0$ (i.e., G is not a number in $\boldsymbol{K}_r$). G is said to be a *true factor* of F if $0 < \deg(G, y) < \deg(F, y)$.

Assume that one knows how to factorize polynomials over $\boldsymbol{K}_0$. The following lemma guarantees the correctness of the factoring algorithm described below.

Lemma 7.5.1. Let $\mathbb{A}$ and F be as above, $c_1, \ldots, c_r$ be r integers,

$$\bar{F} = F|_{y=y-c_1 y_1 - \cdots - c_r y_r},$$

and $\bar{\mathbb{C}}$ be an ascending set in any characteristic series of $\bar{\mathbb{A}} = \mathbb{A} \cup [\bar{F}]$ over $\boldsymbol{K}_0$ with respect to $y \prec y_1 \prec \cdots \prec y_r$. Let $\bar{C}$ be the first polynomial in $\bar{\mathbb{C}}$ and

$$C = \bar{C}|_{y=y+c_1 y_1 + \cdots + c_r y_r}.$$

If $\bar{\mathbb{C}}$ is perfect, then $|\bar{\mathbb{C}}| = r + 1$. If $\bar{\mathbb{C}}$ is moreover irreducible, then the GCD of F and C is irreducible over $\boldsymbol{K}_r$.

Proof. Since $\mathbb{A}$ is irreducible and F is reduced with respect to $\mathbb{A}$,

$$\mathrm{Dim}(\bar{\mathbb{A}}) = \mathrm{Dim}(\mathbb{A} \cup [F]) = 0.$$

If $\bar{\mathbb{C}}$ is perfect, then $\dim(\bar{\mathbb{C}}) = 0$. It follows that $|\bar{\mathbb{C}}| = r + 1$.

Let $\bar{\mathbb{C}}$ be irreducible and $(\eta, \boldsymbol{\eta}) = (\eta, \eta_1, \ldots, \eta_r)$ be any generic zero of $\bar{\mathbb{C}}$; then $(\eta, \boldsymbol{\eta}) \in \mathrm{Zero}(\bar{\mathbb{A}})$. Hence, there exists an irreducible factor $\bar{G}$ of $\bar{F}$ over $\boldsymbol{K}_r$ such that $\bar{G}(\eta, \boldsymbol{\eta}) = 0$; $(\eta, \boldsymbol{\eta})$ is a generic zero of $\mathbb{A} \cup [\bar{G}]$. By Lemma 4.3.1, $\mathrm{prem}(\bar{C}, \mathbb{A} \cup [\bar{G}]) = 0$. It follows that $G = \bar{G}|_{y=y+c_1 y_1 + \cdots + c_r y_r}$ is a divisor of C over $\boldsymbol{K}_r$.

Let $\bar{H}$ be another irreducible factor of $\bar{F}$ that is *distinct* from $\bar{G}$ over $\boldsymbol{K}_r$. Then there exists an η' in some extension field of $\boldsymbol{K}_r$ such that

$$\bar{H}(\eta', \boldsymbol{\eta}) = 0, \quad \bar{G}(\eta', \boldsymbol{\eta}) \neq 0, \quad \forall \boldsymbol{\eta} \in \mathrm{Zero}(\mathbb{A}).$$

We claim that $\mathrm{prem}(\bar{C}, \mathbb{A} \cup [\bar{H}]) \neq 0$. For, otherwise, $C(\eta') = 0$, and one can find an $\boldsymbol{\eta}'$ such that $(\eta', \boldsymbol{\eta}') \in \mathrm{Zero}(\bar{\mathbb{C}}) \subset \mathrm{Zero}(\mathbb{A} \cup [\bar{G}])$. This would lead to a contradiction. Hence, $\bar{H}$ cannot be a divisor of $\bar{C}$ over $\boldsymbol{K}_r$.

Let $\bar{C}$ be factorized as $\bar{C} \doteq \bar{D}\bar{G}$ over $\boldsymbol{K}_r$. Then $\bar{C} - \bar{D}\bar{G} \in \mathrm{sat}(\mathbb{A})$. It remains to be shown that $\bar{G}$ is not a divisor of $\bar{D}$ over $\boldsymbol{K}_r$.

Since $\mathrm{prem}(\bar{G}, \mathbb{A}) \neq 0$, by Lemma 4.3.2 there exists a polynomial $Q \in \boldsymbol{K}_0[y, \boldsymbol{y}]$ such that

$$Q\bar{G} - R \in \mathrm{Ideal}(\mathbb{A}) \subset \mathrm{sat}(\mathbb{A}), \quad \text{where } R = \mathrm{res}(\bar{G}, \mathbb{A}) \neq 0, \ R \in \boldsymbol{K}_0[y],$$

and $Q(\eta, \boldsymbol{\eta}) \neq 0$ for any $(\eta, \boldsymbol{\eta}) \in \mathrm{Zero}(\mathbb{A} \cup [\bar{G}])$. As any zero of $\bar{\mathbb{C}}$ is a zero of $\mathbb{A} \cup [\bar{G}]$, any zero of $\bar{C}$ is also a zero of R. This implies that $\bar{C} \mid R$, so there exists a $T \in \boldsymbol{K}_0[y]$ such that $R = T\bar{C}$. It follows that

$$Q\bar{G} - T\bar{D}\bar{G} \in \mathrm{sat}(\mathbb{A}).$$

Because $\mathrm{sat}(\mathbb{A})$ is prime and $\bar{G} \notin \mathrm{sat}(\mathbb{A})$, $Q - T\bar{D} \in \mathrm{sat}(\mathbb{A})$. Thus, for any $(\eta, \boldsymbol{\eta}) \in \mathrm{Zero}(\mathbb{A} \cup [\bar{G}])$

$$Q(\eta, \boldsymbol{\eta}) - \bar{D}(\eta, \boldsymbol{\eta})T(\eta) = 0.$$

Note that $Q(\eta, \boldsymbol{\eta}) \neq 0$. If $\bar{G}$ is a divisor of $\bar{D}$ over $\boldsymbol{K}_r$, then $\bar{D}(\eta, \boldsymbol{\eta}) = 0$. This is a contradiction. Therefore, $\bar{G} \nmid \bar{D}$ and $\bar{G}$ is the GCD of $\bar{F}$ and $\bar{C}$ over $\boldsymbol{K}_r$. The lemma is proved. □

We continue using the above notations and let $\bar{\mathbb{C}} = [\bar{C}_0, \bar{C}_1, \ldots, \bar{C}_r]$ be a characteristic set of $\bar{\mathbb{A}}$ and $\bar{J} = \prod_{i=1}^{r} \mathrm{ini}(\bar{C}_i)$. Suppose that $\bar{\mathbb{C}}$ is perfect, so $\bar{C}_0 \in \boldsymbol{K}_0[y]$. Take an irreducible factor $\bar{C}$ of $\bar{C}_0$ over $\boldsymbol{K}_0$ which does not divide $\bar{J}$, if any, and compute a GCD G of F and $C = \bar{C}|_{y=y+c_1 y_1+\cdots+c_r y_r}$ over $\boldsymbol{K}_r$. In any case, it would be sufficient if G is a true factor of F over $\boldsymbol{K}_r$. Otherwise, we check whether $\bar{\mathbb{C}}$ is quasilinear. If so, then $[\bar{C}, \mathrm{prem}(\bar{C}_1, \bar{C}), \ldots, \mathrm{prem}(\bar{C}_r, \bar{C})]$ is an irreducible ascending set contained in a characteristic series of $\bar{\mathbb{A}}$. Thus, G is an irreducible factor of F over $\boldsymbol{K}_r$ according to Lemma 7.5.1. So what we need is to get a $\bar{\mathbb{C}}$ which is quasilinear and perfect. The linear transformation $y \leftarrow y - c_1 y_1 - \cdots - c_r y_r$ with random integers c_i is introduced to make $\bar{\mathbb{C}}$ quasilinear.

The GCD of F and C over $\boldsymbol{K}_r$ can be obtained from / as the last polynomial in any characteristic set of $\mathbb{A} \cup \{F, C\}$. Moreover, possible true factors of F may be constructed by computing over $\boldsymbol{K}_r$ the GCDs of F with the irreducible factors of $\bar{J}|_{y=y+c_1 y_1+\cdots+c_r y_r}$. The chance to obtain such factors is higher when $\bar{\mathbb{C}}$ is quasilinear.

There is an important practical issue: the factorization of F over $\boldsymbol{K}_r$ is unique only up to a "constant" factor in $\boldsymbol{K}_r$ which is represented here as a polynomial in $\boldsymbol{u}$ and $\boldsymbol{y}$. The size of each factor of F may be dramatically affected by such a constant. Let G be an irreducible factor of F, which may be assumed, without loss of generality, to be in $\mathbf{Q}[\boldsymbol{u}, \boldsymbol{y}, y]$. In general, $\mathrm{lc}(G, y)$ involves both the variables $\boldsymbol{u}$ and $\boldsymbol{y}$. By using algorithm Norm or NormG, one can normalize G by $\mathbb{A}$ to get another polynomial $G^* \in \mathbf{Q}[\boldsymbol{u}, \boldsymbol{y}, y]$ such that $\mathrm{lc}(G^*, y) \in \mathbf{Q}[\boldsymbol{u}]$ and G^* differs from G only by a factor in $\boldsymbol{K}_r$. In many cases G^* is much simpler than G, but the opposite is also true in many other cases. Heuristic use of normalization of this kind may improve the efficiency of FactorB considerably.

Algorithm FactorB: $F^* \leftarrow \mathrm{FactorB}(F, \mathbb{A})$, Given an irreducible ascending set $\mathbb{A} = [A_1, \ldots, A_r] \subset \boldsymbol{K}_0[\boldsymbol{y}]$ and a polynomial $F \in \boldsymbol{K}_0[\boldsymbol{y}, y]$ irreducible over $\boldsymbol{K}_0$ and reduced with respect to $\mathbb{A}$, this algorithm factorizes F into the product F^* of irreducible factors over $\boldsymbol{K}_r = \boldsymbol{K}_0(\boldsymbol{y})$ with adjoining ascending set $\mathbb{A}$ for $\boldsymbol{y}$.

F1. Set $\mathbb{A}^* \leftarrow [A\colon \mathrm{ldeg}(A) > 1, A \in \mathbb{A}]$. If $\mathbb{A}^* = \emptyset$ or $\deg(F, y) \leq 1$ then return F and the algorithm terminates. Otherwise, let $y_{p_1} \prec \cdots \prec y_{p_s}$ be the leading variables of the polynomials in $\mathbb{A}^*$ and set $\Omega \leftarrow \emptyset$.

F2. Choose a set of integers $[c_1, \ldots, c_s] \notin \Omega$; set $\Omega \leftarrow \Omega \cup \{[c_1, \ldots, c_s]\}$ and $\bar{F} \leftarrow F|_{y=y-c_1 y_{p_1}-\cdots-c_s y_{p_s}}$.
Compute a characteristic set $\bar{\mathbb{C}}$ of $\mathbb{A}^* \cup \{\bar{F}\}$ with respect to the variable ordering $y \prec y_{p_1} \prec \cdots \prec y_{p_s}$. If $|\bar{\mathbb{C}}| \neq s + 1$, then go back to F2. Let $\mathbb{I}$ be the set of all irreducible factors (over $\boldsymbol{K}_0$) of the polynomials in $\mathrm{ini}(\bar{\mathbb{C}})$ and $\mathbb{F}$ the set of those irreducible factors (over $\boldsymbol{K}_0$) of the first polynomial in $\bar{\mathbb{C}}$ which do not divide any polynomial in $\mathbb{I}$.

F3. If $\bar{\mathbb{C}}$ is quasilinear, then go to F4. If $|\mathbb{F}| \leq 1$, then go to F2; else set $\mathbb{I} \leftarrow \mathbb{I} \cup \mathbb{F}$ and $\mathbb{F} \leftarrow \emptyset$.

F4. Set $G \leftarrow F, \mathbb{P}' \leftarrow \emptyset$, and

$$\mathbb{F} \leftarrow \mathbb{F}|_{y=y+c_1 y_{p_1}+\cdots+c_s y_{p_s}},$$
$$\mathbb{I} \leftarrow \mathbb{I}|_{y=y+c_1 y_{p_1}+\cdots+c_s y_{p_s}}.$$

For each $P \in \mathbb{F} \cup \mathbb{I}$ while $\deg(G, y) > 1$, do:
Compute a GCD F_P of G and P over $\boldsymbol{K}_r$ with heuristic normalization. If $0 < \deg(F_P, y) < \deg(G, y)$, then set $G \leftarrow G/F_P$ over $\boldsymbol{K}_r$ and $\mathbb{P} \leftarrow \mathbb{P} \cup \{F_P\}$ when $P \in \mathbb{F}$, or $\mathbb{P}' \leftarrow \mathbb{P}' \cup \{F_P\}$ otherwise.

F5. If $\mathbb{P} \cup \mathbb{P} \neq \emptyset$, then return

$$F^* \leftarrow \prod_{P \in \mathbb{P}} P \cdot \prod_{P \in \mathbb{P}' \cup \{G\}} \mathrm{FactorB}(P, \mathbb{A}^*)$$

and the algorithm terminates. If $\bar{\mathbb{C}}$ is quasilinear and $\mathbb{F} \neq \emptyset$, then return $F^* \leftarrow F$; else go to F2.

The correctness of FactorB follows from Lemma 7.5.1. It is not easy to see whether the algorithm always terminates, i.e., whether a perfect quasilinear characteristic set can be produced in a finite number of steps. Fortunately, the probability of obtaining a quasilinear characteristic set by a random choice of integers $c_1, \ldots, c_s$ in step F2 is 1. This is because in general

$$\deg(\mathrm{prem}(P, Q, x), x) = \deg(Q, x) - 1,$$

while prem is the principal operation in the characteristic set algorithm. So in practice, termination has never been a problem for us.

An immediate variation in FactorB is to compute instead a characteristic series in step F2. The irreducible factors of F are determined from those ascending sets in the series whose irreducibility can be easily verified. The ordering for the variables $y, y_{p_1}, \ldots, y_{p_s}$ may be arbitrary as long as y is arranged with the lowest order. As the purpose of this step is to produce polynomials in $\boldsymbol{K}_0[y]$ by successive elimination of the variables, other elimination methods may be used as well. In fact, algorithm FactorB can be considered as a variant of the method of Trager (1976) on the basis of resultant computation.

The two algorithms described above are of sufficient generality. If the transcendental elements $\boldsymbol{u}$ do not appear in the adjoining polynomials A_i, the factorization can be viewed as performed over the usually called *algebraic-number field* $\mathbf{Q}(\boldsymbol{y})$. If $\boldsymbol{u}$ appear in the A_i, the factorization is performed over the *algebraic-function field* $\boldsymbol{K}_r$. Then the algorithm is relatively slow, mainly because the involvement of $\boldsymbol{u}$ greatly increases the complexity of variable elimination and GCD computation.

Example 7.5.2. During the computation of the irreducible decomposition in Example 7.2.7, several polynomials have to be factorized over algebraic-extension fields. We take one of them as an example: factorize

$$F = 4y_5^2 - 4u_1 y_5 - 4y_5 - 3u_2^2 + u_1^2 + 2u_1 + 1$$

over $\mathbf{Q}(u_1, u_2, y_2)$ with y_2 having adjoining polynomial $A = 4y_2^2 - 3$.

Substituting y_5 in F by $y_5 + y_2$, we have

$$\begin{aligned}\bar{F} &= F|_{y_5=y_5+y_2}\\ &= 4[y_5^2 + (2y_2 - u_1 - 1)y_5 + y_2^2 - (u_1+1)y_2] - 3u_2^2 + u_1^2 + 2u_1 + 1.\end{aligned}$$

A characteristic set of $\{\bar{F}, A\}$ with respect to the ordering $y_5 \prec y_2$ is

$$\mathbb{C} = [C_1, \bar{F} - A],$$

in which C_1 factors over $\mathbf{Q}$ into $(C_0 + 6u_2)(C_0 - 6u_2)$ with

$$C_0 = 4y_5^2 - 4(u_1+1)y_5 - 3u_2^2 + u_1^2 + 2u_1 - 2.$$

Let us take the first factor of C_1 and substitute y_5 back by $y_5 - y_2$. The resulting polynomial is

$$\begin{aligned}D = 4[y_5^2 - (2y_2 + u_1 + 1)y_5 + y_2^2 + (u_1+1)y_2]\\ - 3u_2^2 + 6u_2 + u_1^2 + 2u_1 - 2.\end{aligned}$$

To find a GCD of D and F over $\mathbf{Q}(u_1, u_2, y_2)$, we compute a characteristic set $\bar{\mathbb{C}}$ of $\{D, F, A\}$ with respect to the ordering $y_2 \prec y_5$:

$$\bar{\mathbb{C}} = [A, 4y_2y_5 - 2(u_1+1)y_2 - 3u_2].$$

The second polynomial F_1 in $\bar{\mathbb{C}}$ is a true factor of F over $\mathbf{Q}(u_1, u_2, y_2)$. Removing this factor from F, one obtains the other true factor

$$F_2 = \mathrm{pquo}(F, F_1, y_5) = 4y_2y_5 - 2(u_1+1)y_2 + 3u_2.$$

Therefore, F is factorized as the product $F_1F_2/3$ over $\mathbf{Q}(u_1, u_2, y_2)$.

Remark 7.5.1. Here are some heuristics which may be useful for implementing algebraic-factoring algorithms. The first is a result from algebraic-number theory: Let $A \in \boldsymbol{K}[x]$ and $F \in \boldsymbol{K}[y]$ be two irreducible polynomials of degrees m in x and l in y, respectively. If m and l are relatively prime, then F is always irreducible over the algebraic-extension field $\boldsymbol{K}(x)$ with A as adjoining polynomial for x.

Secondly, let $A \in \boldsymbol{K}[x]$ and $F \in \boldsymbol{K}[y]$ be two polynomials irreducible over $\boldsymbol{K}$, and let $\tilde{A}$ and $\tilde{F}$ be the homogenization of A and F by z with respect to x and y, respectively. Let $\tilde{R} = \mathrm{prem}(\tilde{F}, \tilde{A}, z)$ with $I = \mathrm{lc}(\tilde{A}, z)$ such that $I^q\tilde{F} = \tilde{Q}\tilde{A} + \tilde{R}$ for some integer $q \geq 0$. Then any factorization of $R = \tilde{R}|_{z=1}$ over $\boldsymbol{K}$ divided by I^q is a factorization (not necessarily complete) of F over the algebraic-extension field $\boldsymbol{K}(x)$ with A as adjoining polynomial for x. This is obvious by plunging $z = 1$ into the pseudo-remainder formula. There is more possibility for R to be reducible when $\tilde{R}$ does not contain the variable z.

The homogenization above is not needed if A and F involve a transcendental element. To be precise, let $A \in \boldsymbol{K}[u, x]$ and $F \in \boldsymbol{K}[u, y]$ be two irreducible polynomials with $\deg(F, u) \geq \deg(A, u) > 0$. Let $R = \mathrm{prem}(F, A, u)$ with $I = \mathrm{lc}(A, u)$ such that $I^qF = QA + R$ for some integer $q \geq 0$. Then any factorization of R over $\boldsymbol{K}$ divided by I^q, upon reducing the higher powers of x in each component by A, is a factorization (not necessarily complete) of F over the extension field $\boldsymbol{K}(u, x)$ with u being a transcendental element and A the adjoining polynomial for x.

7.6 Center conditions for certain differential systems

Problem

Consider plane autonomous differential systems of center and focus type

$$dx/dt = y + P(x, y), \quad dy/dt = -x + Q(x, y), \tag{7.6.1}$$

where $P(x, y)$ and $Q(x, y)$ are polynomials beginning with terms of total degree >1 in x and y with indeterminate coefficients $\boldsymbol{u} = (u_1, \dots, u_e)$. As explained in Wang (1991a), one can compute a locally positive polynomial $L(x, y) \in \mathbf{Q}[\boldsymbol{u}, x, y]$ and polynomials $v_3, v_5, \dots, v_{2j+1}, \dots \in \mathbf{Q}[\boldsymbol{u}]$ such that the differential of $L(x, y)$ along the integral curve of (7.6.1) is of the form

$$\frac{dL(x, y)}{dt} = v_3 y^4 + v_5 y^6 + \cdots + v_{2j+1} y^{2j+2} + \cdots,$$

where v_{2j+1} is called the jth *Liapunov constant* of (7.6.1).

The origin, a singular point of (7.6.1), is said to be a *center* for (7.6.1) if and only if

$$v_3 = v_5 = \cdots = v_{2j+1} = \cdots = 0.$$

The necessary and sufficient conditions given in this way require infinitely many equations $v_{2j+1} = 0$, $j = 1, 2, \dots$ in a finite number of indeterminates. The polynomial ideal generated by $v_3, v_5, \dots, v_{2j+1}, \dots$ in $\mathbf{Q}[\boldsymbol{u}]$ has finite bases. Hence for any P and Q of given total degree m there exists an N_m such that v_3, $v_5, \dots, v_{2N_m+1}$ form such a basis, but we do not know any upper bound for N_m.

On the other hand, there are other methods for deriving center conditions. The explicit expressions of the conditions for a number of concrete systems have been obtained. Unfortunately, many of the conditions are erroneous and incomplete. In the next section we show how elimination methods can be used to examine the correctness of the conditions and to establish the relationship among different sets of conditions.

The computation and manipulation of Liapunov constants relate to and are useful for several other problems such as distinguishing between center and focus, searching for higher-order foci and constructing limit cycles (the second part of Hilbert's 16th problem) in the qualitative theory of differential equations. The study of these problems forms an entire subject of mathematics. Some of the treatments require solving polynomial equations, determining whether a polynomial equation follows from a system of polynomial equations and inequations, simplifying a polynomial by using a set of polynomial relations, etc., and thus elimination techniques may have applications therein. They are not discussed here. In this section, we only explain some aspects of the problem with reference to a particular class of cubic different systems.

Kukles' system

In what follows, we present a classical example of 1944 to illustrate the application. The author began investigating this example in 1986; the same example has also been studied by several other researchers since our results were published. However, the problem is still unsolved and the example remains challenging.

Let us consider a class of cubic differential systems, called *Kukles' system*, which is the particular case of (7.6.1) with

$$\begin{aligned} P(x,y) &= 0, \\ Q(x,y) &= a_{20}x^2 + a_{11}xy + a_{02}y^2 + a_{30}x^3 \\ &\quad + a_{21}x^2y + a_{12}xy^2 + a_{03}y^3. \end{aligned} \tag{7.6.2}$$

Kukles (1944) showed that in this case the origin is a center "if and only if" one of the following conditions holds:

$$\left.\begin{aligned} \alpha &= a_{30}a_{11}^2 + a_{21}\lambda = 0, \\ \beta &= (3a_{03}\lambda + \lambda^2 + a_{12}a_{11}^2)a_{21} - 3a_{03}\lambda^2 - a_{12}a_{11}^2\lambda = 0, \\ \gamma &= \lambda + a_{20}a_{11} + a_{21} = 0, \\ \delta &= 9a_{12}a_{11}^2 + 2a_{11}^4 + 9\lambda^2 + 27a_{03}\lambda = 0; \end{aligned}\right\} \tag{K1}$$

$$a_{03} = \alpha = \beta = \gamma = 0; \tag{K2}$$

$$a_{03} = a_{11} = a_{21} = 0; \tag{K3}$$

$$a_{03} = a_{02} = a_{20} = a_{21} = 0, \tag{K4}$$

where $\lambda = a_{02}a_{11} + 3a_{03}$. The above conditions have been commonly recognized and used in standard textbooks (e.g., Nemytskii and Stepanov 1960). Recent research interest and activity on Kukles' system started in the later 1980s when Jin and Wang (1990) discovered, by using the methods of Gröbner bases and characteristic sets, the following example:

$$\begin{aligned} &a_{20} \neq 0, \ \ a_{11} = 0, \ \ a_{02} = -2a_{20}, \ \ a_{30} = -a_{20}^2/3, \\ &\quad a_{21}^2 = a_{20}^4/2, \ \ a_{12} = 0, \ \ a_{03} = -a_{21}/3 \end{aligned} \tag{JW}$$

which is not covered by Kukles' conditions. Our computations suggested that for this example the origin is a center and thus Kukles' conditions are incomplete; the incompleteness was soon confirmed by Christopher and Lloyd (1990). Afterwards, several papers were published to give other examples and to establish the complete conditions. For example, Lloyd and Pearson (1992) together with C. J. Christopher found the following set of conditions:

$$\begin{aligned} \kappa_1 &= 81a_{20}^3a_{02} - 2(18a_{11}^2r - 4a_{11}^4 - 27a_{11}^2a_{20}^2 - 81a_{20}^4) = 0, \\ \kappa_2 &= 9\eta a_{30} + 36a_{11}^2r + 8a_{11}^4 + 90a_{11}^2a_{20}^2 + 243a_{20}^4 = 0, \\ \kappa_3 &= \eta a_{21} - a_{20}a_{11}(27r - 2a_{11}^2 - 9a_{20}^2) = 0, \\ \kappa_4 &= 81a_{20}^2\eta a_{12} + 2a_{11}^2(144a_{11}^2r - 567a_{20}^4 - 270a_{11}^2a_{20}^2 \\ &\quad + 243a_{20}^2r - 32a_{11}^4) = 0, \\ \kappa_5 &= 3\eta a_{03} + a_{11}(a_{02}\eta + 27a_{20}r + 14a_{20}a_{11}^2 + 72a_{20}^3) = 0, \end{aligned} \tag{CLP}$$

where

$$\begin{aligned} \eta &= 16a_{11}^2 + 81a_{20}^2, \\ \kappa_0 &= 162a_{11}^2r^2 - (2a_{11}^2 + 9a_{20}^2)^3 = 0, \\ a_{20}a_{11} &\neq 0. \end{aligned}$$

On the other hand, the incompleteness of Kukles' conditions was already pointed out independently by Cherkas (1978). Cherkas investigated Kukles' system with a different approach and derived the following set of conditions instead of (K1):

$$\begin{aligned}
&\gamma = 0,\\
&\theta_1 = 6a_{20}a_{03} + a_{20}a_{11}a_{02} - a_{21}a_{02} - a_{11}a_{12} - 2a_{30}a_{11} - \tfrac{2}{9}a_{11}^3 = 0,\\
&\theta_2 = 6a_{30}a_{03} - 3a_{20}^2a_{03} + a_{30}a_{11}a_{02} + a_{20}a_{02}a_{21} + a_{20}a_{12}a_{11}\\
&\qquad - a_{21}a_{12} - a_{30}a_{21} - \tfrac{2}{3}a_{11}^2a_{21} = 0,\\
&\theta_3 = a_{30}a_{21}a_{02} - 6a_{20}a_{30}a_{03} + a_{30}a_{11}a_{12} + a_{20}a_{21}a_{12} - \tfrac{2}{3}a_{11}a_{21}^2 = 0,\\
&\theta_4 = a_{30}a_{21}a_{12} - 3a_{30}^2a_{03} - \tfrac{2}{9}a_{21}^3 = 0,
\end{aligned}\tag{C1}$$

which contain the conditions (JW). He also proved that, for $a_{03} = 0$, his conditions coincide with Kukles'.

Since center conditions may be derived by different methods as noted above, among the obtained conditions there are some equivalent or containment relations which cannot be observed without involving heavy computations. For Kukles' system, one can easily verify that the third condition (K3) is contained in both (K1) and (K2), so it is redundant. An irreducible decomposition of (K1) consists of two components of which one is (K3).

To examine the relation between (K1) and (C1), we may compute an irreducible decomposition of the variety defined by (C1). The decomposition has been given in detail as Example 6.2.3.

From (6.2.11) and the decomposition of (K1) into irreducible components, one can see that two components of (C1) coincide with the two components of (K1). The third component of new conditions is given by $\mathbb{V}_2 = 0$. The following examines the relationship between this set of conditions and (CLP).

Let $\mathbb{P}_\kappa = \{\kappa_0, \ldots, \kappa_5\}$ and refer to Example 6.2.3 for w_2, $\mathbb{V}_2$, etc. Computing a characteristic set of $\mathbb{P}_\kappa$ or a triangular series of $[\mathbb{P}_\kappa, \{a_{20}, a_{11}, \eta\}]$ with respect to the ordering $\omega_2 \prec r$, one may find that

$$\mathrm{Zero}(\mathbb{P}_\kappa/a_{20}a_{11}\eta) = \mathrm{Zero}(\mathbb{T}_\kappa/a_{20}a_{11}\eta)$$

with $\mathbb{T}_\kappa = [\bar{T}_1, \ldots, \bar{T}_6]$, where $\bar{T}_1$, $\bar{T}_2$ and $\bar{T}_3$ are the first, the second, and the fourth polynomial in $\mathbb{V}_2$, $\bar{T}_5 = \gamma$, and

$$\begin{aligned}
\bar{T}_4 &= 243a_{20}^3a_{12} + 2(16a_{11}^2 + 27a_{20}^2)a_{11}a_{21} - 4a_{20}(2a_{11}^2 + 9a_{20}^2)a_{11}^2,\\
\bar{T}_6 &= -27a_{20}a_{11}r + 3(2a_{11}^2 + 27a_{20}^2)a_{21} + a_{20}(2a_{11}^2 + 9a_{20}^2)a_{11}.
\end{aligned}$$

It can be verified that $\mathrm{rem}(\bar{T}_4, \mathbb{V}_2) = 0$ and $\mathrm{prem}(\mathbb{V}_2, \mathbb{T}_\kappa) = \{0\}$. Hence,

$$\mathrm{Zero}(\mathbb{V}_2/a_{20}a_{11}\eta) = \mathrm{Zero}([\bar{T}_1, \ldots, \bar{T}_5]/a_{20}a_{11}\eta).$$

Let $\boldsymbol{a}$ stand for $(a_{20}, a_{11}, a_{02}, a_{30}, a_{21}, a_{12}, a_{03})$. It follows that

$$\mathrm{Zero}(\mathbb{V}_2/a_{20}a_{11}\eta) = \{\boldsymbol{a}\colon\ (\boldsymbol{a}, r) \in \mathrm{Zero}(\mathbb{P}_\kappa/a_{20}a_{11}\eta)\}.$$

This shows that the conditions

$$\mathbb{V}_2 = 0, \quad a_{20}a_{11}\eta \neq 0$$

are equivalent to (CLP) with $\eta \neq 0$. Note that $\eta \neq 0$ is implied by $a_{20}a_{11} \neq 0$ over **R**. Therefore, (CLP) is a subset of (C1) and thus a rediscovery of Cherkas' conditions.

$\mathbb{V}_2 = 0$ is simplified to the center conditions (JW) and

$$a_{20} = a_{11} = a_{30} = a_{21} = a_{12} = a_{03} = 0 \tag{7.6.3}$$

when $a_{11} = 0$, and to the conditions (7.6.3) and

$$a_{20} = a_{02} = a_{21} = a_{12} = a_{03} = 0, \quad 9a_{30} + a_{11}^2 = 0 \tag{K0}$$

when $a_{20} = 0$. (7.6.3) is contained in Kukles' conditions (K1), (K2), and (K3), and so is (K0) in (K4). As a consequence, all the center conditions for Kukles' system discovered by Christopher, Lloyd, Pearson, Jin, and the author are already covered by the conditions $\mathbb{V}_2 = 0$. In summary, we have the following.

Theorem 7.6.1. The set of center conditions (C1) holds if and only if one of the following four sets of conditions holds: (K0), (K1), (JW), and (CLP).

Therefore, the three sets of conditions (C1), (K2), and (K4) cover all the known center conditions for Kukles' system.

Our computational approach has opened the way to the independent discovery of the incompleteness of Kukles' conditions and the nontrivial relations among the different sets of center conditions known so far. The derivations for Kukles' system show that this work depends heavily on the systematic use of elimination methods.

Having Cherkas' conditions (C1) does not prevent one from investigating Kukles' system further. This is because there are doubts about Cherkas' method. The author found that some conditions derived by him for other differential systems also appear to be incomplete. The incompleteness has been confirmed by N. G. Lloyd and J. M. Pearson.

Derivation of center conditions

The problem of deriving necessary center conditions can be reduced partially to decomposing large polynomial systems, for which the major computational tools used are elimination techniques based on characteristic sets, Gröbner bases and resultants. The derivation has proved to be thorny and intractable because the occurring polynomials are too large in terms of degree and number of terms to be manageable.

Computationally, one takes a suitable N, forms the polynomial set

$$\mathbb{P}_N = \{v_3, v_5, \ldots, v_{2N+1}\},$$

and simplifies or solves $\mathbb{P}_N = 0$ to obtain the necessary conditions for the origin to be a center. The sufficiency of the conditions, i.e., $\mathbb{P}_N = 0$ implies that $v_{2j+1} = 0$ for all $j > N$, is proved separately by sophisticated mathematical techniques.

Table 2. Liapunov constants for Kukles' system

	v_5	v_7	v_9	v_{11}	v_{13}	v_{15}	v_{17}	v_{19}
Number of terms	13	49	131	292	577	1046	1775	2859
Total degree	4	6	8	10	12	14	16	18
MLIC[a]	2	4	6	9	13	17	22	27

[a] Maximum length of integer coefficients

We have implemented a program called DEMS in Fortran, Scratchpad II, and Maple for computing Liapunov constants from any differential systems of center and focus type. For Kukles' system, the first Liapunov constant is $v_3 = \gamma/3$. To simplify calculations, we replace a_{21} in (7.6.2) by

$$-(3a_{03} + a_{11}a_{02} + a_{11}a_{20}).$$

Then $v_3 = 0$ and the next eight Liapunov constants computed by DEMS may be characterized as in Table 2. These polynomials are made available in Maple format via World Wide Web from http://www-leibniz.imag.fr/ATINF/Dongming.Wang/PEAA/Wang.html. The Kukles problem is reduced partially to simplifying the conditions given by $\mathbb{P}_N = 0$ and examining their relationships with the existing center conditions.

It seems still unknown whether (C1), (K2), and (K4) cover all the center conditions for Kukles' system. According to theorem 4.1 in Lloyd and Pearson (1992) and the result of the previous section, there are no center conditions of positive dimension other than (C1), (K2), and (K4) for Kukles' system. In fact, Lloyd and Pearson conjectured that there are no other center conditions at all. The difficulties of searching for the complete conditions are caused by the involved large-scale polynomial computations. Despite this, one often gets encouraged by seeing some hope to find new conditions when coming to manipulate the polynomials which are large and appear to follow some bizarre yet regular patterns.

From the known center conditions for Kukles' system, one sees that the algebraic variety Zero($\mathbb{P}_N$) should become reducible for a sufficiently big N. So a natural idea is to decompose $\mathbb{P}_N$ into irreducible components. However, elementary application of the previously mentioned elimination algorithms to $\mathbb{P}_N$ would fail due to the size of the polynomials in $\mathbb{P}_N$. The reducibility occurs and thus splitting $\mathbb{P}_N$ into subsystems becomes possible as N increases. When splitting happens, one gets smaller subsystems and thus the involved computations become easier. Unfortunately, the size of v_{2N+1} expands rapidly as N increases. So a big N would cause some problem as well.

We have taken $N = 7$ and made several attempts including interactive elimination to decompose the polynomial set $\mathbb{P}_7$ into irreducible triangular systems without success. Decomposing $\mathbb{P}_7$ and establishing the complete center conditions for Kukles' system are still challenging problems that remain open.

Bibliographic notes

Although we have tried to acknowledge the sources of material and work in the text wherever they are used, it is possible that in some cases credits were forgotten or not properly given to the original authors. We apologize for any inadequate omission and unawareness. Here are some additional notes on history and bibliography, of which some were not provided because of interference or loose relevance with the context, and the others are repeated for emphasis.

General

Elimination theory has been developed in the West since the 18th century. Early methods are attributed to Euler (1984) and É. Bézout, while the best known are the method of Gauss (1873) for sets of linear equations and the dialytic method of Sylvester (1904) for sets of general polynomial equations. The former is fundamental and has been used in many different domains; the latter started at studying algebraic invariants and was further developed as the theory of resultants through the British school: A. Cayley, A. L. Dixon, F. S. Macaulay, and others.

The method of triangularizing sets of linear equations, named after Gauss, was also described in the ancient Chinese collection *Chiu Chang Suan Shu* (Nine chapters on the mathematical art, abbreviated *Chiu Chang* hereafter) which appeared early in the first century and was commentated by Hui Liu in 260 AD. The book *Chiu Chang* was designed by first asking a daily life question and then giving an answer together with a method for deriving the answer. The example of solving the following set of three linear equations, extracted from the eighth chapter (*Fang Chhêng Shu* – the way of calculating by tabulation), is one of the 246 problems included in the book:

$$\begin{cases} 3x + 2y + z = 39, \\ 2x + 3y + z = 34, \\ x + 2y + 3z = 26. \end{cases}$$

The method given in *Chiu Chang* proceeds by first placing the coefficients and constant terms of the equations in a matrix form and then reducing the matrix with column operations to another triangular matrix. The latter represents the equations $36z = 99$, $5y + z = 24$, and $3x + 2y + z = 39$, from which the values of z, y, and x are successively found with ease. See Boyer (1968, pp. 218 f), Needham (1959, pp. 24–28 and 115 f) and van der Waerden (1983, pp. 47–49) for more details.

Fang Chhêng Shu illustrated by 18 problems deals with sets of simultaneous linear equations in an arbitrary number of unknowns, using both positive and negative numbers. The last problem, involving four equations and five unknowns, foreshadows indeterminate equations. The method described in *Chiu Chang* is systematic and effective and has the same algorithmic feature as that proposed by C. F. Gauss in 1826. In view of this fact and the anonymity of *Chiu Chang*, the method was called *China-Gauss elimination* by W.-t. Wu. In fact, it has already been known as *Chinese matrix method* in mathematical history (see Boyer 1968, p. 248). Several of the algorithms described in this book can be considered as generalizations of the China-Gauss elimination.

The most widely known elimination methods of solving simultaneous algebraic equations of high degree and problems about the solvability of such systems are those based on resultants. The exploration of general elimination methods in China is also of long standing. By the 13th century, Chinese algebraists had already developed a method, called *Ssu Yuan Shu* (Four-element process), that can solve sets of polynomial equations of high degree in four variables. Polynomial arithmetic and elimination are among the most important achievements of Chinese ancient mathematics. The methods then developed were used not only for efficient resolution of algebraic equations but also as algebraic tools for systematic treatment of geometric problems.

We conclude these general notes by quoting the following Taoist paradoxes from I-Chi Tsu's preface to Shih-Chieh Chu's *Ssu Yuan Yü Chien* (Precious mirror of the four elements) of 1303 (Needham 1959, p. 47).

"By moving the expressions upwards and downwards, and from side to side, by advancing and retiring, alternating and connecting, by changing, dividing and multiplying, by assuming the unreal for the real and using the imaginary for the true, by employing different signs for positive and negative, by keeping some and eliminating others and then changing the positions of the counting-rods, by attacking from the front or from one side, as shown in the four examples – he finally succeeds in working out the equations and roots in a profound yet natural manner ... "

Chapter 1

Although the material in this chapter was taken from various sources, the reader may find most of the concepts and results in the books by van der Waerden (1950, 1953) and Knuth (1981). The presentation of subresultants is based largely on chap. 7 of Mishra (1993).

Chapters 2–4

The concept and method of characteristic sets were introduced by Ritt (1932, 1950) for differential polynomial ideals. It was W.-t. Wu, who realized the power of Ritt's method in the later 1970s and has considerably refined and developed it for polynomial sets (instead of ideals). In particular, Wu dropped the irreducibility requirement so that characteristic sets of arbitrary polynomial sets can be defined and computed in different senses. Extensive work on the subject has been done by Wu himself (1984, 1986a, 1987, 1989a, 1994), members of his group (MMRC 1987–1996), Chou and Gao (1990b, 1993), Gallo and Mishra (1991), and Wang (1992b, 1995a). The presentation of the characteristic-set method in this book is based on Wang (1989) and Wu (1994).

The elimination algorithms described in Sects. 2.3 and 3.2 root in the elimination theory of Seidenberg (1956a, b). The adaption and refinement were made by the author (Wang 1993). The notion of simple systems is due to Thomas (1937). The decomposition algorithms using SRS in Sects. 2.4 and 3.3 are also proposed by us (Wang 1998), for which the exposition of Mishra (1993, chap. 7) on subresultants has been helpful.

The contents of Sects. 4.1–4.3 come mostly from Wu (1984, 1986a, 1994) and Wang (1993).

Chapter 5

The concept of regular sets was introduced independently by Kalkbrener (1993) under the name of regular chains and by Yang and Zhang (1994) under the name of proper ascending chains. Related work has also been done by Gao and Chou (1993). The algorithm based on SRS for computing regular series is given in Sect. 5.1 and Wang (2000) for the first time, and so are some of the properties about regular systems proved. The inclusion of Sect. 5.2 is motivated by the work of Lazard (1991).

The Gröbner basis method was invented by Buchberger (1965). Most of the material in Sect. 5.3 originates from Buchberger (1985). The history and extensive literature on Gröbner bases are covered by Adams and Loustaunau (1994) and Becker and Weispfenning (1993).

The base of Sect. 5.4 is van der Waerden (1950, chap. XI), Kapur and Lakshman (1992), and Chionh and Goldman (1995), which contain a lot of historical and bibliographic information.

Chapter 6

The presentation of dimension and unmixed decomposition is based partly on the work done by Chou and Gao (1990b, 1993), and Kalkbrener (1993), with some generalizations. Methods for computing prime bases of irreducible ascending sets were suggested by Chou et al. (1990), Wang (1989), Wu (1989b), and Ritt (1950). The technique of using Gröbner bases to construct saturation bases is also contained in Gianni et al. (1988). Irreducible decomposition of algebraic varieties was investigated in Wang (1989, 1992a).

The algorithm of primary ideal decomposition is attributed to Shimoyama and Yokoyama (1996).

Chapter 7

Many researchers have worked on and contributed to automated geometry theorem proving; see Wang (1996b) for a list of 171 references. We ought to mention Wu (1978, 1984, 1986c, 1994) and the work done by his students (Wang and Gao 1987; MMRC 1987–1996), Chou (1988), Kapur (1988), and Kutzler and Stifter (1986), just to name a few. In particular, Chou (1988) contains 512 geometric theorems which were proved by an implementation based on Wu's method and the Gröbner bases method. Zero decompositions were used for geometric theorem proving by Ko (1988), Chou and Gao (1990b), and Wang (1995c).

Automated derivation of unknown relations was initiated by Wu (1986b) and Chou (1987); further work was carried out by Chou and Gao (1990a) and Wang (1995b). The implicitization of parametric objects was investigated by various researchers; see Buchberger (1987) and Li (1989), for background and literature information. The methods of algebraic factorization described in Sect. 7.5 are taken from Hu and Wang (1986) and Wang (1992c). Several other geometric applications of elimination methods can be found in Buchberger (1987), Wang (1995b), and MMRC (1987–1996).

References

Adams, W. W., Loustaunau, P. (1994): An introduction to Gröbner bases. American Mathematical Society, Providence, R.I.

Aubry, P., Lazard, D., Moreno Maza, M. (1999): On the theories of triangular sets. J. Symb. Comput. 28: 105–124.

Becker, T., Weispfenning, V. (1993): Gröbner bases: a computational approach to commutative algebra. Springer, Berlin Heidelberg New York Tokyo (Graduate texts in mathematics, vol. 141).

Boyer, C. B. (1968): A history of mathematics. J. Wiley, New York.

Brown, W. S., Traub, J. F. (1971): On Euclid's algorithm and the theory of subresultants. J. ACM 18: 505–514.

Buchberger, B. (1965): Ein Algorithmus zum Auffinden der Basiselemente des Restklassenringes nach einem nulldimensionalen Polynomideal. Ph.D. thesis, Universität Innsbruck, Innsbruck, Austria.

Buchberger, B. (1985): Gröbner bases: an algorithmic method in polynomial ideal theory. In: Bose, N. K. (ed.): Multidimensional systems theory. Reidel, Dordrecht, pp. 184–232.

Buchberger, B. (1987): Applications of Gröbner bases in non-linear computational geometry. In: Rice, J. R. (ed.): Mathematical aspects of scientific software. Springer, Berlin Heidelberg New York Tokyo, pp. 59–87 (The IMA volumes in mathematics and its applications, vol. 14).

Cherkas, L. A. (1978): Conditions for the equation $yy' = \sum_{i=0}^{3} p_i(x)y^i$ to have a center. Differentsial'nye Uravneniya 14: 1594–1600.

Chionh, E. W., Goldman, R. N. (1995): Elimination and resultants. IEEE Comput. Graphics Appl. 15/1: 69–77; 15/2: 60–69.

Chou, S.-C. (1987): A method for the mechanical derivation of formulas in elementary geometry. J. Automat. Reason. 3: 291–299.

Chou, S.-C. (1988): Mechanical geometry theorem proving. Reidel, Dordrecht.

Chou, S.-C., Gao, X.-S. (1990a): Mechanical formula derivation in elementary geometries. In: Proceedings ISSAC '90, Tokyo, August 20–24, 1990. Association for Computing Machinery, New York, pp. 265–270.

Chou, S.-C., Gao, X.-S. (1990b): Ritt-Wu's decomposition algorithm and geometry theorem proving. In: Stickel, M. E. (ed.): 10th International Conference on Automated Deduction. Springer, Berlin Heidelberg New York Tokyo, pp. 207–220 (Lecture notes in computer science, vol. 449).

Chou, S.-C., Schelter, W. F., Yang, J.-G. (1990): An algorithm for constructing Gröbner bases from characteristic sets and its application to geometry. Algorithmica 5: 147–154.

Christopher, C. J., Lloyd, N. G. (1990): On the paper of Jin and Wang concerning the conditions for a centre in certain cubic systems. Bull. London Math. Soc. 22: 5–12.

Collins, G. E. (1967): Subresultants and reduced polynomial remainder sequences. J. ACM 14: 128–142.

Collins, G. E. (1971): The calculation of multivariate polynomial resultants. J. ACM 18: 515–532.

Cox, D., Little, J., O'Shea, D. (1992): Ideals, varieties, and algorithms. Springer, Berlin Heidelberg New York Tokyo (Undergraduate texts in mathematics).

Dixon, A. L. (1908): The eliminant of three quantics in two independent variables. Proc. London Math. Soc. 6: 468–478.

Euler, L. (1984): Elements of algebra. Translated by J. Hewlett. Reprint of the 5th edition Longman, Orme, and Co., London 1840. Springer, Berlin Heidelberg New York Tokyo.

Gallo, G., Mishra, B. (1991): Efficient algorithms and bounds for Wu-Ritt characteristic sets. In: Mora, T., Traverso, C. (eds.): Effective methods in algebraic geometry Birkhäuser, Boston, pp. 119–142 (Progress in mathematics, vol. 94).

Gao, X.-S., Chou, S.-C. (1992): Solving parametric algebraic systems. In: Proceedings ISSAC '92, Berkeley, July 27–29, 1992. Association for Computing Machinery, New York, pp. 335–341.

Gao, X.-S., Chou, S.-C. (1993): On the dimension of an arbitrary ascending chain. Chin. Sci. Bull. 38: 799–804.

Gauss, C. F. (1873): Werke, Band IV. Herausgegeben von der Königlichen Gesellschaft der Wissenschaft zu Göttingen.

Gianni, P., Trager, B. M., Zacharias, G. (1988): Gröbner bases and primary decomposition of polynomial ideals. J. Symb. Comput. 6: 149–167.

Hartshorne, R. (1977): Algebraic geometry. Springer, Berlin New York.

Hilbert, D. (1901): Mathematische Probleme. Arch. Math. Phys. (3) 1: 44–63; 213–237.

Hu, S., Wang, D. (1986): Fast factorization of polynomials over rational number field or its extension fields. Kexue Tongbao 31: 150–156.

Jacobson, N. (1974): Basic algebra I. Freeman, San Francisco.

Jin, X., Wang, D. (1990): On the conditions of Kukles for the existence of a centre. Bull. London Math. Soc. 22: 1–4.

Kalkbrener, M. (1993): A generalized Euclidean algorithm for computing triangular representations of algebraic varieties. J. Symb. Comput. 15: 143–167.

Kalkbrener, M. (1994): Prime decompositions of radicals in polynomial rings. J. Symb. Comput. 18: 365–372.

Kapur, D. (1988): A refutational approach to geometry theorem proving. Artif. Intell. 37: 61–93.

Kapur, D., Lakshman, Y. N. (1992): Elimination methods: an introduction. In: Donald, B. R., Kapur, D., Mundy, J. L. (eds.): Symbolic and numerical computation for artificial intelligence. Academic Press, London, pp. 45–87.

Kapur, D., Saxena, T. (1995): Comparison of various multivariate resultant formulations. In: Proceedings ISSAC '95, Montreal, July 10–12, 1995. Association for Computing Machinery, New York, pp. 187–194.

Knuth, D. E. (1981): The art of computer programming, vol. 2, 2nd ed. Addison-Wesley, Reading, Mass.

Ko, H.-P. (1988): Geometry theorem proving by decomposition of quasi-algebraic sets: an application of the Ritt-Wu principle. Artif. Intell. 37: 95–122.

Ko, H.-P., Hussain, M. A. (1985): ALGE-prover: an algebraic geometry theorem proving software. Tech. Rep. 85CRD139, General Electric Company, Schenectady, N.Y.

Kukles, I. S. (1944): Sur les conditions nécessaires et suffisantes pour l'existence d'un centre. Dokl. Akad. Nauk 42: 160–163.

Kusche, K., Kutzler, B., Stifter, S. (1987): Implementation of a geometry theorem proving package in Scratchpad II. In: Davenport, J. H. (ed.): EUROCAL '87. Springer, Berlin Heidelberg New York Tokyo, pp. 246–257 (Lecture notes in computer science, vol. 378).

Kutzler, B., Stifter, S. (1986): On the application of Buchberger's algorithm to automated geometry theorem proving. J. Symb. Comput. 2: 389–397.

Lazard, D. (1981): Résolution des systèmes d'équations algébriques. Theor. Comput. Sci. 15: 77–110.

Lazard, D. (1991): A new method for solving algebraic systems of positive dimension. Discr. Appl. Math. 33: 147–160.

Li, Z. (1989a): Determinant polynomial sequences. Chin. Sci. Bull. 34: 1595–1599.

Li, Z. (1989b): Automatic implicitization of parametric objects. Math. Mech. Res. Preprints 4: 54–62.

Lloyd, N. G., Pearson, J. M. (1992): Computing centre conditions for certain cubic systems. J. Comput. Appl. Math. 40: 323–336.

Loos, R. (1983): Generalized polynomial remainder sequences. In: Buchberger, B., Collins, G. E., Loos, R. (eds.): Computer algebra: symbolic and algebraic computation, 2nd edn. Springer, Wien New York, pp. 115–137.

Macaulay, F. S. (1921): Note on the resultant of a number of polynomials of the same degree. Proc. London Math. Soc. 21: 14–21.

Macaulay, F. S. (1964): The algebraic theory of modular systems. Stechert-Hafner Service Agency, New York London.

Mishra, B. (1993): Algorithmic algebra. Springer, Berlin Heidelberg New York Tokyo (Texts and monographs in computer science).

MMRC (1987–1996): Mathematics-Mechanization Research Preprints, nos. 1–14. Academia Sinica, China.

Needham, J. (1959): Science and civilisation in China, vol. 3. Cambridge University Press, Cambridge.

Nemytskii, V. V., Stepanov, V. V. (1960): Qualitative theory of differential equations. Princeton University Press, Princeton, N.J.

Ritt, J. F. (1932): Differential equations from the algebraic standpoint. American Mathematical Society, New York.

Ritt, J. F. (1950): Differential algebra. American Mathematical Society, New York.

Seidenberg, A. (1956a): Some remarks on Hilbert's Nullstellensatz. Arch. Math. 7: 235–240.

Seidenberg, A. (1956b): An elimination theory for differential algebra. Univ. California Publ. Math. (N.S.) 3/2: 31–66.

Shimoyama, T., Yokoyama, K. (1996): Localization and primary decomposition of polynomial ideals. J. Symb. Comput. 22: 247–277.

Sylvester, J. J. (1904): The collected mathematical papers, vol. I. Cambridge University Press, Cambridge.

Thomas, J. M. (1937): Differential systems. American Mathematical Society, New York.

Thomas, J. M. (1946): Division sequence. Duke Math. J. 13: 459–469.

Trager, B. M. (1976): Algebraic factoring and rational function integration. In: Proceedings SYMSAC '76, Yorktown Heights, August 10–12, 1976. Association for Computing Machinery, New York, pp. 219–226.

van der Waerden, B. L. (1950): Modern algebra, vol. II. Frederick Ungar, New York.
van der Waerden, B. L. (1953): Modern algebra, vol. I. Frederick Ungar, New York.
van der Waerden, B. L. (1983): Geometry and algebra in ancient civilizations. Springer, Berlin Heidelberg New York Tokyo.
Wang, D. (1987): Mechanical approach for polynomial set and its related fields. Ph.D. thesis, Academia Sinica, Beijing, People's Republic of China (in Chinese).
Wang, D. (1989): Characteristic sets and zero structure of polynomial sets. Lecture notes, RISC Linz.
Wang, D. (1991a): Mechanical manipulation for a class of differential systems. J. Symb. Comput. 12: 233–254.
Wang, D. (1991b): On the parallelization of characteristic-set-based algorithms. In: Zima, H. P. (ed.): Parallel computation. Springer, Berlin Heidelberg New York Tokyo, pp. 338–349 (Lecture notes in computer science, vol. 591).
Wang, D. (1992a): Irreducible decomposition of algebraic varieties via characteristic sets and Gröbner bases. Comput. Aided Geom. Des. 9: 471–484.
Wang, D. (1992b): A strategy for speeding-up the computation of characteristic sets. In: Havel, I., Koubek, V. (eds.): Mathematical foundations of computer science 1992. Springer, Berlin Heidelberg New York Tokyo, pp. 504–510 (Lecture notes in computer science, vol. 629).
Wang, D. (1992c): A method for factorizing multivariate polynomials over successive algebraic extension fields. Preprint, RISC Linz.
Wang, D. (1993): An elimination method for polynomial systems. J. Symb. Comput. 16: 83–114.
Wang, D. (1994): Algebraic factoring and geometry theorem proving. In: Bundy, A. (ed.): Automated deduction – CADE-12. Springer, Berlin Heidelberg New York Tokyo, pp. 386–400 (Lecture notes in computer science, vol. 814).
Wang, D. (1995a): An implementation of the characteristic set method in Maple. In: Pfalzgraf, J., Wang, D. (eds.): Automated practical reasoning: algebraic approaches. Springer, Wien New York, pp. 187–201 (Texts and monographs in symbolic computation).
Wang, D. (1995b): Reasoning about geometric problems using an elimination method. In: Pfalzgraf, J., Wang, D. (eds.): Automated practical reasoning: algebraic approaches. Springer, Wien New York, pp. 147–185 (Texts and monographs in symbolic computation).
Wang, D. (1995c): Elimination procedures for mechanical theorem proving in geometry. Ann. Math. Artif. Intell. 13: 1–24.
Wang, D. (1996a): GEOTHER: a geometry theorem prover. In: McRobbie, M., Slaney, J. K. (eds.): Automated deduction – CADE-13. Springer, Berlin Heidelberg New York Tokyo, pp. 166–170 (Lecture notes in computer science, vol. 1104).
Wang, D. (1996b): Geometry machines: from AI to SMC. In: Calmet, J., Campbell, J. A., Pfalzgraf, J. (eds.): Artificial intelligence and symbolic computation. Springer, Berlin Heidelberg New York Tokyo, pp. 213–239 (Lecture notes in computer science, vol. 1138).
Wang, D. (1998): Decomposing polynomial systems into simple systems. J. Symb. Comput. 25: 295–314.
Wang, D. (2000): Computing triangular systems and regular systems. J. Symb. Comput. 30: 221–236.
Wang, D., Gao, X.-S. (1987): Geometry theorems proved mechanically using Wu's method – part on Euclidean geometry. Math. Mech. Res. Preprints 2: 75–106.

Wu, W.-t. (1978): On the decision problem and the mechanization of theorem-proving in elementary geometry. Sci. Sin. 21: 159–172.

Wu, W.-t. (1984): Basic principles of mechanical theorem proving in elementary geometries. J. Syst. Sci. Math. Sci. 4: 207–235.

Wu, W.-t. (1986a): On zeros of algebraic equations – an application of Ritt principle. Kexue Tongbao 31: 1–5.

Wu, W.-t. (1986b): A mechanization method of geometry and its applications I: distances, areas and volumes. J. Syst. Sci. Math. Sci. 6: 204–216.

Wu, W.-t. (1986c): On reducibility problem in mechanical theorem proving of elementary geometries. Chin. Q. J. Math. 2: 1–20.

Wu, W.-t. (1987): A zero structure theorem for polynomial equations-solving. Math. Mech. Res. Preprints 1: 2–12.

Wu, W.-t. (1989a): Some remarks on characteristic-set formation. Math. Mech. Res. Preprints 3: 27–29.

Wu, W.-t. (1989b): On the generic zero and Chow basis of an irreducible ascending set. Math. Mech. Res. Preprints 4: 1–21.

Wu, W.-t. (1990): On a projection theorem of quasi-varieties in elimination theory. Chin. Ann. Math. (Ser. B) 11: 220–226.

Wu, W.-t. (1994): Mechanical theorem proving in geometries: basic principles. Translated from the Chinese by X. Jin and D. Wang. Springer, Wien New York (Texts and monographs in symbolic computation).

Yang, L., Zhang, J.-Z. (1994): Searching dependency between algebraic equations: an algorithm applied to automated reasoning. In: Johnson, J., McKee, S., Vella, A. (eds.): Artificial intelligence in mathematics. Oxford University Press, Oxford, pp. 147–156.

Subject index